Post-combustion Carbon Dioxide Capture Materials

Inorganic Materials Series

Series editors:
Duncan W. Bruce, *University of York, UK*
Dermot O'Hare, *University of Oxford, UK*
Richard I. Walton, *University of Warwick, UK*

Titles in the Series:

How to obtain future titles on publication:
A standing order plan is available for this series. A standing order will bring delivery of each new volume immediately on publication.

For further information please contact:
Book Sales Department, Royal Society of Chemistry, Thomas Graham House, Science Park, Milton Road, Cambridge, CB4 0WF, UK
Telephone: +44 (0)1223 420066, Fax: +44 (0)1223 420247,
Email: booksales@rsc.org
Visit our website at www.rsc.org/books

Post-combustion Carbon Dioxide Capture Materials

Edited by

Qiang Wang
Beijing Forestry University, P. R. China
Email: qiangwang@bjfu.edu.cn

ROYAL SOCIETY OF CHEMISTRY

Inorganic Materials Series No. 2

Print ISBN: 978-1-78801-109-9
PDF ISBN: 978-1-78801-335-2
EPUB ISBN: 978-1-78801-545-5
Print ISSN: 2472-3819
Electronic ISSN: 2472-3827

A catalogue record for this book is available from the British Library

The Royal Society of Chemistry is a charity, registered in England and Wales, Number 207890, and a company incorporated in England by Royal Charter (Registered No. RC000524), registered office: Burlington House, Piccadilly, London W1J 0BA, UK, Telephone: +44 (0) 20 7437 8656.

For further information see our web site at www.rsc.org

Printed in the United Kingdom by CPI Group (UK) Ltd, Croydon, CR0 4YY, UK

Preface

Carbon dioxide (CO_2) is one of the major greenhouse gases that contribute to global warming and anthropogenic climate change. Unfortunately, it is predicted that the atmospheric CO_2 concentration will continue to increase in the next few decades because fossil fuels will still be the dominant energy source. In recent years, worldwide efforts have been made to reduce the CO_2 emissions, among which capturing using solid adsorbents/sorbents has attracted intense attention from both academia and industry. I have been working on CO_2 capture materials since 2009, and have witnessed nearly one-thousand papers per year being published in this field. With this rapid development, I believe that it is was necessary to edit a book to summarize all the important progresses made with each type of CO_2 capture material. Professor Dermot O'Hare, University of Oxford suggested that this could be part of the Royal Society of Chemistry's Inorganic Materials Series. He suggested two books entitled *Pre-combustion Carbon Dioxide Capture Materials* and *Post-combustion Carbon Dioxide Capture Materials*. I am confident that these works will benefit advanced undergraduates, postgraduates and researchers working in both academia and industry on this topic.

This book is organized into five chapters and focuses mainly on the low-temperature CO_2 adsorbents used for post-combustion CO_2 capture from flue gases. This book aims to present the full picture of various post-combustion CO_2 capture materials including carbon, zeolite and silica, metal organic frameworks, and alkali metal carbonate, *etc*. The discussion of each type of adsorbent starts with the fundamental mechanism for CO_2 capture, followed by the preparation and modification of the materials, and their capture capacity, kinetics, and recycling stability, *etc*. The application status of the above

Inorganic Materials Series No. 2
Post-combustion Carbon Dioxide Capture Materials
Edited by Qiang Wang

Published by the Royal Society of Chemistry, www.rsc.org

mentioned materials for post-combustion CO_2 capture is presented in the fifth chapter.

The editor thanks all the contributors to this book, particularly Professor Jin Zhou (Shandong University of Technology, China), Professor Wei Xing (China University of Petroleum, China), Professor Guillaume Laugel (Université Pierre et Marie Curie, France), Professor Benoît Louis (Université de Strasbourg, France), Professor Wha-Seung Ahn (Inha University, South Korea), Professor Hirofumi Kanoh (Chiba University, Japan), Professor Rajender Gupta (University of Alberta, Canada), and all the students and researchers involved in each chapter.

Also, I would like to express special acknowledgements to Professor Duncan Bruce (University of York, UK), Professor Dermot O'Hare (University of Oxford, UK), and Professor Richard Walton (University of Warwick, UK), who accepted and supported this project, and to Connor Sheppard, Leanne Marle, Sylvia Pegg, and Robin Driscoll for all their support during the editing of this book. Finally, I thank the Royal Society of Chemistry for supporting this edition.

Qiang Wang

Contents

Inorganic Materials Series No. 2
Post-combustion Carbon Dioxide Capture Materials
Edited by Qiang Wang

Published by the Royal Society of Chemistry, www.rsc.org

CHAPTER 1

Carbon-based CO_2 Adsorbents

JIN ZHOU*[a], XUAN WANG[a] AND WEI XING*[b]

[a]Shandong University of Technology, School of Chemistry and Chemical Engineering, Zibo, 255000, PR China; [b]China University of Petroleum, School of Science, State Key Laboratory of Heavy Oil Processing, Qingdao, 266580, PR China
*E-mail: zhoujin@sdut.edu.cn, xingwei@upc.edu.cn

1.1 Introduction

Carbon dioxide (CO_2) has been recognized to be the biggest driver of global warming, which is one of the most serious problems that our world is facing.[1–3] Furthermore, CO_2 is an important source of C1 chemical engineering, and could be converted into high-value chemical products *via* chemical,[2,3] photochemical,[4,5] or electrochemical processes.[6,7] However, the efficient capture of CO_2 is essential for these processes. So, there is an urgent need to develop CO_2 capture and storage (CCS) technologies. The basic concept of CCS is to capture CO_2 from emissions without releasing it into the atmosphere. CCS can be classified as post-combustion, pre-combustion, and oxy-fuel combustion technologies. Among the current CCS technologies, post-combustion capture, a technology for capturing CO_2 from post-combustion emission gases (*e.g.*, flue gas from power plants) is the most easily applied technology for existing emission sources.

Inorganic Materials Series No. 2
Post-combustion Carbon Dioxide Capture Materials
Edited by Qiang Wang

Published by the Royal Society of Chemistry, www.rsc.org

In general, post-combustion capture technologies include chemical absorption, dry adsorption, membrane-based technologies, and cryogenic technologies. Currently, chemical absorption is the most applicable technology for CO_2 capture in power plants, but this technology suffers from several drawbacks. The biggest challenge in applying a chemical absorption process for post-combustion is how to reduce the heat of regeneration. Another problem is the release of hazardous byproducts. For these reasons, dry adsorption using solid adsorbents is considered to be promising for the capture of post-combustion CO_2.

The dry adsorption technique is a process of selective adsorption of CO_2 from post-combustion gases using solid adsorbents, which has advantages such as a simple device, easy operation, it is environmentally friendly and has a high energy efficiency. When evaluating solid adsorbents, it is important to consider their surface area, apparent density, pore size and volume, feasibility of regeneration, stability, abundance and sustainability. CO_2 capture by solid adsorbents mainly relies on the mechanism of physical adsorption that is also interfered with by some weak interactions between CO_2 and the adsorbent's surface (*i.e.* hydrogen bonding or electric quadrupole interactions). Due to the main contribution of van der Waals forces to the physical adsorption, materials that possess a developed microporous texture are preferred for CO_2 capture. Nowadays, many kinds of solid adsorbent materials with porous textures, such as porous carbonaceous materials,[8,9] zeolites,[10] zeolitic imidazolate frameworks (ZIFs),[11] metal–organic frameworks (MOFs),[12] covalent organic frameworks (COFs),[13] and porous coordination polymers (PCPs),[14] have been investigated, and show excellent CO_2 capture performances.

Among these solid adsorbents, porous carbonaceous materials have been studied intensively because of their desirable physical and chemical properties, such as low cost, variety of form (powder, fibers, aerogels, composites, sheets, monoliths, tubes, *etc.*), ease of processability, controllable porosity (adjustable pore size and its distribution, high specific surface area and pore volume), and tailored surface chemistry (O, N, S, P, F or other heteroatom doping). They also possess some other advantages, particularly for adsorption applications: (1) carbon materials have excellent stability especially in hot and humid environments; (2) gas sorption on carbon materials is not moisture-sensitive because the surface is usually hydrophobic; (3) the energy consumption of regeneration is low due to the nature of physical adsorption; (4) the adsorption/desorption temperatures are always below 373 K; (5) these materials can be used at atmospheric pressure.

In this chapter, we summarize the recent research progress made in developing carbon-based sorbents for post-combustion CO_2 capture. Specifically, this chapter will provide overviews of (1) porous carbons, (2) graphene-based porous materials, (3) carbon nanotubes, (4) carbon-based hybrid sorbents, and (5) important factors influencing CO_2 uptake over carbon adsorbents.

1.2 Porous Carbons

Porous carbons have been extensively studied in the field of CO_2 capture. In order to control the pore structure in carbon materials, a variety of preparation methods have been developed and certain successes have been achieved. Herein, we summarize typical preparation methods for porous carbon adsorbents, including chemical activation, physical activation, metal ion activation, templating methods, and the combined method of templating and activation. In each section, synthesis principles, carbon precursors, pore structures, as well as their CO_2 adsorption performance, are discussed.

1.2.1 Chemical Activation

Activated carbon is the oldest and most widely used carbon material. Generally, the production routes of activated carbons are divided into physical activation and chemical activation. In chemical activation, the carbon precursor is mixed/impregnated with an activating agent (such as KOH, H_3PO_4, $ZnCl_2$, K_2CO_3, *etc.*), then the precursor is simultaneously carbonized and activated at an elevated temperature (from 400 to 900 °C) and under an inert atmosphere (usually N_2 or Ar). In physical activation, the carbon precursor is usually pre-carbonized at temperatures over 500 °C in an inert atmosphere to remove non-carbon species, followed by etching by an oxidizing gas (such as CO_2, steam, and air) at a higher temperature (from 700 to 1200 °C). Comparatively, chemical activation needs a lower temperature and shorter activation time, and generally results in a higher specific surface area and more uniform pore size distribution (PSD), while physical activation is simple and does not require chemical agents and repeated washing procedures to remove the inorganic residues after activation. The structure of activated carbons, containing the surface area, pore size and its distribution, and surface chemistry, *etc.*, strongly depends on the activation conditions, activating agents as well as the carbon precursors used.

1.2.1.1 KOH as an Activating Agent

KOH is the most common activating agent in chemical activation. Jaroniec *et al.* treated a commercial carbon sorbent (Ambersorb 563) with five of the most commonly-used activating agents, including CO_2, H_2O, NH_3, KOH, and $ZnCl_2$, and compared their activating power for the evolution of microporosity responsible for CO_2 adsorption.[15] N_2 adsorption analysis showed that the investigated activating agents enlarged microporosity and consequently surface area and pore volume of the carbons in the following order: KOH > CO_2 > NH_3 > H_2O > $ZnCl_2$. It was shown that KOH activation yielded the highest volume of micropores and small micropores. Besides, the CO_2 uptake for the KOH-activated sample was the highest, indicating that KOH activation appears to be the most effective to obtain carbon adsorbents for CO_2 capture.

However, the mechanism of KOH activation has not been totally understood due to the complexity of this process. In a previous review about KOH activated carbon materials for energy storage, Wang *et al.* suggested that KOH activation is a synergistic process of chemical activation, physical activation, and carbon lattice expansion by metallic K intercalation.[16] Firstly, the potassium species serving as chemical activating reagents vigorously etch the carbon framework by the redox reactions shown in eqn (1.1)–(1.3). Secondly, H_2O (from dehydration of carbon precursors or eqn (1.4)) and CO_2 (eqn (1.5) and (1.6)) produced *in situ* in the activation system further develop the porosity through the gasification of carbon, namely physical activation (eqn (1.7) and (1.8)). Meanwhile, the produced metallic K intercalates into the lattices of the carbon matrix, responsible for both stabilization and widening of the interlayer spacing (Figure 1.1). After removal of the intercalated metallic K and other K compounds by washing, the expanded carbon lattices cannot return to their previous non-porous structure and thus create a narrow microporosity and large specific surface areas.

$$6KOH + 2C \rightarrow 2K + 2H_2 + 2K_2CO_3 \tag{1.1}$$

$$K_2CO_3 + 2C \rightarrow 2K + 3CO \tag{1.2}$$

$$C + K_2O \rightarrow 2K + CO \tag{1.3}$$

$$2KOH \rightarrow K_2O + 3H_2O \tag{1.4}$$

$$CO + H_2O \rightarrow CO_2 + H_2 \tag{1.5}$$

$$K_2CO_3 \rightarrow K_2O + CO_2 \tag{1.6}$$

$$C + H_2O \rightarrow CO + H_2 \tag{1.7}$$

$$C + CO_2 \rightarrow 2CO \tag{1.8}$$

Figure 1.1 Activation mechanism by the penetration of metallic K into the carbon lattices. (a) Carbon lattices, (b) metallic K intercalated in the carbon lattices, (c) activated carbon.

Generally, the raw materials of KOH activation could be classified into non-renewable fossil-based materials and renewable biomass resources. As shown in Table 1.1, various fossil-based precursors, such as petrol coke,[17,18] pitch,[19] and synthetic polymers,[20–22] have been used as precursors for the preparation of porous carbon adsorbents. Wahby *et al.* prepared a series of carbon molecular sieves (CMS) from petroleum pitch using KOH as the activating agent. Depending on the type of petroleum residue and the conforming step applied, the prepared CMS possessed a well-defined pore size

Table 1.1 Porous carbons prepared by KOH activation of fossil-based resources for CO_2 capture.

Precursor	Surface area ($m^2\,g^{-1}$)	Pore volume ($cm^3\,g^{-1}$)	Surface chemistry	CO_2 uptake (25 °C, 1 bar, $mmol\,g^{-1}$)	CO_2–N_2 selectivity	Reference
Petroleum pitch	3100	1.40	—	4.09	2.8[a]	23
Petroleum pitch	2895	1.42	—	5.23	—	24
Coal–pitch mixture	1044	0.50	—	4.00	5.94[a]	26
Petroleum coke	1445	0.58	N-doped	4.57	22[c]	18
Petroleum coke	1745	—	—	3.45	—	25
Phenolic resin	2400	1.07	—	4.60	—	27
Waste ion-exchange resin	828	0.42	—	1.85	—	28
Vinylidene chloride	2151	0.90	—	3.97	12.59[c]	29
Styrene divinylbenzene resin	3870	2.07	—	4.75	—	30
Polypyrrole	1700	0.88	N-doped	3.90	—	20
Polypyrrole–graphene composite	1360	0.59	N-doped	4.30	16[a]	31
Polyaniline	1091	0.61	N-doped	4.30	8[a]	21
Polyacrylonitrile	2231	1.16	N-doped	4.50	—	22
Polyurethane	1516	0.64	N-doped	4.33	12[a]	32
Urea furfural resin	1013	0.53	N-doped	4.70	—	33
1,3-Bis(cyanomethyl imidazolium) chloride	1317	0.59	N-doped	5.39	62[b]	34
Polyimine	1561	0.75	N-doped	3.10	47[a]	35,36
Polythiophene–graphene composite	1567	—	S-doped	4.50	51[c]	37
Melamine-doped phenolic resins	1286	0.54	N-doped	4.04	52.9[b]	38

[a]The selectivity calculated by the ratio of CO_2–N_2 sorption capacity.
[b]The selectivity based on ideal adsorbed solution theory (IAST).
[c]The selectivity of Henry's law.

(0.35–0.7 nm), together with a very large surface area up to 3100 $m^2 g^{-1}$, thus exhibited a high CO_2 adsorption capacity up to 4.09 mmol g^{-1} at 1 bar and 25 °C.[23] After further optimizing the process parameters, including the nature of the petroleum residue, the KOH–pitch ratio, the mesophase content, the temperature and time of activation, the CO_2 uptake could increase to 5.23 mmol g^{-1} at 1 bar and 25 °C.[24] The activated carbons by the KOH activation of petroleum coke possessed high surface areas, over 3000 $m^2 g^{-1}$, and a high CO_2 capacity, over 15 wt% at 1 bar.[25] High-resolution analysis of N_2 sorption isotherms concluded that the micropores smaller than 1 nm played a critical role in CO_2 capture under ambient conditions due to the high-density filling of CO_2 in these small pores.[25] Using petroleum coke as the precursor, Yang *et al.* prepared N-doped porous carbons by combining ammoxidation with KOH activation.[18] The sample prepared under mild conditions (a low temperature of 650 °C and a low KOH–precursor ratio of 2) showed the highest CO_2 uptake of 4.57 mmol g^{-1} at 25 °C and 1 bar, while the CO_2/N_2 selectivity and CO_2 heats of adsorption of the sorbent were 22 and 37 kJ mol^{-1}, respectively. The high CO_2 capture capacity was attributed to the synergetic effect of N-doping and high narrow microporosity, while the latter was suggested to contribute more.[18]

Besides petroleum coke and pitch, synthetic polymers, such as phenolic resins,[27] poly(vinylidene chloride),[29] styrene-divinylbenzene resin,[30] polypyrrole,[20] polyaniline,[21] polyurethane,[32] urea furfural resins,[33] polyacrylonitrile,[22] and polyimine,[37,38] have also been widely used as precursors for the preparation of porous carbon adsorbents. For instance, Jaroniec *et al.* prepared activated carbon spheres by direct KOH activation of phenolic resin spheres obtained by a modified Stöber method.[27] Due to the small micropore (<0.8 nm) and large specific surface areas, the prepared porous carbon spheres exhibited superior CO_2 uptakes reaching 4.6 and 8.9 mmol g^{-1} at 23 and 0 °C under 1 bar, respectively.

Many studies have reported that N-doping could significantly improve the CO_2 capture performance of porous carbons, especial CO_2 uptake at low partial pressure and the selectivity of CO_2-to-N_2 (Table 1.1). Activation/carbonization of nitrogen-containing synthetic polymers is an important approach to preparing N-doped porous carbons. In 2010, Lu *et al.* reported a facile and rapid preparation of N-doped porous carbon monoliths using a basic amino acid as both the catalyst and nitrogen source.[39] This monolithic carbon directly pyrolyzed at 500 °C exhibited a CO_2 adsorption capacity of 3.13 mmol g^{-1} at room temperature. KOH activation of nitrogen-containing polymers could result in highly developed microporosity at the same time maintaining the N-doping. Sevilla *et al.* prepared highly porous N-doped carbon by using polypyrrole (PPy) as the carbon precursor and KOH as the activating agent.[20] The mildly activated carbon (KOH–PPy = 2) showed a larger CO_2 uptake than the severely activated ones (KOH–PPy = 4) as a result of two important characteristics for the mildly activated samples: (a) a high-level N-doping (up to 10.1 wt% N) identified as main pyridonic-N and a small proportion of pyridinic-N groups, and (b) a narrower micropore size. Moreover, these polypyrrole-based carbons showed a high adsorption rate for the capture of CO_2, more than 95% of the CO_2 being adsorbed in 2 min with a high kinetic CO_2–N_2 selectivity of 12. Other nitrogen-containing polymers, such as polyacrylonitrile,[22] polyaniline,[21] polyurethane,[32] urea furfural resin,[33] and polyimine,[35,36] have also been used as precursors and the corresponding N-doped porous carbons showed CO_2 uptakes of 4.50, 4.30, 4.33, 4.70, and 3.10 mmol g^{-1} at 25 °C and 1 bar, respectively.

From Table 1.1, we find that the CO_2 uptakes for most of the activated carbons are lower than 5.0 mmol g^{-1}. Recently, Sethia and Syari prepared a series of strictly microporous N-doped activated porous carbons by using a nitrogen-containing ionic liquid of 1,3-bis(cyanomethyl imidazolium) chloride as a precursor and KOH as the activating agent.[34] The optimized material exhibited an extraordinary CO_2 uptake of 5.39 mmol g^{-1} at 25 °C and 1 bar, which is the highest uptake reported so far for activated carbons. They concluded that both nitrogen content and ultramicropores played important roles in the CO_2 capture, with the latter being predominant.[34]

Besides the N-doped porous carbons, S-doped porous carbons have also been prepared by the KOH activation of sulfur-containing synthetic polymers and are applied in CO_2 capture. For instance, S-doped microporous carbon materials can be prepared by the KOH activation of polythiophene–graphene composite.[37] This material displayed a high CO_2 uptake of 4.5 mmol g^{-1} at 25 °C and 1 bar, as well as an impressive CO_2 adsorption selectivity over N_2 (51), CH_4 (12), and H_2 (214) based on the initial slopes method of Henry's law.[40–42]

Fossil-derived materials, such as petroleum coke, pitch, coal, and synthesized polymers are non-renewable and expensive. Considering the demand for a huge amount of solid CO_2 sorbents, the development of low-cost carbon materials from renewable and sustainable sources is urgent.[43] Biomass materials are renewable, easily available and very cheap, which are promising raw materials for the preparation of carbon adsorbents. As shown in Table 1.2, there have been over 40 kinds of biomass employed as precursors for porous carbon adsorbents, and most of them are obtained *via* chemical or physical activation, especially KOH activation. The specific surface areas and pore volumes of the biomass-based carbons were tuned with controlled carbonization and/or activation processes. The biomass used previously could be roughly divided into three categories: (1) polysaccharides, such as sucrose,[44] cellulose,[45] chitosan,[46] starch,[47,48] *etc.*, (2) raw biomass or waste biomass, such as nutshells,[49,50] wood residues,[51] Jujun grass,[52] *Enteromorpha prolifera*,[53] sugar cane bagasse,[54] granular bamboo,[55] leaves,[56] and others, and (3) microorganisms, including fungi[57] and yeast.[58]

Polysaccharides could be converted into carbonaceous materials *via* a hydrothermal treatment approach using water as the carbonization medium under a self-generated pressure. In 2011, Sevilla and Fuertes reported a series of porous carbons prepared by the KOH activation of hydrothermally carbonized polysaccharides (starch and cellulose) and biomass (sawdust).[48]

Table 1.2 Biomass-based porous carbons prepared by KOH activation for CO_2 capture.

Biomass	Activating agents	Surface area ($m^2 g^{-1}$)	Pore volume ($cm^3 g^{-1}$)	Doping	CO_2 capacity at 25 °C and 1 bar ($mmol g^{-1}$)	Reference
Sugarcane bagasse	None	388			1.67	54
Palm shell	None	753	—		2.3	60
Chitosan–poly-benzoxazine	None	679	0.37		5.72	66
Carboxymethyl-cellulose	Hydro-thermal	1063	0.70	N-doped	2.84	67
Palm shell	KOH	1890	0.82		4.40	61
Peanut shell	KOH	1790	0.77	N-doped	1.54[a]	50
Sawdust hydrochar	KOH	2370	1.15		4.80	48
Celtuce leaves	KOH	3404	1.88		4.36	56
Bamboo	KOH	2332	1.00		4.50	55
Enteromorpha prolifera	KOH	481	0.37	N-doped	1.40[b]	53
Coffee ground	KOH	2785	1.36		2.81	68
Chicken feather	KOH	1610	0.86	N-doped	6.50[c]	69
Coconut shell	KOH	1535	0.60	N-doped	5.00	49
Jujun grass	KOH	1512	0.74		5.00	52
Gelatin	KOH	1636	0.51	N-doped	3.80	47
Wheat flour	KOH	1438	0.65		3.48	59
Pine cone shell	KOH	1033			4.80	70
Popcorn	KOH	1489	0.71		4.60	71
Microalgae	KOH	1745	0.89	N-doped	4.30	64
Broussonetia papyrifera bark	KOH	1759	0.92		4.45	72
Coconut shell	KOH	1593	0.68	N-doped	4.47	62
Burnt wood	KOH	1554	0.7		5.00	73
Oil palm shell	KOH	1630	0.92		2.70	74
Black locust	KOH	2511	1.16	N-doped	5.05	75
Fish scales	KOH	3206	2.29	N-doped	3.89	76
Fungi	KOH	2264	0.92		3.30	57
Yeast	KOH	1348	0.67	N-doped	4.77	58
Sucrose	KOH	1745	0.9	N-doped	4.30	44
Chitosan	KOH	970	0.46	N-doped	5.00	77
Granular bamboo	KOH	2641	1.25		4.5	55
Cellulose	NaOH	615	0.64	N-doped	4.99	45
Chitosan	K_2CO_3	1381	0.57	N-doped	3.86	46
Corncobs	NH_3	1154	0.57	N-doped	2.81	78
Palm shell	NH_3	889	0.47	N-doped	1.70	79
Banana peel	CO_2	1426	0.83	N-doped	2.70	80
Coconut shell	CO_2	1327	0.65		3.90	81
Olive stones	CO_2	1215	0.51		3.10	82
Grass cuttings	CO_2	841	0.38		1.45[d]	83
Cotton stalk	Steam	439	0.19	N-doped	2.25	84

[a]At 25 °C and 0.15 bar.
[b]At 2 °C, N_2–CO_2 = 85 : 15.
[c]At 0 °C and 1 bar.
[d]At 0 °C and 0.1 bar.

The prepared porous carbons were used as CO_2 adsorbents. Although the porous carbons prepared at a ratio of KOH–precursor = 2 have a considerably low-level pore development than those obtained at a ratio of KOH–precursor = 4, they exhibited significantly better CO_2 capture capacities. This is mainly due to the presence of a larger number of narrow micropores (<1 nm) in the mildly activated carbons. The sawdust-based carbon activated at 600 °C using KOH–precursor = 2 gave the highest CO_2 uptake of 4.8 mmol g^{-1} (21.2 wt%) at 25 °C and 1 atm, which is still among the highest ever reported for activated carbons now.

Microporous carbons were prepared from wheat flour (mainly starch) *via* a combined process of pre-carbonization and post-KOH activation.[59] The pore texture of the wheat-flour-based carbons was significantly improved by the post-chemical activation of KOH and varied with the KOH–carbon ratio. By increasing the KOH–carbon ratio up to 3, narrow micropores smaller than 0.8 nm were developed primarily, and the corresponding activated carbon possessed a moderate surface area but the highest volume of pores smaller than 0.8 nm, thus delivering the highest CO_2 adsorption capacities of 5.70 and 3.48 mmol g^{-1} at 0 and 25 °C, respectively. The experimental results confirmed that CO_2 adsorption uptake at ambient conditions was significantly dependent on the volume of narrow micropores with a pore size of less than 0.8 nm rather than the total volume or specific surface area.

Biomass waste is not only abundant, and sustainably renewable, but is also more cost-effective as carbon precursors. A wide range of biomass waste has been used to prepare porous carbon materials and showed excellent CO_2 sorption capacities. Shell is one of them. As early as 2004, Tan and Ani reported carbon molecular sieves by direct carbonization of palm shell waste at 600–1000 °C.[60] When the material was pyrolyzed at 600 °C, the micropore surface area was 753 m^2 g^{-1}; it showed CO_2 adsorption near 2.3 mmol g^{-1} at 25 °C. When the pyrolysis temperature increased to 1000 °C, the micropore surface area and the pore volume decreased; however, the selectivity toward CO_2 increased.[60] After further KOH activation at 600 °C, the CO_2 uptake of palm-shell-based porous carbon significantly increased from 2.7 mmol g^{-1} to 4.4 mmol g^{-1} at 25 °C and 1 bar, due to the much more developed microporosity of the activated carbons compared to the non-activated ones.[61] Hu and his co-workers prepared two series of nitrogen-doped porous carbons from coconut shells.[50,62] One type was prepared by urea modification and KOH activation, and the other was prepared by a successive process of pre-oxidization by H_2O_2, ammoxidation, and KOH activation. These carbons were found to exhibit very high CO_2 uptakes at 1 bar, almost 5.0 mmol g^{-1} and 4.47 mmol g^{-1} at 25 °C, 1 bar, respectively. The pre-oxidization by H_2O_2 increased oxygen-containing surface groups, which were expected to increase the amount of nitrogen incorporated into the carbon product in the subsequent ammoxidation process. The resulting carbons possessed a much higher nitrogen content and a narrower microporosity than the control sample without H_2O_2

pre-treatment. These results made coconut shell an attractive precursor for carbon adsorbents.

Shahkarami *et al.* performed a comprehensive study on the effects of the nature of precursors and carbonization conditions on the physical and chemical properties of activated carbon and the influence of these factors on CO_2 adsorption capacity, selectivity, and stability of the produced activated carbon in a fixed-bed reactor and mixed feed stream of $N_2/CO_2/O_2$.[63] In that work, three types of abundant feedstocks were used: agricultural waste (wheat straw and flax straw), forest residue (sawdust and willow ring), and animal manure (poultry litter). The selected precursors were carbonized *via* both fast and slow pyrolysis processes and were converted to porous carbons after KOH activation. Slow-pyrolysis-based activated carbon had a lower surface area and total pore volume but higher CO_2 adsorption capacity in the presence of N_2. Sawdust-based activated carbon synthesized from the slow pyrolysis possessed the largest ultramicropore volume of 0.36 $cm^3\ g^{-1}$, and the highest CO_2 adsorption capacity (78.1 mg g^{-1}) in N_2 but low selectivity (2.8) over O_2 because of the oxygen functional groups on the surface. The ultramicropores and surface chemistry of adsorbents were found to be much more important than particle size, total pore volume, and internal surface area of the adsorbents.

Besides terrestrial raw biomass, marine raw biomass, such as microalgae[64] and *Enteromorpha prolifera*,[53] were also converted into carbons. For instance, a sugar-rich microalgae (*Chlorococcum* sp.) was used to prepare N-doped activated carbons through hydrothermal carbonization and a subsequent activation of KOH or NH_3.[64] Although the NH_3-activated carbon possessed a much higher nitrogen content than the KOH-activated ones, the latter exhibited higher CO_2 sorption capacities than the former, indicating the main contribution of narrow microporosity on the CO_2 capacities. Ma *et al.* have successfully prepared highly porous N-doped carbon monoliths by using binary H_3PO_4–HNO_3 mixed acid as a co-activating agent and sodium alginate, a marine biopolymer, as a carbon precursor.[65] The SA-2N–P carbon prepared at a volume ratio of HNO_3–H_3PO_4 = 2 showed high CO_2 adsorption capacities of 8.99 mmol g^{-1} at 0 °C and 4.57 mmol g^{-1} at 25 °C, along with a high CO_2 capacity of 1.51 mmol g^{-1} at 25 °C and 0.15 bar.

Microorganisms are another promising kind of carbon precursor. Wang *et al.* reported a series of porous carbons with narrow microporosities by KOH activation of pre-carbonized fungi (Agaricus).[57] A moderate CO_2 uptake of 5.5 mmol g^{-1} and a high CO_2–N_2 selectivity of 27.3 at 0 °C, 1 bar were obtained. Similarly, Shen *et al.* found that yeast-based carbon was a promising adsorbent.[58] Hierarchical microporous carbon with a specific surface area of 1348 $m^2\ g^{-1}$ and a pore volume of 0.67 $cm^3\ g^{-1}$ was prepared by KOH activation of yeast. This type of carbon material showed a high CO_2 uptake of 4.77 mmol g^{-1} and a fast adsorption rate with an equilibrium time less than 10 min at 25 °C.

1.2.1.2 Other Chemicals as Activating Agents

Other activating agents, such as $ZnCl_2$,[90,91] NaOH,[85] K_2CO_3,[38,86] H_3PO_4,[92] $NaNH_2$,[94] *etc.* have also been used to prepare porous carbon for CO_2 capture. $ZnCl_2$ is the second most commonly used activating agent in chemical activation.[96–98] However, there are only a few reports on $ZnCl_2$ activated carbons for CO_2 capture, as shown in Table 1.3, which may be due to its relatively undeveloped porosity as a result of the low activating power of $ZnCl_2$.[90,91] For example, $ZnCl_2$-activated polypyrrole gave a specific surface area of 1283 $m^2\ g^{-1}$ and pore volume of 0.46 $cm^3\ g^{-1}$, much lower than those of KOH-activated polypyrrole (1700 $m^2\ g^{-1}$ and 0.88 $cm^3\ g^{-1}$).[20,91] The as-prepared N-doped carbon exhibited a moderate CO_2 uptake of 3.80 mmol g^{-1} at 25 °C and 1 bar when the activation temperature was 600 °C.[91] Although the $ZnCl_2$ activated carbons showed lower CO_2 uptakes compared to the KOH activated carbons, the $ZnCl_2$ activation usually gave a higher carbon yield and higher carbon density thus could exhibit some advantage in volumetric CO_2 capacities.

Using K_2CO_3 as an activating agent, Fan *et al.* activated chitosan into N-doped microporous carbon.[46] By changing the weight ratio of K_2CO_3–chitosan and the activation temperature, the porosity and nitrogen content

Table 1.3 Activated carbons prepared by other chemical activations.

Precursor	Activating agents	S_{BET} ($m^2\ g^{-1}$)	Pore volume ($cm^3\ g^{-1}$)	CO_2 uptake[a] (mmol g^{-1})	CO_2–N_2 selectivity	Reference
Polypyrrole	NaOH	1005	0.53	3.73	—	85
Polyaniline	K_2CO_3	3670	1.70	7.60[d]	—	86
Resorcinol resin	Potassium oxalate	2130	1.10	4.7	—	87
m-Aminophenol–formaldehyde resin	K^+	664	—	3.90	50[c]	88
Carboxylic phenolic resins	Cs^+	1312	0.67	5.20	18[c]	89
Polyfuran	$ZnCl_2$	1123	0.64	3.46	27.8[b]	90
Polypyrrole	$ZnCl_2$	1283	0.46	3.80	—	91
Eucalyptus wood	H_3PO_4	2079	1.29	3.22	—	92
Pine cone	H_3PO_4	1470	—	1.91	—	93
Mesoporous carbon	$NaNH_2$	1917	1	3.66	16.2[c]	94
RF polymer	NH_3	1458	0.65	4.54	—	95

[a]CO_2 capacities at 25 °C and 1 bar.
[b]Selectivity based on the ideal adsorbed solution theory (IAST).
[c]The selectivity of Henry's law.
[d]CO_2 capacity at 0 °C and 1 bar.

of the prepared carbons could be tuned in the range of 1180–2567 $m^2 g^{-1}$ and 1.29–6.02 wt%, respectively. The sample prepared at 635 °C with a K_2CO_3-chitosan ratio of 2 showed a CO_2 uptake of 3.86 mmol g^{-1} at 25 °C, 1 atm, five consecutive recyclabilities, and a good CO_2–N_2 selectivity of *ca.* 21. Silvestre-Albero prepared a series of activated carbons by pre-carbonization of PANI at different temperatures and post-activation of KOH or K_2CO_3.[86] They studied the activating effects of KOH and K_2CO_3. It was found that carbonization temperature significantly influenced the porosity of the prepared carbons when using KOH as an activating agent, while K_2CO_3 mainly produced microporosity, independent of the carbonization temperature. The highest CO_2 uptake of these carbons was 7.60 mmol g^{-1} at 1 bar and 0 °C.

In the last decade, ammonia (NH_3) treatment has been used to introduce nitrogen surface groups into the carbon framework. Several authors have reported ammonia-treated porous carbons for CO_2 capture.[79,99] Pevida observed that nitrogen groups were successfully introduced into carbon frameworks after a NH_3 treatment, and the nitrogen content was proportional to the oxygen content of the pristine porous carbons.[99] Due to the incorporated basic nitrogen groups, the ammoxidized carbons showed enhanced CO_2 uptakes at high adsorption temperatures. However, the low reaction efficiency between the NH_3 and the carbon resulted in a relatively low N-doping level. Recently, Geng *et al.* reported a NH_3-assisted activation process in which NH_3 played the roles of activating agent and nitrogen source at the same time.[78] When carbonizing a corncob in a NH_3 atmosphere, the nitrogen could be easily incorporated into a carbon framework. The N content increased as the activation temperature increased, reaching a high level of ~12 wt% at 800 °C, along with an increase in the specific surface area and pore volume. Especially, the nitrogen was mainly incorporated in the form of phenyl amine and pyridinic N groups, which were very efficient for CO_2 capture. These carbons showed a moderate CO_2 uptake of 2.81 mmol g^{-1}, but superior IAST CO_2–N_2 selectivity up to 82. Similarly, Hu *et al.* prepared a N-doped hierarchical porous carbon by carbonization of a cellulose aerogel under a NH_3 atmosphere.[45] This N-doped carbon aerogel exhibited a N-doping of 4.62 wt%, and a high CO_2 adsorption capacity of 4.99 mmol g^{-1} at 25 °C and 1 atm.

Most of the other activating agents usually show a lower activating ability than KOH. For example, KOH-activated polypyrrole carbons show a much higher specific surface area than the NaOH-activated ones under the same activation conditions (*e.g.*, 2940 $m^2 g^{-1}$ *vs.* 1453 $m^2 g^{-1}$, activation conditions: the weight ratio activating reagent–polypyrrole = 2, and activation temperature = 700 °C).[20,85] Interestingly, $NaNH_2$, a strong base commonly used in organic synthesis, was reported to exhibit a more powerful activation ability compared to NaOH and KOH. At an activating reagent–carbon weight ratio of 2 and an activation temperature of 550 °C, $NaNH_2$ could activate a much higher microporosity, especially more small micropores, than KOH and NaOH, resulting in a higher CO_2 uptake of 3.66 mmol g^{-1} at 25 °C and 1 bar, verifying the superiority of $NaNH_2$ in the activation under relatively moderate conditions.[94]

1.2.2 Physical Activation

The so-called physical activation is the partial gasification of the carbon framework with CO_2, steam, and air, or a combination of these, at high temperatures (from 700 to 1200 °C) as shown in eqn (1.9) to (1.12). In view of porosity development, the most important variables in the gasification process are the activating agents, the final burn-off ratio, and the inorganic impurities. Rodríguez-Reinoso and Molina-Sabio comprehensively studied the evolution in the porosity of several series of activated carbons prepared by physical activation of lignocellulosic materials (uncatalyzed and iron-catalyzed) in CO_2 or in a water–nitrogen mixture.[100] They found that at the initial stage of CO_2 activation, the micropore and macropore volume increase coincided with the proceeding burn-off. Further burn-off with CO_2 opened and enlarged the micropores of the char with even a shift to meso- and macropores, the ablation of the exterior of the particle being very important at high burn-offs over 50%.

The final activated carbon had a well-developed micro- and macroporosity, with a relatively small portion of mesoporosity. The Fe-catalyzed CO_2 gasification was initially very fast even at a lower gasification temperature and declined at an increasing burn-off due to the deactivation of the Fe catalyst. Steam activation produced a more selective attack of the carbon framework and a more uniform widening of porosity. It developed a large number of micropores, but fewer macropores, compared to CO_2 activation.

$$C + CO_2 \rightarrow 2CO \tag{1.9}$$

$$C + H_2O \rightarrow CO + H_2 \tag{1.10}$$

$$C + O_2 \rightarrow CO \tag{1.11}$$

$$C + O_2 \rightarrow CO_2 \tag{1.12}$$

1.2.2.1 CO_2 as an Activating Agent

Mietek Jaroniec *et al.* prepared a series of porous carbon spheres by a combination of pre-carbonization and subsequent CO_2 activation of phenolic resin spheres obtained by a modified Stöber method.[101] The activated carbon spheres had diameters from 200 to 420 nm, a high surface area (from 730 to 2930 $m^2\ g^{-1}$), abundant narrow micropores (<1 nm, pore volume of 0.28 to 1.12 $cm^3\ g^{-1}$), thus exhibited very high CO_2 adsorption capacities: 8.05 and 4.55 mmol g^{-1} at 1 bar and two temperatures, 0 and 25 °C, respectively.[101] They further reported a series of N-doped activated carbon spheres.[102] These materials were prepared by one-pot hydrothermal synthesis in the presence of resorcinol-formaldehyde resin as a carbon precursor and ethylenediamine (EDA) as both a base catalyst and nitrogen source, followed by carbonization in a N_2 atmosphere and CO_2 activation. The N-doping level and the particle size can be tuned by varying the EDA amount in the range of 2.80–7.20% and 50–1200 nm, respectively. These activated carbon spheres exhibited a high surface area and volume of micropores smaller than 1 nm, reaching up to 1224 $m^2 g^{-1}$ and 0.36 $cm^3 g^{-1}$, and could capture 6.2 and 4.1 mmol g^{-1} of CO_2 at 0 °C and 25 °C. Obviously, the CO_2 uptake of the N-doped CO_2-activated carbon sphere was lower than those of non-doped or KOH-activated ones[101] due its relative lower surface area and ultramicropores, indicating the primary importance of small micropores.

Two biomass materials – olive stones and almond shells – have been directly activated by CO_2 to prepare porous carbon adsorbents for CO_2 capture.[82] The effect of holding time on overall yield, pore texture, and CO_2 uptake, was studied. As the holding time was increased, the overall yield decreased and more porosity was developed, thus resulting in an increase of CO_2 uptake. The olive-stone-based carbons showed a higher specific surface area and pore volume than the almond-shell-based ones, indicating that the olive stone was more easily activated by CO_2. These carbons showed a CO_2 uptake up to 3.10 mmol g^{-1} at 120 kPa and 25 °C, and 0.6–1.1 mmol g^{-1} at 15 kPa and 25–50 °C.

1.2.2.2 Steam as an Activating Agent

Steam was also used to activate both synthetic polymers and biomass into highly porous carbons. N-doped activated carbons were prepared by steam activation of melamine-modified phenol-formaldehyde resins.[103] As the activation temperature increased, the specific surface area significantly increased from 382 to 1439 $m^2 g^{-1}$. The carbon activated at 850 °C could adsorb 6.71 mmol g^{-1} CO_2 at 0 °C and 1 atm, and released 93.6% of the CO_2 adsorbed when the temperature increased to 50 °C. Young-Jung Heo and Soo-Jin Park prepared cellulose-fiber-based ultramicroporous carbons by steam activation.[104] The steam activation was found to have a strong influence on the development of new pores and the expansion of pore sizes and to be effective in developing optimal micropores for CO_2 adsorption. The porous carbon prepared by pre-carbonization at 800 °C and further activation with steam exhibited a high CO_2 adsorption capacity of 3.78 mmol g^{-1} at 25 °C and 1 bar, as well as an impressive Henry's law selectivity of 47.1 for CO_2 over N_2.

In short, researchers have synthesized a large number of porous carbon materials from a wide range of raw materials and activating agents for CO_2 capture. The porosity and chemical surface properties varied with factors including raw materials, the pre-treatment process, activating agents, and activation temperature, *etc.* Particularly, many biomass materials, classified as polysaccharides, raw biomass and microorganisms here, are employed to successfully fabricate porous carbon adsorbents. Biomass-based carbon materials have shown high performances in CO_2 capture and will play a more important role in post-combustion CO_2 capture in the future. However, considering the huge emissions of carbon dioxide, there will be an enormous demand for solid CO_2 sorbents, and biomass materials that are easy to mass-produce and are widely-collected should be the focus of studies. KOH is still the most common activating agent, but it is highly corrosive. Additionally, the KOH dosage in the current literature is too high (*e.g.*, 200 wt% of the carbon precursor). Therefore, the development of new activating methods is also needed.

1.2.3 Metal Ion Activation

It has been widely accepted that the small micropores in carbon frameworks, especially ultramicropores smaller than 0.8 nm, are the most efficient for CO_2 capture.[105–107] Chemical or physical activation methods are still widely used in the preparation of highly microporous carbon materials. In order to get a high capture capacity of CO_2, severe activation conditions, like excess activating agents and high burn-offs, are usually employed to increase the ultramicroporosity. In this way, an inhomogeneous activation is inevitable, resulting in a wide pore size distribution (PSD), low carbon yield and enhanced cost. Most activated carbon materials show a wide pore size distribution (PSD) and contain substantial supermicropores (larger than 1.0 nm), resulting in low CO_2 uptake, especially at a low CO_2 pressure (*e.g.*, CO_2 capture from flue gas). For instance, Zhao *et al.* have reported that porous carbon prepared by the direct carbonization of poly(vinylidene chloride) gave a slightly higher CO_2 uptake than the further KOH-activated ones at 25 °C and 1 bar, because KOH activation generated some large micropores or small mesopores in the carbon framework, and then resulted in the reduction of interactions between CO_2 molecules and the pore walls.[29]

Ultramicroporosity is desired for a high CO_2 capture performance. Recently, we reported a new activating strategy for strictly ultramicroporous carbons by using phenolic resins as platforms, that could be called "single-ion activation".[49,50,108] In this process, mono-dispersed alkali ions are introduced *via* an acid–base reaction between an alkali hydroxide and the acidic groups (–OH or –COOH) of phenolic resins (Figure 1.2). After a direct carbonization of the prepared phenolic salts of resin, strictly ultramicroporous carbons are obtained. The mono-dispersed ions in the form of $-O^-M^+$ or $-COO^-M^+$ serve as activating agents, and produce a homogeneous activation effect, resulting in highly developed and narrowly-distributed ultramicropores.

This strategy was first used to prepare N-doped microporous carbons (NMCs).[88] As shown in Figure 1.2, potassium phenolate of *meta*-aminophenol-formaldehyde resin was firstly prepared by the aqueous poly-condensation of *meta*-aminophenol and formaldehyde using equimolar KOH as both a catalyst precursor and a base. During the polycondensation reaction, the K^+ ion can exchange with H^+ in the hydroxyl groups, leading to the single

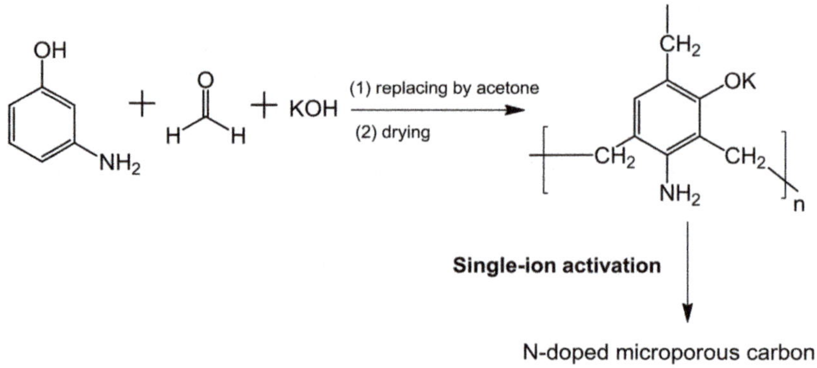

Figure 1.2 An illustration of single-ion activation.

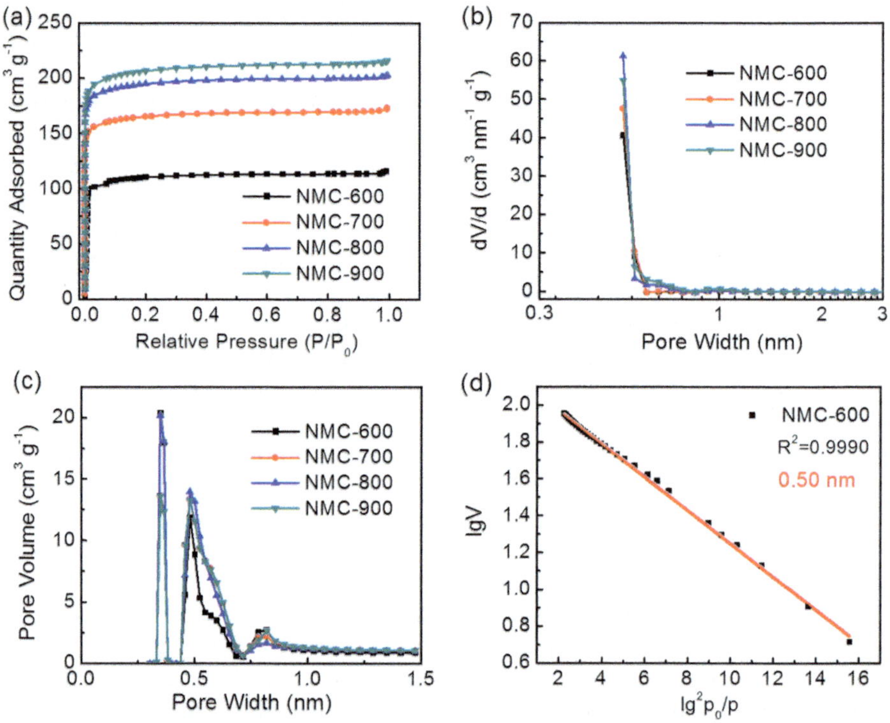

Figure 1.3 (a) Nitrogen sorption isotherms, (b) QSDFT pore size distributions derived from N_2 sorption, (c) NLDFT pore size distributions derived from CO_2 sorption, (d) D–R plots for CO_2 sorption by NMC-600. Reproduced from ref. 88 with permission from the Royal Society of Chemistry.

dispersion of K^+ ions in the bulk of the phenolic resin. After a simple carbonization process, N-doped ultramicroporous carbons were obtained.

From Figure 1.3, it could be seen that all the NMC samples exhibited a standard sorption isotherm of type I with a very narrow knee at a very low relative pressure (p/p^0 < 0.02) and an absence of apparent adsorption increment in the relative pressures over 0.02, indicating that the porous carbons had a narrow microporosity and that no mesoporosity was present. The specific surface area of NMCs gradually increased from 272 $m^2\ g^{-1}$ to 664 $m^2\ g^{-1}$ as the activation temperature increased from 600 °C to 900 °C. Dubinin–Radushkevich (D–R) plots of CO_2 sorption on NMCs exhibited well-defined linear shapes, showing that the porosity of NMCs was strictly made up of uniform ultramicropres.[107,109] The average sizes of these ultramicropores calculated according to the slope of the D–R plots gradually increased from 0.50 to 0.58 nm as the carbonization temperature increased from 600 °C to 900 °C.

The carbon prepared at 600 °C (NMC-600) exhibited a high CO_2 uptake of 3.90 mmol g^{-1} at 25 °C and 1 bar, although this carbon possessed a very low specific surface area of 272 $m^2\ g^{-1}$. Besides, this carbon gave an outstanding CO_2 uptake of 1.67 mmol g^{-1} mmol g^{-1} at 25 °C and 0.15 bar, and an impressive CO_2-over-N_2 selectivity of 50 based on Henry's law. At 75 °C and 1 bar, the carbon prepared at 800 °C (NMC-800) still showed a very high CO_2 uptake of 2.0 mmol g^{-1}. The excellent performance of NMCs for CO_2 capture indicated that these carbons preferably interact with CO_2 molecules based on the combined contributions of the extremely small micropores *ca.* 0.5 nm and the polar surface caused by the N, O-doping.

The strategy of single-ion activation could be applied to finely tune the sizes of ultramicropores by using different alkali metal ions (Li^+, Na^+, K^+, Rb^+, Cs^+) as activating agents (Figure 1.4).[89] All the samples presented a

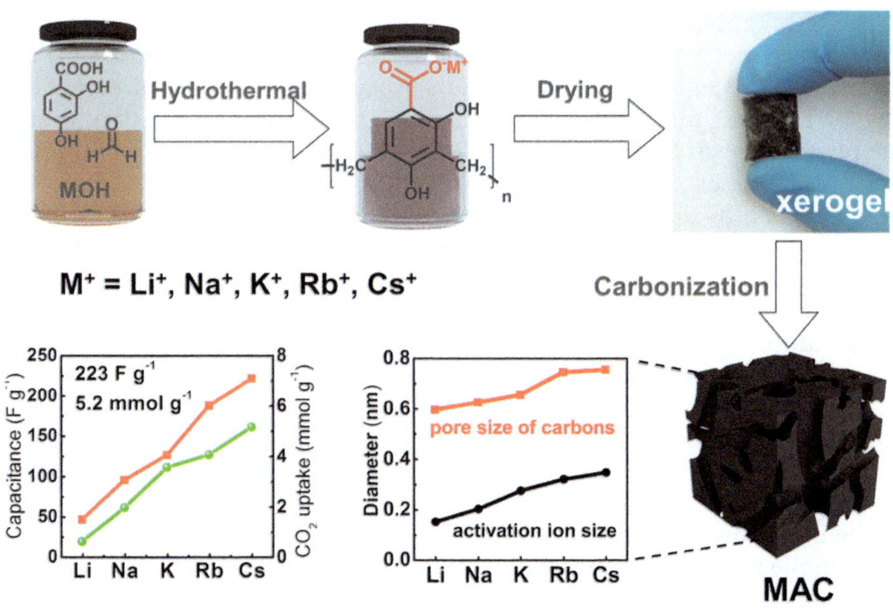

Figure 1.4 Schematic diagram of the synthesis of MAC. Reproduced from ref. 89, with permission from John Wiley and Sons, Copyright 2016 Wiley-VCH Verlag GmbH & Co. KGaA, Weinheim.

standard type I N_2 adsorption isotherm, showing a narrow micropore size distribution. The pore volume and the apparent surface areas gradually increased from 0.07 to 0.70 $cm^3\ g^{-1}$ and 111 to 1312 $m^2\ g^{-1}$, respectively, while varying the activating ion from Li^+ to Cs^+, thereby revealing that the activation power significantly increased from Li^+ to Cs^+. All the samples, except Li^+-activated carbon (LiAC), exhibited well-defined linear D–R plots (correlation coefficient $R^2 > 0.995$), indicating the existence of strictly uniform ultramicropores (Figure 1.5). Interestingly, a systematic widening of micropores took place; the pore size gradually increased from 0.60 nm up to 0.76 nm as the activation ions varied from Li^+ to Cs^+. In other words, the ultramicropore size was finely tuned at sub-angstrom level by simply varying the activating ions.

When applying for CO_2 capture, the Cs^+-activated carbon (CsAC) showed a superior CO_2 adsorption capacity of 5.20 mmol g^{-1} at 25 °C and 1.0 bar, high CO_2-to-N_2 selectivity and full regenerability for four consecutive cycles (Figure 1.6). Furthermore, the resulting carbons present ultrahigh capacitances of up to 223 F g^{-1} or 205 F cm^{-3} in an ionic liquid electrolyte. The outstanding performance of these carbons is due to their uniform pore size, which is finely tuned to adapt to the dimensions of the CO_2 molecule and the ions of ionic liquids. The CO_2 capacities and specific capacitances reported here were higher than for most traditional activated carbons. More importantly, the dosage of activation agents used herein was only 30 wt% of the carbon precursors as counted in KOH, much lower than that used in traditional chemical activation,[89] confirming the high efficiency of metal ion activation.

The mechanism of metal ion activation was systematically studied by thermogravimetry-mass spectrometry (TG-MS) analysis and *in-situ* X-ray diffraction (XRD) (Figure 1.7). The pyrolysis process of the potassium salt of carboxylic phenolic resin (PR-COOK) was much more complicated than that of the phenolic resin carboxylic acid (PR-COOH) due to the activation of metal ions. Three gases, including H_2O (*m*/*z* 18), CO (*m*/*z* 28), and CO_2 (*m*/*z* 44), were detected by mass spectrometry. *In-situ* XRD patterns of PR-COOK indicated that the PR-COOK began to decompose into K_2CO_3 (PDF card 71-1466) at around 200 °C, and then transformed into K_2O at 600 °C, along with two steps of weight loss and the release of CO_2 and CO. When the carbonization temperature increased to 800 °C, new peaks at 10.4° and 16.5° appeared, which is attributed to K (100) (PDF card 01-0500) and KC_8 (100) (PDF card 04-0221), and large amounts of CO were detected (Figure 1.7). These facts are strong evidence that metallic potassium was produced by the reduction of K_2O and was further intercalated into the carbon layers to form graphite intercalation-like compounds.

The mechanism of single-ion activation was suggested as: first, alkali salts of phenolic resins decomposed into alkali carbonate (M_2CO_3), H_2O and CO_2 below 400 °C. Second, alkali oxides (M_2O) were produced by the thermal decomposition of M_2CO_3 or the redox reaction between M_2CO_3 and C. Third, framework carbon atoms were etched by M_2O *via* the vigorous redox reaction

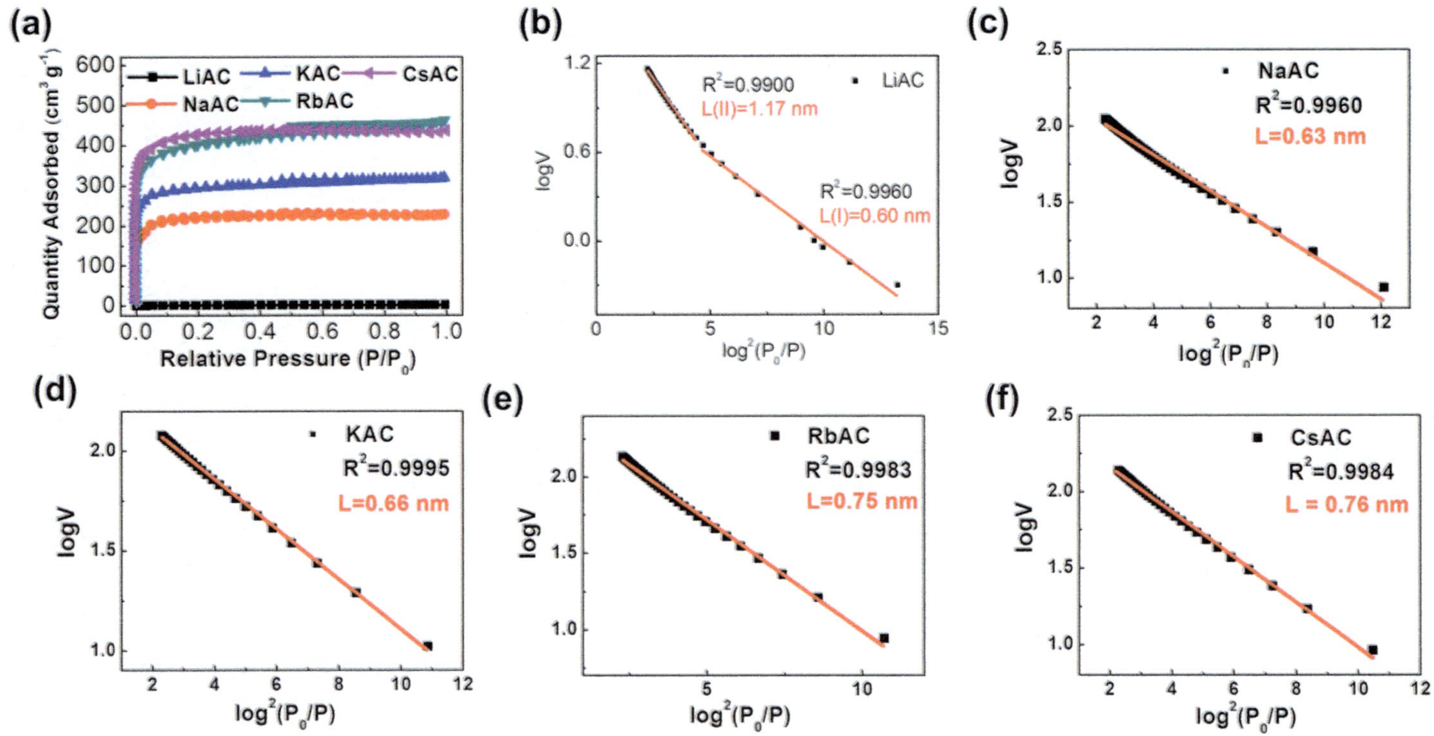

Figure 1.5 (a) Nitrogen adsorption isotherms; analysis of the narrow microporosity by Dubinin–Radushkevich equations for: (b) LiAC, (c) NaAC, (d) KAC, (e) RbAC and (f) CsAC. Adapted from ref. 89, with permission from John Wiley and Sons, Copyright 2016 Wiley-VCH Verlag GmbH & Co. KGaA, Weinheim.

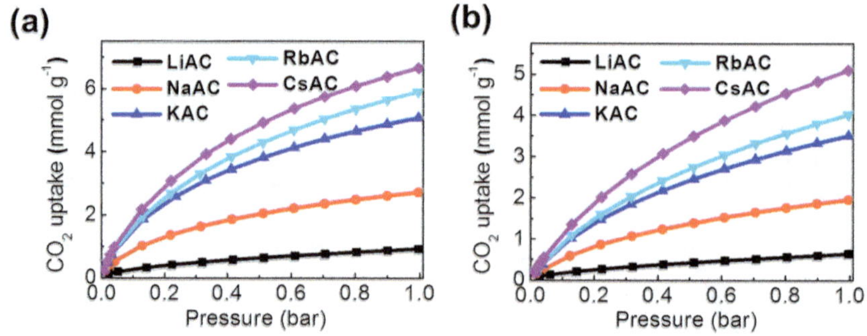

Figure 1.6 (a) CO_2 adsorption isotherms at 0 °C and 1.0 bar; (b) CO_2 adsorption isotherms at 25 °C and 1.0 bar. Adapted from ref. 89, with permission from John Wiley and Sons, Copyright 2016 Wiley-VCH Verlag GmbH & Co. KGaA, Weinheim.

of $M_2O + C \rightarrow 2M + CO$, then the produced metallic alkali intercalated into the lattices of the carbon matrix. The interlayer spacing will be primarily determined by the size of the metal intercalate, and this spacing (*i.e.*, pore size) will be systematically widened with increasing metal ion size from Li^+ to Cs^+. In a word, the activation is similar to traditional KOH activation, but much more efficient.

This strategy was further applied for the synthesis of microporous carbon spheres for the following reasons: (1) the developed ultramicroporosity offered a large space for CO_2 adsorption; (2) the small size of the carbon spheres ensured a short diffusion distance of the CO_2 molecule; (3) the gaps between carbon spheres allowed a rapid flow of flue gas.[110] In that work, potassium salts of resorcinol-formaldehyde (RF) resin spheres were prepared by the reaction of KOH and pristine RF resin spheres, and followed by carbonization into microporous carbon materials (denoted as CS-x, where x stands for the weight ratio of KOH–RF) (Figure 1.8). When the prepared resins were mixed with KOH, the –OH groups of the RF resins immediately reacted with KOH *via* a fast acid–base reaction and were changed into mono-dispersed K^+ in the form of –OK groups. Based on a simplified molecular structure of RF resins, the KOH–RF resins' weight ratios of 0.50, 0.75, 1, 1.5 and 2 are roughly equivalent to the KOH–OH molar ratios of 0.50, 0.75, 1, 1.5 and 2. In the cases of CS-0.5, CS-0.75 and CS-0.75, the activating agents exclusively existed as –OK groups, while in the cases of CS-1.5 and CS-2, the KOH was in excess (Figure 1.8).

After activation, the majority of the carbon particles retained, roughly, their spherical shape even when the weight ratio of KOH to RF resins was up to 2 (Figure 1.9). The detailed microscopy morphology and the porosity of the prepared carbon materials were closely related to KOH dosage (Figures 1.9 and 1.10). The CS-0.5, CS-0.75 and CS-1 samples showed regular spherical shapes, a smooth carbon surface, and an exclusively narrow microporosity (<0.8 nm), indicating that the K^+ activation almost occurred in the interiors

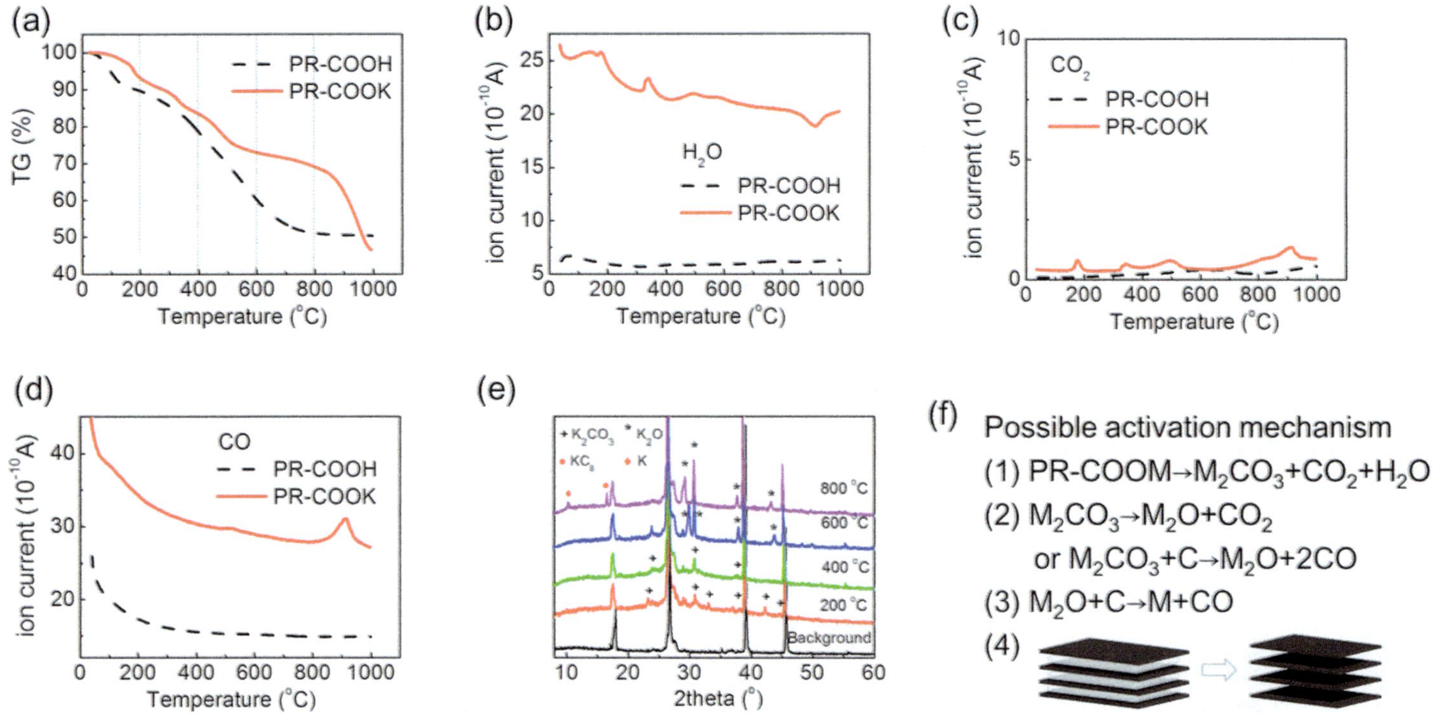

Figure 1.7 (a) TG curves, MS responses of evolved gases in TG-MS analysis, (b) H_2O, (c) CO_2, and (d) CO, (e) *in-situ* XRD patterns (CuKa) during the carbonization of PR-COOK, and (f) possible activation mechanism. Reprinted from ref. 89, with permission from John Wiley and Sons, Copyright 2016 Wiley-VCH Verlag GmbH & Co. KGaA, Weinheim.

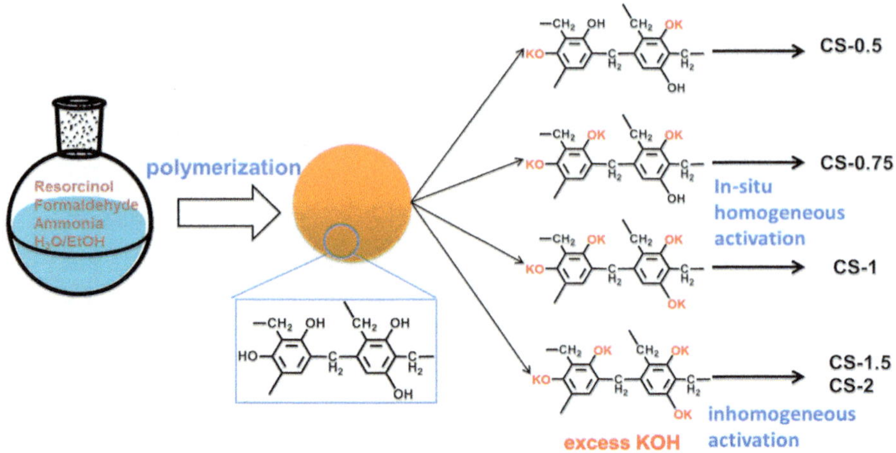

Figure 1.8 Illustration for preparation of the porous carbon spheres. Reprinted from X. Wang, J. Zhou, W. Xing, B. Liu, J. Zhang, H. Lin, H. Cui and S. Zhuo, Resorcinol–formaldehyde resin-based porous carbon spheres with high CO_2 capture capacities, *J. Energy Chem.*, **26**, 1007–1013, Copyright 2017, with permission from Elsevier.

of the carbon mono-spheres. In comparison, some defects and roughness were observed on the surface of the CS-1.5 and CS-2 carbon particles, and a widening of PSD took place, indicating a non-homogeneous activation. As the KOH dosage increases, the surface areas and volumes of the micropores first increase and then decrease, and CS-1 possesses the largest specific surface area, micropore surface area and volume, up to 1235 $m^2\ g^{-1}$, 1084 $m^2\ g^{-1}$ and 0.57 $cm^3\ g^{-1}$, respectively.

The as-prepared carbon spheres exhibited excellent CO_2 capture capacities ranging from 6.55–7.34 mmol g^{-1} at 0 °C and 4.27–4.83 mmol g^{-1} at 25 °C and 1 bar, and the highest values belonged to CS-1, which is attributed to its largest micropore volume with narrow micropore size. The value of 4.83 mmol g^{-1} is higher than those for lots of phenolic-resin-based porous carbon materials under the same conditions, including porous carbon spheres prepared by traditional chemical or physical activation, such as CO_2-activated carbon spheres,[101,107] N-rich microporous carbon spheres,[102] and KOH-activated carbon spheres.[27]

Overall, single-ion activation has been used to develop and adjust ultramicroporosity. This strategy has exhibited many merits, such as a low demand for activating agents, a high activating efficiency, easy operation, and it reaches high CO_2 sorption capacities; it is worthy of further study.

1.2.4 Templating Method

Historically, the templating method was first reported by Knox and co-workers in 1986, who demonstrated the synthesis of graphitic porous carbons for liquid chromatography separation by impregnation of spherical porous

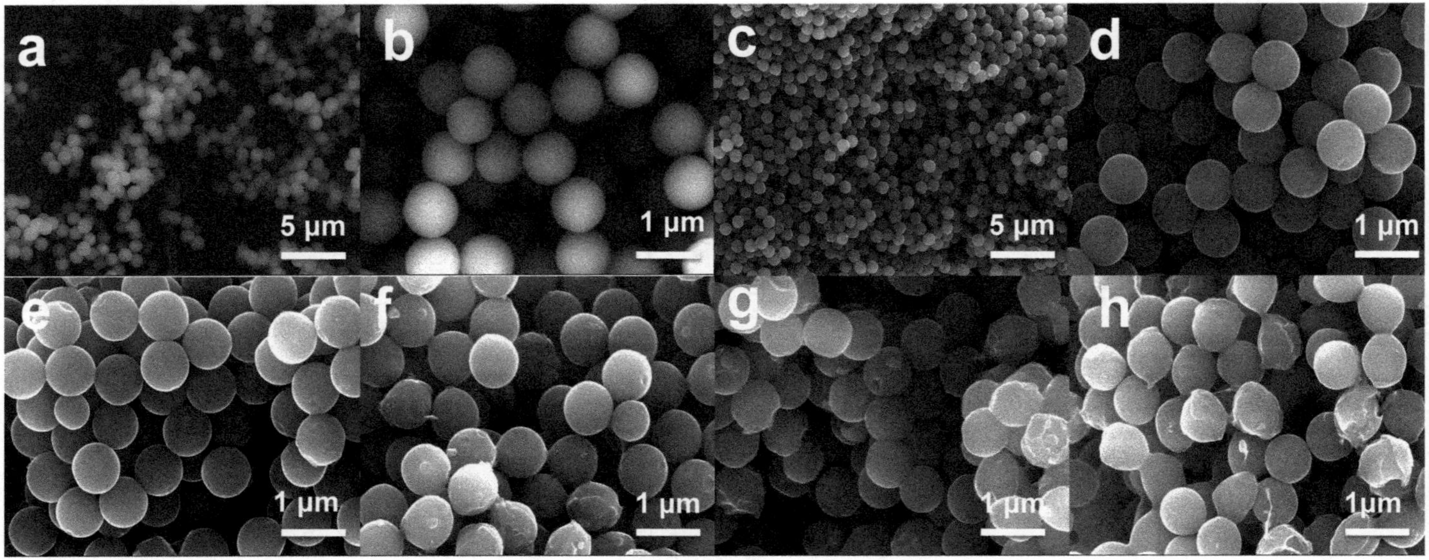

Figure 1.9 Microscopic morphology of CS-*x*. SEM images: (a, b) RF resin spheres, (c, d) CS-0.5, (e) CS-0.75, (f) CS-1, (g) CS-1.5 and (h) CS-2. Reprinted from X. Wang, J. Zhou, W. Xing, B. Liu, J. Zhang, H. Lin, H. Cui and S. Zhuo, Resorcinol–formaldehyde resin-based porous carbon spheres with high CO_2 capture capacities, *J. Energy Chem.*, **26**, 1007–1013, Copyright 2017, with permission from Elsevier.

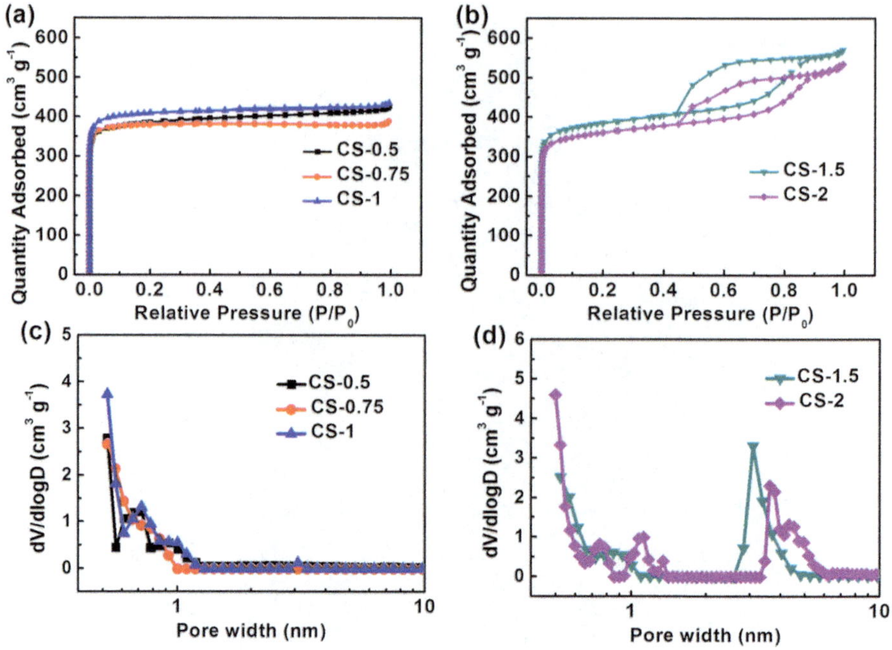

Figure 1.10 N_2 sorption measurements for CS-*x* materials (a, b) N_2 sorption isotherms, (c, d) PSDs plots calculated by QSDFT model. Reprinted from X. Wang, J. Zhou, W. Xing, B. Liu, J. Zhang, H. Lin, H. Cui and S. Zhuo, Resorcinol–formaldehyde resin-based porous carbon spheres with high CO_2 capture capacities, *J. Energy Chem.*, **26**, 1007–1013, Copyright 2017, with permission from Elsevier.

silica gel particles with phenolic resin and subsequent carbonization and removal of silica.[111] Since then, this method has received extensive attention and various types of template carbons are synthesized. The resulting carbon synthesized by the templating method, called a templated carbon, possesses a relatively narrow PSD and controlled architecture. The templated carbonization method permits one to control the carbon structure in terms of various aspects, such as pore structure, specific surface area, microscopic morphology and graphitizability, which makes this method very attractive.[112,113]

Hard templates and soft templates are the main kinds of templates used as molds to form templated carbons. The hard template method includes the following steps: (a) synthesis of a suitable porous template; (b) introduction of a suitable carbon precursor into the template pores using the method of wet impregnation, chemical vapor deposition or a combination of both methods; (c) polymerization and carbonization of the carbon precursor; and (d) removal of the inorganic template.[114,115] Following these steps, porous carbon with a specific pore structure is formed. Compared with the hard template, the soft template is a kind of surfactant, which has a strong interaction with the carbon source, and mesoporous carbons with different structures can be obtained through the soft template method. This method possesses good controllability and operability; as a result, it has very good application prospects. The mechanisms of the soft template method include a liquid crystal template mechanism, a synergistic assembly mechanism, a "rod micellar" mechanism and so on; these mechanisms have been widely recognized.[116]

Since the porous carbon replicates the morphology of the template, selecting a suitable template to synthesize the carbon with a specific porous structure is the most important step. Template carbons with different pore structures, such as ordered porous carbon, disordered porous carbon and hierarchical porous carbons, can be prepared by different templates, such as porous silica,[117–120] zeolites,[121] metal–organic frameworks (MOFs),[122] nanoparticles (*e.g.*, MgO[123] and SiO_2[124]), and surfactant micelles.[125–131] These templates will be described in detail in the following sections.

1.2.4.1 Porous Silica as a Hard Template

Mesoporous molecular sieve SBA-15 has a large surface area, uniform pore diameter distribution, adjustable pore size and wall thickness, and high hydrothermal stability. Therefore, SBA-15 is an excellent template to prepare ordered porous carbons.[120,132] Wang *et al.* designed a new method of using a one-step "surfactant-assisted" nanocasting route to synthesize highly ordered mesoporous graphitic carbon. They selected ordered mesoporous silica SBA-15 maintaining its triblock copolymer surfactant P123 ($EO_{20}PO_{70}EO_{20}$) as the hard template and natural organic soybean oil (SBO) as a carbon precursor. Finally, the ordered mesoporous graphitic carbon was synthesized and named as MGC-1 (Figure 1.11).[133] The main advantage of this method is that the carbon precursor SBO can easily infiltrate the

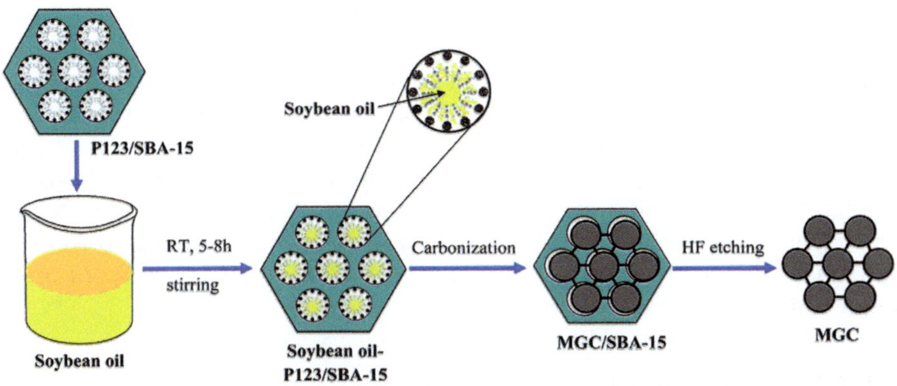

Figure 1.11 Schematic illustration of the synthesis process of ordered mesoporous graphitic carbon by a "surfactant-assisted" nanocasting route. Reproduced from ref. 133, http://dx.doi.org/10.1038/srep26673, © The Authors. Published under the terms of the CC BY 4.0 licence, https://creativecommons.org/licenses/by/4.0/.

mesopore channels of the silica template with the help of P123 to reach a high filling level, resulting in an enhanced yield of graphitic carbon materials. Because of the improved structural ordering, the mesoporous carbon, after amine modification, could adsorb more CO_2 (57.2 mg g^{-1}) compared with amine-functionalized carbon prepared without the assistance of surfactant (31.3 mg g^{-1}).[133]

In order to clarify the influence of porosity and surface chemistry on carbon dioxide capture, Sanchez–Sanchez *et al.* synthesized ordered mesoporous carbons with high surface areas and pore volumes by the hard template method, in which 3-aminobenzoic acid acted as both carbon precursor and nitrogen source, and SBA-15 acted as the hard template. The synthesis was accomplished in the presence of different H_3PO_4 concentrations. The surface chemistry and the porous texture of the carbons could be easily modulated by varying the H_3PO_4 concentration and carbonization temperature. Finally, they found that with H_3PO_4 concentrations higher than 50 wt%, the pyrolysis mechanism of the thermally polymerized precursor changed and both the structural order and mesopore arrangement declined. On the contrary, low H_3PO_4 concentrations led to mesoporous carbon materials with a very narrow PSD and favored nitrogen retention at the carbon surface. By comparing the amount of CO_2 adsorption of different samples under different conditions, they also proved that the CO_2 adsorption capacity depends on the narrow micropore volume and is almost unaffected by the surface chemistry at 0 °C and 1 bar; the largest CO_2 uptake of 5.04 mmol g^{-1} was obtained for the carbon with the most developed ultramicroporous structure. On the other hand, the nitrogen functional groups exerted a beneficial influence on CO_2 capture, while oxygen and phosphorus functionalities exerted a negative influence at 25 or 50 °C and 1 bar. Moreover, among the N-containing groups, the pyrrolic N groups provoked the largest improvements in CO_2 adsorption.[134]

In addition to SBA-15, there are also other porous silica templates used to prepare ordered mesoporous carbons. For instance, Tiwari prepared oxygen enriched carbon adsorbents with epoxy resin as the precursor and MCM-41 as a template.[135] The sample prepared at 700 °C shows the highest capacity of CO_2 adsorption of 0.65 mmol g^{-1} at 30 °C at a CO_2 concentration of 12.5%. Zhao *et al.* reported nitrogen-doped mesoporous carbons, which were prepared by a nanocasting route using tri-continuous mesoporous silica IBN-9 as a hard template and *p*-diaminobenzene as a nitrogen-containing precursor.[136] One of the carbons possessed a very high nitrogen doping concentration (13 wt%) and exhibited an excellent performance for CO_2 adsorption at about 4.50 mmol g^{-1} at 298 K and 1 bar.

1.2.4.2 Zeolite as a Hard Template

Zeolites are highly crystalline aluminosilicate materials that possess uniform sub-nanometer-sized pores. The pore channel apertures vary between 0.3 and 1.5 nm. As a result, zeolite-templated carbons (ZTCs) possess highly ordered microporosity and a large surface area, which is helpful in enhancing the capabilities of CO_2 capture. With these advantages, ZTCs are especially attractive and many groups have prepared different types of ordered microporous carbons using various zeolites as templates to study the capabilities of CO_2 capture.[137,138]

Xia *et al.* synthesized zeolite-templated, high-surface-area, microporous, N-doped carbons, which exhibit CO_2 uptake capacities of up to 6.9 mmol g^{-1} at 273 K at 1 bar and 4.4 mmol g^{-1} at 298 K at 1 bar, along with CO_2 adsorption energies of 36 kJ mol^{-1} to 20 kJ mol^{-1}.[139] Combined with their ease of preparation, regeneration stability, excellent recyclability and high selectivity for CO_2, the N-doped zeolite-templated carbons are suitable for CO_2 capture. They further successfully synthesized structurally well-ordered sulfur-doped microporous carbon materials for the first time.[121] The sulfur-modified ordered microporous carbons were obtained by infiltration of the carbon precursors (2-thiophenemethanol) into the pores of zeolite EMC-2 *via* impregnation combined with a CVD method. They found that S-doped microporous carbons with well-resolved XRD patterns arising from a high level of zeolite-type structural ordering can be nanocasted from zeolite EMC-2. The carbon materials possessed a high surface area with varied sulfur content depending on the preparation conditions. They found that both the presence of S functional groups and the textural properties play important roles in the capacity of CO_2 adsorption. These carbons also showed a very high CO_2 adsorption energy up to 59 kJ mol^{-1}.[121]

1.2.4.3 Porous Organic Frameworks as Self-templates

Porous organic frameworks (POFs), such as metal–organic frameworks (MOFs), zeolite imidazole frameworks (ZIFs) and covalent organic frameworks (COFs), have been intensively investigated for CO_2 capture application recently.[42,140,141] MOFs generally consist of two main components: organic ligands and metal ions. The structures of MOFs change with the types of inorganic metal ions and organic ligands and can be functionally modified by a variety of methods. ZIFs are a newly emerging class of porous crystals with extended three-dimensional structures constructed from tetrahedral metal ions (*e.g.*, Zn, Co) bridged by imidazolate. Another kind of porous material that has emerged recently, COFs possess a long-range ordered network composed of covalent connections by the light elements (C, N, O, B). Recently, these porous organic frameworks have been used to prepare porous carbons and related nanostructured functional materials.[142] Those carbons have also been quickly developed and used in CO_2 capture.

Generally, there are two routes to fabricate porous carbons from these porous organic frameworks. One is using the porous organic frameworks as sacrificial templates with the incorporation of an additional carbon source. Xu and his co-workers used the classical MOF-5 as a sacrificial template and precursor with furfuryl alcohol as the additional carbon source to prepare porous carbons, which is the first report of MOF-derived porous carbons.[143] The resultant porous carbon obtained at 1000 °C exhibited a high specific surface area up to 2872 m^2 g^{-1}. The as-synthesized porous carbon exhibited a high H_2 uptake, reaching 2.6 wt% at 77 K and 1 bar, much higher than that of raw MOF-5 under the same conditions. Moreover, this material gave a good electrochemical performance as an electrode material for supercapacitors with a capacitance of 258 F g^{-1} at a current density of 0.05 A g^{-1}.[143] They further investigated the effect of carbonization temperature on the porosity of the resultant porous carbons.[144] The specific surface area for MOF-5 templated porous carbons obtained at carbonization temperatures from 530 to 1000 °C is in the range from 1140 to 3040 m^2 g^{-1}, and the dependence of surface area on carbonization temperature shows a "V" shape.

Wang prepared N-doped porous carbons using ZIF-70 with a high porosity and large pore size as the model system and polyethyleneimine (PEI) molecules as the nitrogen source. Many PEIs can be readily incorporated into the framework pores of ZIF-70. Moreover, the ZIF-70 framework has many N elements, which can be partly doped into a carbon network during the carbonization process. Finally, a series of N-rich porous carbon materials was successfully fabricated by carbonization of polyamine-incorporated MOFs, which had an excellent CO_2 capture capacity of 4.86 mmol g^{-1} at 1 bar and 25 °C.[145] Pachfule *et al.* reported a series of porous carbon materials by using isoreticular zeolitic imidazolate frameworks (IRZIFs).[122] A series of carbon materials named as C-68, C-69 and C-70 was prepared using ZIF-68, ZIF-69 and ZIF-70 as templates and furfuryl alcohol as the additional carbon source under a carbonization temperature of 1000 °C. Among these carbons, the resultant C-70 synthesized using ZIF-70 as the template shows the most developed porosity with the highest specific surface area of 1510 m^2 g^{-1} due to the more porous ZIF-70 template. The porous carbons C-70, C-68 and C-69 show CO_2 uptake capacities (273 K and 1 atm) of 5.45, 4.98 and 4.54 mmol g^{-1} for C-70, C-68 and C-69, respectively. Also, the C-70 carbon shows the highest H_2 uptake capacity (77 K and 1 atm) of 2.37 wt% among these carbons. Zhang *et al.* reported a new strategy of pre-introducing an extra carbon source (furfuryl alcohol) into a PAF (porous aromatic framework) followed by carbonization, which afforded a new microporous carbon material with a small pore size of 0.54 nm, thus facilitating a high CO_2 uptake capacity of about 4.1 mmol g^{-1} at 295 K and 1 bar.[146]

The other route is direct carbonization in which the POFs serve as both self-templates and carbon precursors at the same time without any additives. Srinivas *et al.* reported a new type of hierarchical porous carbon from the controlled carbonization of Zn-MOF-5, Zn-MOF-74, and Al-MIL-53 (Figure 1.12).[147] The authors found that an increase in the size of the MOF-5 precursor could effectively enhance the specific surface area of the resultant porous carbons. The simultaneously enhanced surface area and pore volume in HPC5b2 were attributed to the large, millimeter-sized crystallites of the MOF precursor, that delivered more void space due to the extended pores within the crystals after carbonization, compared with the smaller crystals in HPC5b1. HPC5b2-1100, obtained by carbonization of millimeter-sized MOF-5 at 1100 °C showed a very high specific surface area and

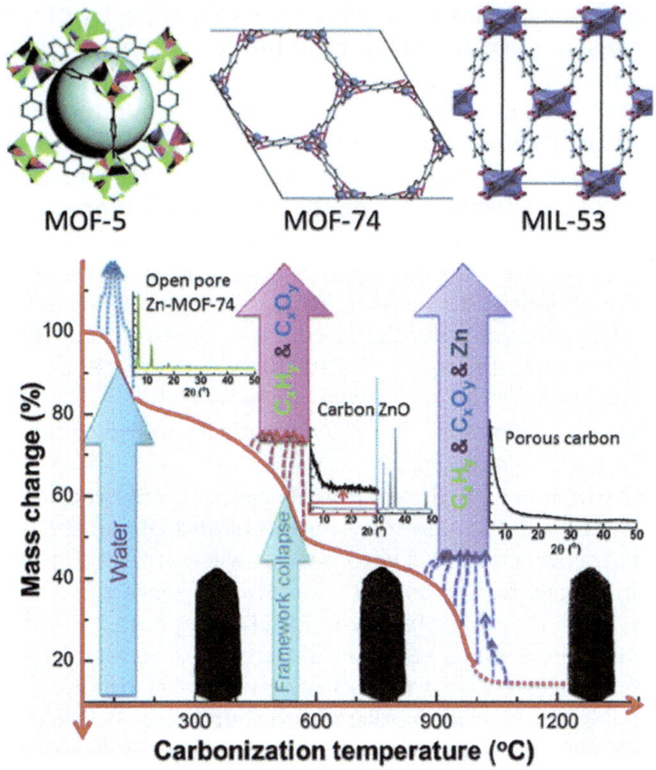

Figure 1.12 Schematic representation of MOF-5, MOF-74 and MIL-53 framework structures. Bottom: mechanism involved in the carbonization process of Zn-MOF-74. The plot represents the MOF mass change *vs.* carbonization temperature. Reproduced from ref. 147 with permission from the Royal Society of Chemistry.

total pore volume, up to 2734 $m^2 g^{-1}$ and 5.23 $cm^3 g^{-1}$, respectively. Through simultaneous thermogravimetric analysis and mass spectroscopy (H_2, H_2O, CO, O_2, CO_2, and carbonaceous gases), the carbonization process and mechanism for all of the MOFs were investigated. Taking Zn-MOF-74 as an example, as shown in Figure 1.12, the first mass-loss, below 150 °C, was due to the evaporation of adsorbed/terminal water molecules within the pores and the second mass-loss between 400 °C and 600 °C was attributed to framework decomposition leading to a major release of carbon containing gaseous products (mostly CO_2, CO, C_6H_6 and a small amount of H_2 and C_xH_y hydrocarbon mixtures) and formation of ZnO occurs. The third mass-loss starting at 900 °C was due to a further release of CO_2 and CO with Zn through the reduction of ZnO by carbon *via* ZnO + C → Zn(g) + CO. In most cases, the CO_2 uptakes in these MOFs-derived carbons were higher than in their MOF precursor. A high CO_2 uptake, over 27 mmol g^{-1} (119 wt%) at 30 bar and 27 °C, is obtained, which is one of the largest reported in the literature for porous carbons. However, the CO_2 capacity of these carbons at ambient pressure was only about 3 mmol g^{-1} due to their relatively low ultramicroporosity.

Ding *et al.* reported a series of N-doped porous carbon monoliths by direct carbonization of IRMOF-3, an isoreticular MOF-3.[148] The ligand (2-amino-benzene-1,4-dicarboxylic acid) served as the nitrogen source. Meanwhile, the ZnO nanoclusters formed during the carbonization process acted as the *in-situ* activator and self-template, which was the key to the transformation of the microporous structure to the meso-macroporous structure. With an increase of the carbonization temperature, the specific surface area, micropore surface area and volume calculated using the t-plot method simultaneously increased. Compared to the pristine IRMOF-3, IRMOF-3/800 carbonized at 800 °C exhibited a much higher CO_2 capacity (3.99 *vs.* 2.32 mmol g^{-1} at 1 bar and 1.64 *vs.* 0.70 mmol g^{-1} at 0.15 bar) and CO_2–N_2 selectivity (22.7 *vs.* 7.4 at 0 °C and 38.8 *vs.* 8.6 at 25 °C).

Ma *et al.* synthesized a N-doped porous carbon using zeolitic imidazolate framework-8 (ZIF-8) both as a solid template and precursor. The resultant carbons were further modified with NH_3. It was confirmed that absorption of CO_2 was enhanced by ammonia treatment. This is attributed to the presence of increased N-containing functional groups and a specific surface area during modifications.[149] Similar to that, Zhao reported the design and synthesis of a series of new B, N-containing cross-linked polymers with well-defined chemical structures and high thermal stabilities. Further, they were efficiently converted into B/N co-doped porous carbons in high yields under pyrolysis treatment. The prepared carbons showed excellent CO_2 uptakes (3.25 mmol g^{-1} at 273 K and 1 atm) with respect to their relatively low surface areas, and high selectivity of CO_2–CH_4 with a ratio of more than 5 : 1 at 298 K.[150]

Unambiguously, MOFs, ZIFs and COFs have been emerging as very promising templates and/or precursors to prepare highly porous carbons that show great potential in gas storage and electrical applications. However, the CO_2 capacities of MOFs (ZIF, COFs)-derived carbons are at a moderate level among the porous carbons. Further exploration is needed, such as how to introduce high microporosity and heteroatom-doping, the development of low-cost MOFs (ZIFs, COFs) and their large-scale synthesis methods, *etc.*

1.2.4.4 Carbide Lattices as Self-templates

Carbide-derived carbons (CDCs) are synthesized by selective extraction of a metal and metalloid from carbide precursors. High-temperature chlorination is the most widely used method for CDC synthesis (eqn (1.13), Figure 1.13).[151,152] In general, the starting carbide precursor templates the initial pore size of the CDCs, and then any range of average pore sizes from less than 1 nm to greater than 10 nm can be achieved by changing the chlorination temperature.[153] From this perspective, this preparation could be considered as a crystal lattice templating process. The most well-known application

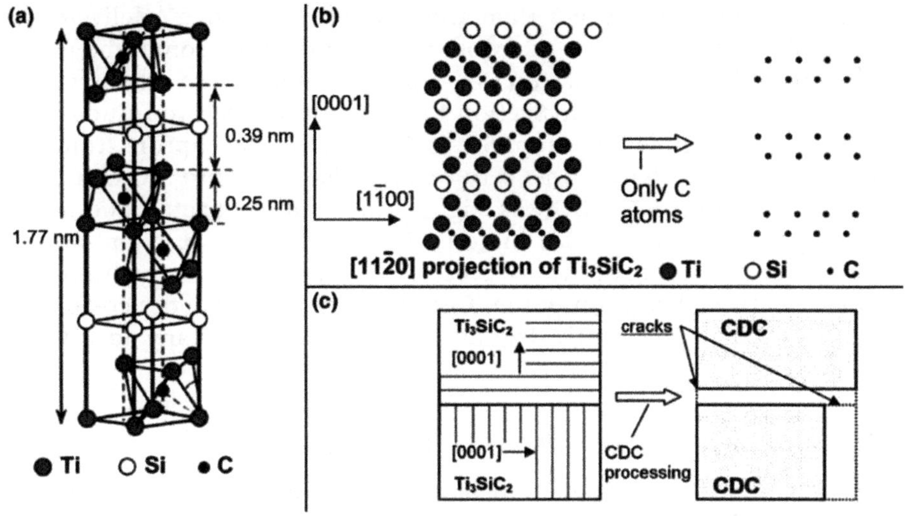

Figure 1.13 (a) Unit cell of Ti_3SiC_2. (b) Schematic of the lattice structure of Ti_3SiC_2 with and without Ti and Si atoms shown. (c) Proposed model for generation of cracks between the grains. Contraction of CDC along the [0 0 0 1] direction of the former Ti_3SiC_2 is suggested. Reprinted from G. N. Yushin, E. N. Hoffman, A. Nikitin, H. Ye, M. W. Barsoum and Y. Gogotsi, Synthesis of nanoporous carbide-derived carbon by chlorination of titanium silicon carbide, *Carbon*, **43**, 2075–2082, Copyright 2005, with permission from Elsevier.

of CDCs is as a supercapacitor, which was led by Prof. Gogotsi.[154,155] These carbons have also shown great promise in H_2 storage,[156–158] and CH_4 storage,[159] as well as CO_2 capture.[106,159–163]

$$MC + x/2Cl_2 \rightarrow MCl_x + C \quad (1.13)$$

The TiC-CDC produced at a chlorination temperature of 800 °C showed a good H_2 capacity of 2.55 wt%, which was further improved to 3.0 wt% after a post-treatment of H_2 annealing.[157] Gogotsi *et al.* prepared a large number of CDCs with various pore sizes by using different starting carbides (B4C, ZrC, TiC and SiC) and chlorination temperatures to explore the effects of pore size and volume on the H_2 sorption and heat of adsorption.[156,157] They found that H_2 storage was dominated by small pores, and thus not directly connected to total specific surface area. A clear dependence on pore size was also suggested – the smaller pores increased both the heat of adsorption and the H_2 uptake.[156]

As far as we know, the first study of CDCs for CO_2 adsorption was reported by Presser and his co-workers in 2011.[106] The Micro-TiC-CDC sample prepared under the combined conditions of chlorination at 700 °C and post-H_2 annealing at 600 °C showed a mean pore size of 0.66 nm and exhibited a very high sorption capacity for CO_2 of up to 7.1 mmol g^{-1} at 0 °C and 1 bar.

The highest CO_2 capacity at 0.1 bar and 0 °C is 1.62 mmol g^{-1} delivered by the sample prepared at 600 °C. More importantly, they comprehensively studied the effect of pore size on CO_2 capacities at different sorption pressures. This will be detailed in Section 1.6.1. Bhatia and Nguyen studied the potential of carbon derived from SiC (SiC-CDC) for CO_2 capture from flue gas.[161] They found that N_2 was very weakly adsorbed onto SiC-CDC, and H_2O adsorption on SiC-CDC was highly kinetically restricted due to the strong hydrophobicity of this material inherited from the diamond-like structure (sp^3 bonding) of the SiC precursor. This feature is very promising for the selective adsorption of CO_2 from real flue gas since the co-adsorption of H_2O can significantly degrade the performance of any adsorbents for CO_2 capture.[161]

1.2.4.5 *Triblock Copolymer as a Soft Template*

Wei *et al.* synthesized nitrogen-doped ordered mesoporous carbon materials by using a self-assembly process with dicyandiamide as a nitrogen source, soluble resol as a carbon source and a triblock copolymer (F127) as a soft template. Figure 1.14 shows the synthesis process of this ordered N-doped mesoporous carbon. In this synthesis, resol molecules can bridge the template F127 and dicyandiamide *via* hydrogen bonding and electrostatic

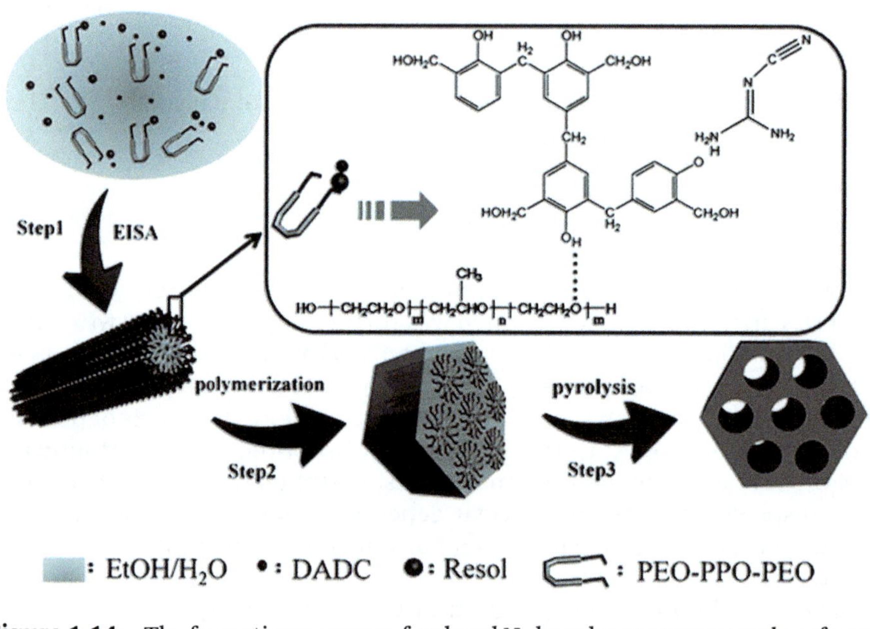

Figure 1.14 The formation process of ordered N-doped mesoporous carbon from a one-pot assembly method using dicyandiamide (DCDA) as a nitrogen source. Reproduced from ref. 131 with permission from John Wiley and Sons, Copyright 2012 WILEY-VCH Verlag GmbH & Co. KGaA, Weinheim.

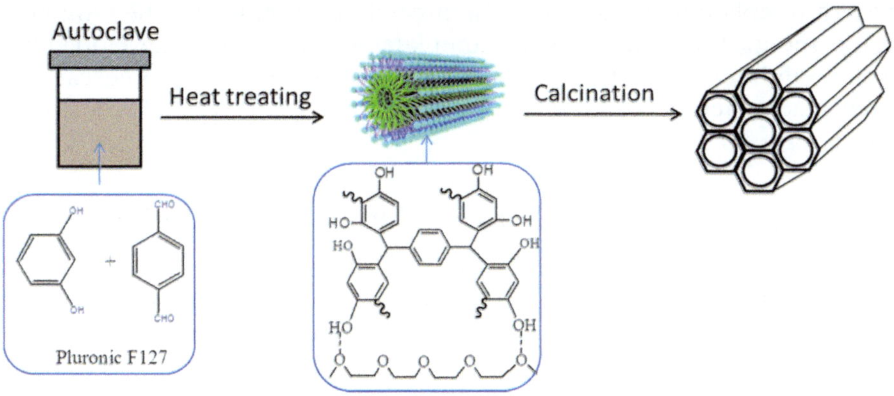

Figure 1.15 Schematic of the one-pot facile synthesis of ordered mesoporous carbons. Reproduced from ref. 127 with permission from the Royal Society of Chemistry.

interactions. The obtained N-doped ordered mesoporous carbons possess tunable mesostructures and pore sizes (3.1–17.6 nm), high surface areas (494–586 m^2 g^{-1}), and high N contents (up to 13.1 wt%). Ascribed to the unique feature of a large surface area and high N content, the carbon materials show a good CO_2 adsorption capacity of 2.8–3.2 mmol g^{-1} at 298 K and 1.0 bar.[131]

Zhang *et al.* reported a facile one-pot melting-assisted solvent-free method to develop ordered mesoporous carbons (Figure 1.15).[127] This method only involved a simple thermal treatment procedure and avoided the time-consuming process of evaporation-induced self-assembly (EISA). Using this method, non-doped and N-doped mesoporous carbons can be synthesized. The non-doped carbon possessed a well-developed hierarchical pore texture with a high surface area of 748 m^{-2} g^{-1} and exhibited a CO_2 capacity of 2.73 mmol g^{-1} at 75 °C and 1 bar. Due to the large amounts of nitrogen in the carbon matrix (11.67%), the N-doped mesoporous carbon exhibited excellent CO_2 capture properties, with a CO_2 capacity of 1.66 mmol g^{-1} as well as an excellent IAST selectivity of CO_2-to-N_2, up to 240.7 at 75 °C and 1 bar. Wu *et al.* reported the consecutive incorporation of carbon nitride and nitrogen-containing functionalities into FDU-type ordered mesoporous carbon materials by using a simple post-synthetic route with melamine as a nitrogen source and F127 as the soft template. The nitrogen-enriched mesoporous carbons deliver promising properties for CO_2 capture with the high heats of adsorption and capacity (6–9 mmol g^{-1} at 10 bar), well retained or even promoted at lower pressures (<3.5 mbar at 303 K).[128]

In summary, the templating method allows us to effectively fabricate porous carbons with specific porous structures and surface chemical properties, which are suitable for CO_2 adsorption. However, there are also some weaknesses. For instance, hard templates are usually prepared by a complicated production process, but are sacrificed after carbonization. Although the soft template method avoids the procedure of removing the templates after carbonization, only a few soft templates are successfully developed due to the requirement of thermal stability. Moreover, the templated carbons usually lack microporosity, which is the really efficient type of porosity for CO_2 capture.

1.2.5 Combined Method of Templating and Activation

Recently, hierarchical porous carbons (HPCs) have attracted much attention. Hierarchical porous carbons possess complex network structures formed by two or more levels of pores, which include mainly micro-mesopores and micro-meso-macropores and so on. These carbons have shown a better performance in many application areas than carbons with a single porosity, on account of the improved mass transport facilitated by the large-sized pores and the high surface area and pore volume from the micro-/mesopores. For CO_2 capture, the combination of micropores and mesopores can simultaneously promote the adsorption and diffusion of CO_2 molecules.

To date, several approaches have been proposed for the synthesis of hierarchical porous carbons. One of the main techniques is a dual templating strategy, in which two templates with dimensions at different length scales are combined to generate multimodal pores. Either two hard templates[164] or a combination of hard and soft templates[165,166] can be employed in this technique. However, the biggest shortcoming of this strategy is its low efficiency in developing microporosity, especially ultramicroporosity, which is crucial for gas adsorption. Another commonly-used method to synthesize hierarchical porous carbons is a combination of a templating method and activation, in which mesopores and micropores are generated by the two different mechanisms respectively.[167] This combined technique could be further classified into a two-step process and a one-step process.

1.2.5.1 *Two-step Process*

As mentioned above, a combination of a templating method and activation is one of the most commonly-used routes to prepare HPCs. This method usually contains two steps: firstly, the construction of mesopores by a template method; then, physical activation or chemical activation of the templated carbons to generate micropores. Many groups have synthesized various hierarchical porous carbons by choosing appropriate templates, carbon precursors and activation conditions.[22,168,169]

Marta Sevilla *et al.* designed and prepared hierarchical porous carbons through the KOH activation of two ordered mesoporous carbons of CMK-3 and CMK-8, which were respectively obtained by using hexagonal SBA-15 and cubic KIT-6 ordered mesostructured silica as hard templates.[98] The process of KOH activation was carried out at different temperatures in the 600–800 °C range. Textural characterization of these activated carbons showed that they had a dual porosity made up of mesopores derived from the templated carbons and micropores generated during the chemical activation step. As a result of the activation process, there was an increase in the surface area and pore volume from 1020 $m^2\ g^{-1}$ and 0.91 $cm^3\ g^{-1}$ for the CMK-8 carbon to a maximum of 2660 $m^2\ g^{-1}$ and 1.38 $cm^3\ g^{-1}$ for the sample activated at 800 °C (KOH–CMK-8 mass ratio of 4). By comparison, various samples exhibited similar CO_2 uptake capacities of around 3.2 mmol g^{-1} at 298 K no matter what type of templated carbon was used as a precursor. Through analysis of the CO_2 adsorption quantities of different porous carbons, they found that CO_2 capture capacity depends on the presence of narrow micropores (<1 nm) rather than on the surface area or pore volume of the porous carbons. Furthermore, it was found that these porous carbons exhibited a high CO_2 adsorption rate, a good selectivity for CO_2–N_2 separation and they can be easily regenerated.[98]

Luis *et al.* demonstrated a novel strategy to synthesize hierarchical porous carbons with large pore volumes, high surface areas, and more importantly, tunable micro/meso/macro porosities. The approach is based on a combination of ice templating, colloidal silica and CO_2 activation for generating interconnected macro-, meso-, and microporosity, respectively (Figure 1.16). During the synthesis, the range, size and extent of porosity can be easily controlled with different conditions. The synthesis was simple, green, reproducible, and used widely available and inexpensive starting materials, all of which made the process highly scalable. It produced a good CO_2 capture capacity of 4.2 mmol g^{-1}. More importantly, for dilute CO_2 (10% CO_2–90% N_2 gas mixture), the sorption capacity measured under dry and

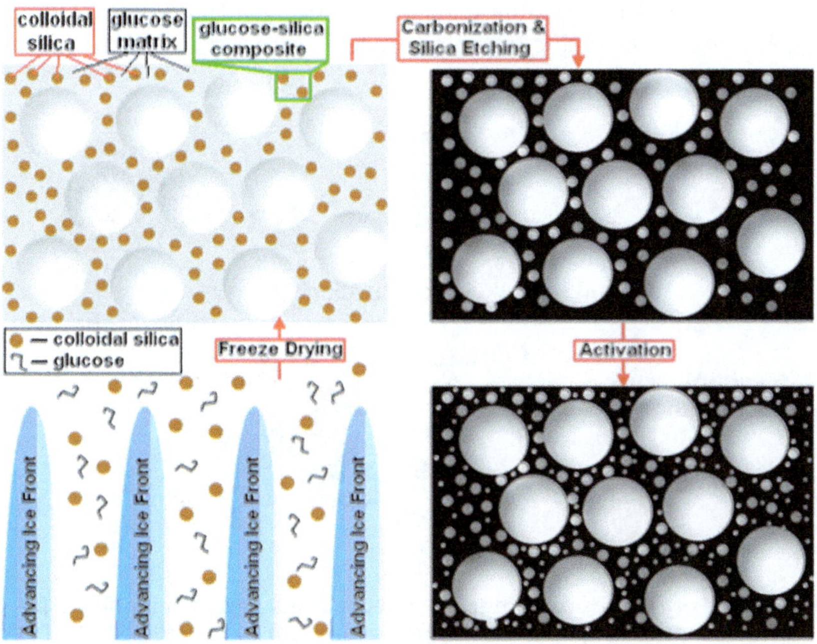

Figure 1.16 Schematic showing the combination of ice templating, colloidal silica and CO_2 activation. Reproduced from ref. 167 with permission from the Royal Society of Chemistry.

moist conditions was the same, which was hardly affected by the presence of moisture.[167]

Many studies have proved that nitrogen-doping is an effective strategy to enhance the CO_2 adsorption capacities of porous carbon adsorbents. However, to achieve a high doping level of nitrogen and a significant porosity in the carbon material simultaneously remains a challenge. Li *et al.* explored a facile approach to construct three-dimensional macroporous nitrogen-doped carbons with periodic ordered macropores and further etched micropores on the frameworks by KOH activation.[169] These hierarchical porous nitrogen-doped carbons were partly graphitic and possessed relatively high nitrogen contents of about 14.45%. The CO_2 adsorption capacities of these samples were about 2.69 mmol g^{-1} at 298 K and 3.82 mmol g^{-1} at 273 K at 1 bar with an extraordinarily high CO_2–N_2 selectivity of 134 calculated from the IAST method. Such an unprecedented CO_2–N_2 selectivity was largely associated with the unusually high N content and the partially graphitic framework of this material.

1.2.5.2 One-step Process

The one-step method is simple and practicable, and is an important way to synthesize hierarchical porous carbons. Hao *et al.* designed a series of poly(benzoxazine-*co*-resol)-based porous carbon monoliths with multiple-length-scale porosity (macro-, meso-, and micropores), a nitrogen-containing framework (polar surface), and remarkable mechanical strength (Figure 1.17).[170] Interconnected mesopores (cubic Im3m symmetry) were formed by the assembly

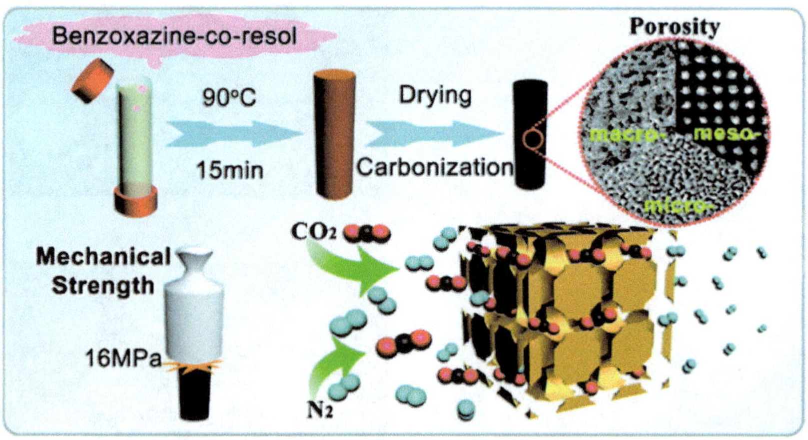

Figure 1.17 Poly(benzoxazine-*co*-resol)-based porous carbon monoliths for CO_2 capture. Reprinted with permission from G.-P. Hao, W.-C. Li, D. Qian, G.-H. Wang, W.-P. Zhang, T. Zhang, A.-Q. Wang, F. Schueth, H.-J. Bongard and A.-H. Lu, *J. Am. Chem. Soc.*, 2011, **133**, 11378, Copyright 2011 American Chemical Society.

of poly(benzoxazine-*co*-resol) under the direction of surfactant Pluronic F127, while co-continuous macropores were formed due to polymerization-induced phase separation. Herein, organic amines were used as both catalyst and nitrogen source. The nature of the organic amines strongly influences the assembly of the mesostructure. It is reported that protic organic bases, such as DMA (dimethylamine), EDA (ethylenediamine), and DAH (1,6-diaminohexane), favored the formation of ordered mesopores in carbon monoliths. In contrast, aprotic organic bases (dimethylamine and triethylamine) resulted in the formation of microporous carbons due to the lack of hydrogen bonds between these amines and F127 molecules. HCM-DAH-1 carbon gave a CO_2 adsorption capacity of 3.3 mmol g^{-1} at 0 °C and 780 mm Hg, a high CO_2–N_2 selectivity of 28, and high isosteric heats of adsorption of 21.1–35.9 kJ mol^{-1}, indicating that the prepared carbon preferentially interacted with CO_2. Further CO_2 activation at 900 °C significantly increased the specific surface area and newly generated micropores with a size of 0.9–1.3 nm; as a result, the CO_2 capacities increased to 4.9 mmol g^{-1}. However, the selectivities of CO_2-to-N_2 and isosteric heats of adsorption decreased after the additional activation, which may be due to the widening of the microporosity.

Liu *et al.* reported the one-step synthesis of nitrogen-doped graphene-like meso-macroporous carbons as highly efficient and selective adsorbents for CO_2 capture. Graphene-like meso-macroporous carbons (GMCs) with high nitrogen contents and controllable nitrogen sites were synthesized by one-step carbonization of dicyandiamide or melamine with glucose. The unique layered structure and the presence of abundant meso-macropores in the GMCs largely enhance the degree of exposure and accessibility of anchored nitrogen sites to CO_2, which provides the GMCs with an excellent performance for the selective capture of CO_2 (about 101 calculated according to the IAST).[171]

Tian *et al.* reported a method of direct carbonization of algae biomass to generate hierarchical porous carbons. Algae biomass is a fantastic precursor for fabricating porous carbon due to its natural composition. *Enteromorpha prolifera*, which was selected as a carbon source, was converted into carbons with evident hierarchical micro-mesoporous structures through a simple direct pyrolysis of freeze-dried algae. The formation process for hierarchical pores was preliminary discussed. The surface areas of the obtained carbons were only close to 450 m^2 g^{-1} with 70% areas inherited from mesopores, resulting in high density materials with the major pore sizes at around 1.5 and 5.0 nm. Moreover, the carbons show a CO_2 uptake of 6.48 mmol g^{-1} at 20 bar and 25 °C.[172]

In this section, we have shown that researchers have fabricated an enormous number of porous carbons as CO_2 adsorbents. As we have reviewed, the porosity (*e.g.*, single- or multi-level, narrow-distributed or wide-distributed, ordered or disordered, microporous or mesoporous, *etc.*) and surface chemistries (pristine or doped) could be engineered well. In general, their synthesis strategies could be roughly categorized into three types: activation, templating, and a combination of both. The CO_2 sorption capacities of these materials could be comparable to those of other well-known porous solid adsorbents, such as MOFs, ZIFs, zeolites, and others. Porous carbons have been one of the most promising candidates for CO_2 adsorbents, and will be more and more important in the field of CO_2 capture.

1.3 Graphene-based Porous Materials

Graphene, the world's first two-dimensional (2D) material, has captured the attention of scientists, researchers, and industry worldwide. In simple terms, graphene is a thin layer of an sp^2-bonded carbon sheet. This material was first isolated and characterized in 2004 by Andre Geim and Konstantin Novoselov at the University of Manchester. In 2010, they received the Nobel prize in physics for their work.

As we know, graphene has many advantages, such as a high specific surface area, superior electron mobility, easy self-assembly into three-dimensional (3D) macroscopic materials with controlled microstructures, high mechanical, chemical, thermal, and electrochemical stabilities, *etc*. With these amazingly attractive properties, graphene already exhibits tremendous promise in many applications: nanoelectronics,[173] supercapacitors,[174] fuel cells,[175] batteries,[176] solar cells,[177] water filtration,[178] gas sorption, separation and storage,[179,180] and sensing,[181] *etc*. In recent years, there has also been an enormous effort to explore graphene and its oxygen-containing derivative, graphene oxide (GO), for CO_2 capture applications. The latest progress with both the experimental and theoretical research into graphene-based CO_2 adsorbents has been summarized in some recent reviews.[182,183] In this section, we mainly focus on the experimental reports of graphene-based porous materials for CO_2 capture.

As gas adsorption is highly dependent on the pore texture of the adsorbents, it is highly desirable that the 2D graphene sheets are engineered further in specific ways to obtain a tailored porosity with a high surface area and pore volume. Accordingly, several techniques have been adapted to develop graphene-based porous adsorbents for CO_2 capture.

1.3.1 Graphene-based Adsorbents by Chemical Activation

Chemical activation with KOH is an efficient technique to introduce pore texture into graphene. In 2011, Zhu *et al.* first reported the KOH activation of microwave exfoliated GO (MEGO).[174] The KOH activation generated a much developed porosity in the product carbon. The specific surface area of the activated MEGO could be controlled by the ratio of KOH–MEGO. The optimized activated MEGO had a well-defined micro-mesopore-sized distribution with a very high BET surface area and pore volume, up to 3100 $m^2\ g^{-1}$ and 2.14 $cm^3\ g^{-1}$, respectively. This activated MEGO carbon could provide a large and accessible surface area for charge accommodation, and, therefore, exhibited an outstanding capacitive performance in organic and ionic electrolytes to achieve high gravimetric energy densities.

In 2012, Srinivas *et al.* reported the synthesis of a range of high surface area graphene-oxide-derived carbons (denoted as GODCs) and their applications toward carbon capture and methane storage.[184] GODCs are prepared by chemical activation with KOH from thermally exfoliated GO and solvothermally reduced GO precursors. The authors explored the effects of activation temperature and KOH concentration upon the development of porosity in the resultant carbons. At low activation temperatures and KOH concentrations, the activation mainly generated pore development in the microporous region, while increasing the activation temperature and KOH concentration resulted in a large proportion of pore development in both the microporous and mesoporous regions (>2 nm). The optimum activation conditions were found to be 800 °C and a GO–KOH ratio of 1 : 9 to obtain a porous carbon (GODCsol-800) with the maximum specific surface area and total pore volume of 1900 $m^2\ g^{-1}$ and 1.65 $cm^3\ g^{-1}$, respectively. The results of CO_2 sorption displayed that the high-pressure CO_2 adsorption trend is more or less linearly dependent on the BET surface area, total pore volume, and average pore size, while the low pressure (1 bar) adsorption behavior is almost independent of the surface area, but is closely related to the narrow pore size distribution in the micropore region. The optimized sample of GODCsol-800 exhibited a high CO_2 adsorption capacity of 72.1 wt% at 300 K and 20 bar, as well as a high methane adsorption capacity of 17.5 wt% at 300 K and 35 bar.

1.3.2 Graphene-based Adsorbents by Physical Activation

Physical activation with steam and CO_2 has also been employed to fabricate porous graphene-based materials for CO_2 capture. Sui *et al.* developed porous graphene-based carbons through physical activation of graphene aerogels using steam as the activating agent.[185] The activation temperature plays a critical role in determining the BET surface area and pore volume of the resultant carbons, and the optimal activation temperature to obtain the highest BET surface area and pore volume is 850 °C. A low activation temperature (750 °C) was less efficient at enhancing the porosity due to the slow reaction rate between graphene aerogel particles and steam. Meanwhile, above 850 °C, the samples also showed a decreased specific surface area and pore volume due to the high burn-off of graphene sheets, thus resulting in the destruction of the porous structure and a lower yield. The steam-activated graphene aerogel exhibited a high specific surface area (830–1230 $m^2\ g^{-1}$), an abundant large pore volume (2.2–3.6 $cm^3\ g^{-1}$), and excellent thermal stability. The optimized SAGA-850 showed a CO_2 adsorption capacity of 2.45 mmol g^{-1} at 1 bar and 273 K, much higher than that of the non-activated GA sample (1.45 mmol g^{-1}) under the same conditions.

Chowdhury and Balasubramanian prepared graphene-based porous carbons though CO_2 activation using reduced graphene oxide as a precursor (Figure 1.18).[186] By increasing the activation temperature of CO_2, the specific surface area, micropore pore volume and total pore volume increase, resulting in an improvement of CO_2 uptake. Specifically, the adsorbent material

Figure 1.18 Illustration of the CO_2 activated reduced graphene oxide (rGO). Reprinted with permission from S. Chowdhury and R. Balasubramanian, *Ind. Eng. Chem. Res.*, 2016, **55**, 7906, Copyright 2016 American Chemical Society.

obtained at an activation temperature of 950 °C (*i.e.*, a-RGO-950) exhibited the largest specific surface area (above 1300 $m^2\ g^{-1}$), the highest pore volume (over 1 $cm^3\ g^{-1}$), and a well-defined bimodal micro-mesoporous structure. This adsorbent material displayed a good gravimetric CO_2 uptake (3.36 and 2.45 mmol g^{-1} at 0 °C, 1 bar and 25 °C, 1 bar, respectively), rapid adsorption kinetics, as well as stable and readily reversible adsorption–desorption cycling behavior at room temperature. Moreover, a-RGO-950 exhibited excellent Henry law CO_2–N_2 selectivities of 162 and 253 under conditions pertinent to CO_2 capture from the dry flue gas steam of a coal-fired (75% N_2 and 15% CO_2) and natural-gas-fired (80% N_2 and 5% CO_2) power plant, respectively.

Similarly, Xia *et al.* prepared a series of hierarchical porous graphene-based carbons (HPGCs) by CO_2 activation of graphite oxide.[187] HPGC-850, which was prepared by 2 hours of CO_2 activation at 850 °C, possessed the highest specific surface area and micropore volume, thus exhibited the highest CO_2 sorption capacity of 1.76 mmol g^{-1} at 274 K and 1 bar, as well as the highest H_2 sorption capacity of 3.76 mmol g^{-1} at 77 K and 1 bar.

1.3.3 Graphene-based Adsorbents by Other Techniques

Other techniques, like hydrogen-induced exfoliation,[188] and covalent functionalization,[189] are also used to introduce pore structures into graphene materials for CO_2 capture. For instance, porous graphene materials prepared *via* hydrogen-induced exfoliation of graphite oxide exhibited a maximum sorption capacity of 21.6 mmol g^{-1} at 11 bar and 25 °C.[188] The physical adsorption nature of CO_2 in the prepared graphene material was confirmed using a Fourier transform infrared spectroscopy (FTIR) study. Beyond the normal hydroxyl (3435 cm^{-1}), carboxyl (1726 cm^{-1}) and carbonyl (1173 cm^{-1}) peaks, a new peak was observed at 2324 cm^{-1} in the IR spectra. This peak corresponds to the asymmetric stretching of CO_2, implying physisorption of CO_2 onto graphene sheets.

Chowdhury reported the CO_2 capture performance of thermally treated graphene oxides.[190] As the thermal treatment temperature increased, the pore texture of the graphene sheets became more developed. GPN-800 treated at

800 °C processed the highest specific surface area (484 $m^2 g^{-1}$) and micropore volume (0.094 $cm^3 g^{-1}$), resulting in the highest CO_2 uptake of 2.9 mmol g^{-1} at 0 °C and 1 bar. Kumar designed two pillared porous graphene frameworks (PGFs) by linking reduced graphene oxide layers with 1,4-diethynylbenzene (PGF-1) and 4,4′-diethynylbiphenyl (PGF-2) *via* a C–C coupling reaction.[189] Both frameworks show high CO_2 uptakes of 112 wt% for PGF-1 and 60 wt% for PGF-2 at 195 K and 0.85 atm.

Besides, graphene has been widely used to fabricate composite materials as a carrier support due to its 2D structure with a high surface area. Lu's group reported the preparation of porous carbon nanosheets (PCNs) with precisely tunable thicknesses in which GO played the role of shape-directing agent. The resorcinol-formaldehyde resins grew *in situ* on the GO sheets due to the bridging effect of asparagine and were converted into porous carbon sheets. The thickness of the carbon sheets was tuned from 20 to 200 nm according to the mass ratio of the resin–GO. At 25 °C and 1 bar, the maximum CO_2 uptakes of PCN-9.9, PCN-17, and PCN-71 with thicknesses of 9.9, 17, and 71 nm, respectively, were 2.02, 2.36, and 2.88 mmol g^{-1}. Moreover, these porous carbon sheets showed a good ability to separate CO_2 from simulated flue gas (a water-saturated CO_2/N_2 stream) under dynamic conditions; the CO_2 capacity reached 0.28 mmol g^{-1} at a CO_2 concentration of 4 v%. PCN-17 could stably work for 200 cycles in total under a CO_2–N_2 gas stream of 14 : 86 v% (Figure 1.19). Kim and colleagues reported various N- or S-doped porous carbons by using graphene/polypyrrole,[31] rGO/polyaniline,[192] rGO/polyindole,[193] and rGO/polythiophene[37] as carbon precursors and KOH as an activating agent. The graphene in the carbon precursor is believed to increase the contact area between the KOH activator and carbon precursors. As a result, the synthesized carbons featured a high surface area, a large pore volume, and developed microporosity, and captured a large amount of CO_2 under ambient conditions (>4 mmol g^{-1}), a value much higher than those of the porous graphene materials discussed earlier.

Hybrid CO_2 adsorbents, like PEI/graphene,[194] polyaniline/graphene,[195] LDH/graphene,[196] *etc.*, have also been reported to exhibit excellent CO_2 capture performances, in which the supporting effect of the graphene ensures the high loading content of the active materials, and improves the stability and CO_2 capacities. These hybrid materials will be briefly discussed in Section 1.5. Clearly, various graphene-based materials have been investigated in the CO_2 capture field. The present studies have demonstrated the promise of graphene-based materials for selectively capturing and isolating CO_2 from flue gas. However, this investigation is still in its infancy, and further research is needed.

1.4 Carbon Nanotubes

Carbon nanotubes (CNTs) are allotropes of carbon with a cylindrical nanostructure. These materials were first discovered by Ijima in 1991 as minority by-products of fullerene synthesis. According to the number of carbon walls,

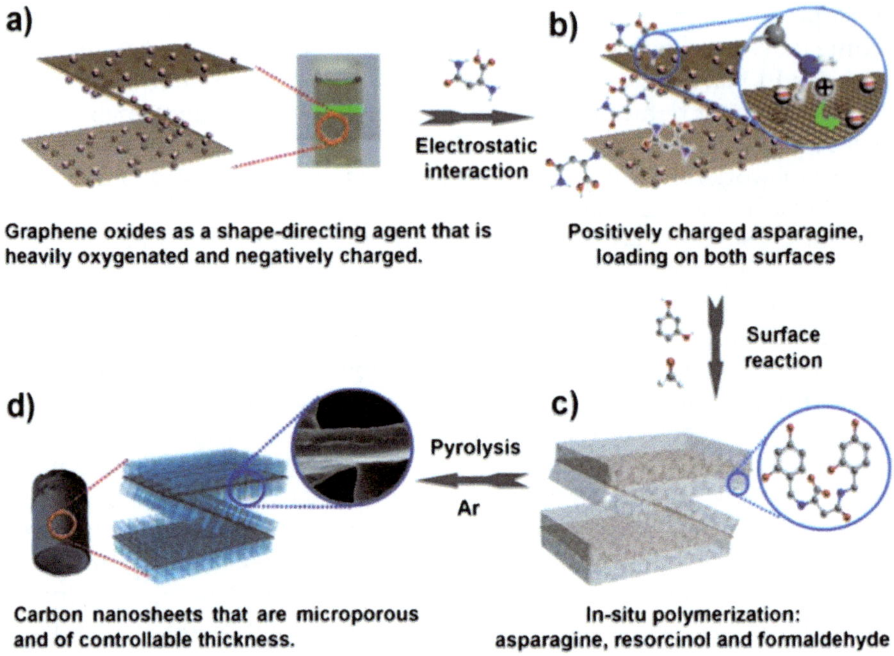

Figure 1.19 Schematic of the formation of the PCNs. (a) The negatively charged GO sheet; the picture (right) shows its Tyndall phenomenon. (b) Positively charged amino acids, equally dispersed at the molecular level on both surfaces of the GO. (c) *In situ* co-polymerization of pre-adsorbed asparagine, resorcinol and formaldehyde. (d) The polymer layer transformed into a microporous carbon layer during pyrolysis in Ar. Reproduced from ref. 191 with permission from The Royal Society of Chemistry.

CNTs are categorized as single-walled nanotubes (SWNTs) and multi-walled nanotubes (MWNTs). The inner diameter of CNTs can vary from approximately 1 nm for SWCNTs to over 10 nm for MWCNTs. Considering the uniform inner diameter and almost defect-free wall of CNTs, the interaction of gas molecules with the inner pores of the CNTs can be described by a smooth potential energy surface. Recent computational simulations have proved that the smoothness of the inner pores make gas molecules diffuse and transport rapidly through the CNTs. For example, the transport rates of light gases, such as H_2, CH_4, and N_2, in CNTs are orders of magnitude faster than in microporous materials with comparable pore sizes.[197,198] This feature makes CNTs ideal candidates for the selective sorption and separation of gases, such as CO_2 capture from flue gas.

Lu *et al.* carried out a comparative study of CO_2 capture by CNTs with inner diameters <10 nm, activated carbons, and zeolites.[199] Raw CNTs showed a CO_2 sorption amount of 22.7 mg g^{-1} under an influent CO_2 concentration of 10% at 25 °C, which is slightly less than activated carbons (24.9 mg g^{-1})

but larger than zeolites (19.0 mg g^{-1}) under the same conditions. After modification with 3-aminopropyltriethoxysilane, the modified CNTs showed an enhanced adsorption capacity of CO_2 of up to 40.9 mg g^{-1} and 96.3 mg g^{-1} at CO_2 concentrations of 10% and 50%, respectively. The mechanism of CO_2 adsorption on these adsorbents appears to be mainly attributable to physical force.

Most studies of pure CNTs for CO_2 capture are simulations, including quantum calculations (*i.e.*, density functional theory (DFT) and *ab initio* simulations),[200–202] grand canonical Monte Carlo (GCMC) simulations,[203,204] and molecular dynamics (MD) simulations.[205] DFT calculations predicted that zigzag SWNTs possessed stronger bonds with CO_2 molecules compared to armchair and chiral ones, and N-doping could increase the binding energies between zigzag SWNTs and CO_2, suggesting that zigzag SWNTs show greater promise as a means of CO_2 capture.[200] Based on Monte Carlo simulations, Liu and Bhatia found that increasing the nanotube diameter from 1.36 nm (10, 10) to 2.03 (15, 15) led to enhanced CO_2 capacity at 5 bar due to the increase of accessible volume, while the change in chirality had a negligible effect.[205] Furthermore, the efficiency of CO_2 capture from flue gas can be significantly improved by pre-adsorbing small clusters of water in carbon nanotubes.[205] Rahimi *et al.* simulated adding a positive charge to the carbon nanotubes, which caused a significant increase in adsorption (up to 35% at a pressure of 1.88 bar), while the gas adsorption decreased by up to 15% on negatively charged carbon nanotubes.[206] The increase or decrease of adsorption upon charging was attributed to the change in potential energy for the interactions between the individual CO_2 molecules and the nanotubes.

Owing to their extraordinary thermal conductivity, mechanical, and electrical properties, carbon nanotubes find applications as additives to various structural materials. Jin *et al.* fabricated carbon composite monoliths by CO_2 activation of a composite of a commercial phenolic resin and carbon nanotubes.[207] CNTs in the composite monoliths played an important role in improving pore structures and the CO_2 capacity of CNT-modified CCMs. CNT-modified carbon monolith samples were much more reactive with CO_2 during activation; they exhibited significant burn-off within a much shorter activation time, thus achieving a well-developed porosity, especially a narrow microporosity. CPD-30 with a burn-off of 25.5 wt% gave the highest CO_2 uptake of 3.5 mmol g^{-1} at 298 K and 1 atm, while CPD-15 with a burn-off of 16.7 wt% had the highest CO_2 uptake of at 1.1 mmol g^{-1} at 298 K and 0.15 atm, more than twice that of the resin alone, Res-60 (only 0.55 mmol g^{-1}). Mixed matrix membranes (MMMs) based on CNTs dispersed inside a polymer matrix have been well-developed and characterized for CO_2 separation measurements. For instance, Ahmad *et al.* fabricated MMMs from cellulose acetate with β-cyclodextrin-functionalized MWCNTs (MWCNTs-F) using a wet phase inversion technique.[208] The results showed that MMMs with 0.1 wt% loadings of functionalized MWCNTs demonstrated enhanced permeance and selectivity towards the separation of CO_2/nitrogen (N_2), as well as outstanding mechanical properties with a high tensile strength of 16 MPa. This behavior results from the enhanced compatibility of MWCNTs-F within the polymer chain segments, which increased the sufficient interlayer spacing.

As discussed, CNTs have attracted much attention in the field of CO_2 capture. These materials could directly serve as adsorbents, as well as supporters/additives for the fabrication of hybrid adsorbents. However, the studies reported have mainly focused on simulations; more experimental studies should be performed in the future.

1.5 Carbon-based Hybrid Adsorbents

CO_2 sorption on pristine carbon materials is essentially "physisorption", which is very sensitive to temperature and pressure, and is usually low-selective. Generally, the sorption capacities of CO_2 on carbon materials will significantly decrease as the sorption temperature increases or the sorption pressure decreases. For example, the CO_2 uptakes of the previously reported porous carbons at 0.15 bar and 25 °C (typical CO_2 partial pressure in flue gas) are usually lower than 1.5 mmol g^{-1}. Reducing the pore size to extremely small micropores (<0.5 nm) and introducing polar surface chemistry can elevate the CO_2 uptake to about 2 mmol g^{-1} at low pressure or high temperature.[88] However, this will lead to a serious obstruction to gas diffusion kinetics. Synthesis of carbon-based hybrid sorbents is an effective strategy to enhance CO_2 capture performance. Mesoporous carbons, carbon nanotubes, and graphene are commonly used because these materials possess larger pore sizes, an opened pore texture and a high surface area, ensuring a high mass loading and good dispersion of the incorporated components. Herein, carbon-based hybrid sorbents are classified into carbon–organic hybrid sorbents and carbon–inorganic ones according to the nature of the incorporated component.

1.5.1 Carbon–Organic Hybrid Adsorbents

Organic amines are the most commonly used functional components for carbon–organic hybrid sorbents. Various amines, such as monoamines, diamines, triamines, polyamines and amine-containing polymers, have been studied for this purpose. In the last few years, several excellent review papers have been published and provide a good perspective towards amine-functionalized solid adsorbents for CO_2 capture and separation.[209–211] Generally, there are two ways to incorporate the amines: (1) impregnation, and (2) grafting.

With impregnation, the amine compounds are dispersed inside the pores of the carbon support, and thereby, they produce an enhanced CO_2 capture performance relative to the mass loading of the bulk amines. In theory, the more amines that are loaded, the higher the CO_2 capture capacities of the amine-impregnated carbon materials. The dispersion of the incorporated amines is also important. The excess loading of amines may result in a decline of the amine utilization ratio due to a lengthening of the diffusion distance and the agglomeration of the amine molecules.[212,213]

Polyethyleneimine (PEI) is most commonly used due to its high CO_2 uptake/release capacity and good stability. For example, Wang *et al.* developed highly efficient CO_2 sorbents by loading PEI on mesoporous carbons.[212] As the PEI loading increased, the utilization ratio of the PEI first increased and then decreased, and the optimal PEI loading is determined to be 65 wt% with a CO_2 sorption capacity of 4.82 mmol g^{-1} in 15% CO_2/N_2 at 75 °C.[212] These sorbents could work over a wide CO_2 concentration range from 5% to 80%. Furthermore, moisture was found to have a promoting effect on the sorption separation of CO_2. This group also reported a sorbent of PEI-impregnated millimeter-sized mesoporous carbon spheres for post-combustion CO_2 capture (Figure 1.20).[214] Under the optimal conditions of using a high molecular weight PEI and polyethylene glycol (PEG) loading of 20 wt%, the adsorbent exhibited a high equilibrium adsorption capacity of 187.5 mg g^{-1} (4.26 mmol g^{-1}) for 15% CO_2 at 75 °C at a relative humidity of 60%. Noticeably, these adsorbents could be regenerated by a novel electric swing adsorption (ESA) operation due to the good electrical conductivity of the carbon support.

With grafting, the amine groups are chemically bound to the carbon supports. In 2011, Houshmand *et al.* reviewed this strategy well.[209] In their

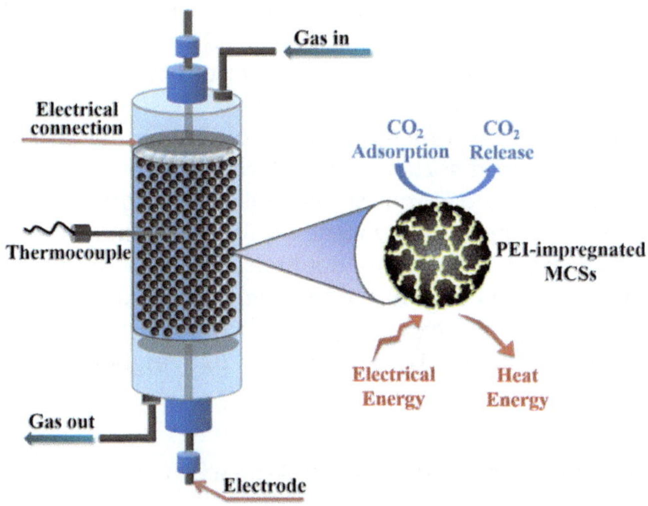

Figure 1.20 Schematic of the adsorption column for electric swing adsorption. Reprinted from M. Wang, L. Yao, J. Wang, Z. Zhang, W. Qiao, D. Long and L. Ling, Adsorption and regeneration study of polyethylenimine-impregnated millimeter-sized mesoporous carbon spheres for post-combustion CO_2 capture, *Appl. Energy*, **168**, 282–290, Copyright 2016, with permission from Elsevier.

review, the methods used for grafting are roughly categorized into amination, silylation with aminosilanes, nitration followed by reduction, anchoring diamines/polyamines, anchoring halogenated amines, surface-initiated polymerization of ethyleneimine and its derivatives, and plasma treatment *etc.* To compare, the interactions between the grafted amines and carbon supports are strong covalent bonds, while the interactions between the impregnated amines and carbon supports are mainly weak interactions, such as dipole–dipole interactions, van der Waals forces, or hydrogen bonding. Therefore, amine groups incorporated by grafting are more stable and do not desorb during regeneration.[209] Nevertheless, grafting suffers the shortcoming of a complicated synthesis process. In addition, with both strategies, the porosities of the carbon supports should be tailored to be suitable for the amine compounds selected.

Besides the commonly used mesoporous carbon materials, graphene and carbon nanotubes have also been used as supports. For instance, Liu *et al.* prepared hybrid sorbents by covalent grafting of PEI on hydroxylated 3D graphene (HG).[196] The optimal HG-PEI hybrid sorbent exhibited a high CO_2 adsorption capacity up to 4.13 mmol g^{-1} at 1 atm of dry CO_2. PEI-impregnated graphene-silica sheets also showed an excellent CO_2 adsorption capacity (171 mg g^{-1} at 75 °C/100 kPa) and good cycle stability (>20 cycles).[194] Lu *et al.* grafted 3-aminopropyltriethoxysilane (APTS) onto MWCNTs, and studied the CO_2 capture behaviors of the amine-loaded MWCNTs.[215,216] The adsorption capacity of CNT(APTS) was significantly influenced by the presence of water vapor and reached a maximum of 2.45 mmol g^{-1} at 2.2% water vapor.[215] In addition to the high surface area and opened pore texture, which promotes diffusion of CO_2 to the active adsorption sites, the high thermal conductivity of graphene and CNTs allowed a fast transfer of heat and avoided degradation of the organic amines.

Other organics, such as ionic liquids and quinones, are also used to fabricate carbon-based hybrid sorbents. Tamilarasan reported that polymerized ionic liquid functionalized graphene showed an adsorption capacity that was 22% higher than graphene, while ionic liquid functionalization improved it only by 2%.[217] Wang *et al.* developed a hydroquinone (quinone)-functionalized hierarchical porous carbon sorbent *via* the Friedel–Crafts reaction.[218] The hydroquinone-grafted carbon gave a CO_2 adsorption capacity of 5.41 mmol g^{-1} at 1 bar and a CO_2–N_2 selectivity of 26.5, higher than those of pristine porous carbon.

1.5.2 Carbon–Inorganic Hybrid Adsorbents

Fabrication of carbon–inorganic hybrid materials is another approach to enhance CO_2 uptake. Various carbon materials, such as porous carbon materials,[219] carbon nanotubes,[220] and graphene and its derivative,[221] are used as solid supports. The inorganic sorbents are mostly metal compounds, including metal coordination polymers (MOFs,[222,223] ZIFs[224]), metal oxides (MgO,[221] CaO,[225] ZnO[226]), alkali metal carbonates,[227] layered double hydroxides

(LDHs),[228] *etc.* According to a recent review by Prof. Wang, these carbon-inorganic hybrid sorbents could be classified into intermediate-temperature (200–400 °C) and high-temperature (>400 °C) solid CO_2 sorbents, and have been summarized well.[229]

In summary, a large number of carbon-based hybrid sorbents have been studied. Due to the strong affinity of incorporated components to CO_2, these sorbents have shown high CO_2 capture capacities at high sorption temperatures or low CO_2 concentrations. Carbon supports have been proved to have significant promoting effects on CO_2 capture, which could be mainly because: (1) the high surface area and pore volume of carbon supports improve the dispersion of active materials while ensuring a high mass loading; (2) the opened pore texture facilitates the diffusion of CO_2 molecules to the active adsorption sites; (3) the high dispersion also guarantees a high adsorption efficiency and good cyclic stability.

1.6 Effect of Carbon Structure on CO_2 Adsorption

A large number of studies have shown that there are two main strategies to improve the CO_2 adsorption capacity of carbons. One simple strategy is to introduce polarity by chemical doping in the carbon framework to enhance its affinity with CO_2. Another approach that has been investigated to enhance CO_2 capture capacities is to tailor pore structures to make them suitable for CO_2 adsorption. In this section, we pay attention to studies of the effect of carbon structure on CO_2 adsorption, which are of use for the development of high-performance carbon-based adsorbents.

1.6.1 Pore Size Effect

1.6.1.1 Analysis of Porosity

The IUPAC (International Union for Pure and Applied Chemistry) has proposed a classification of pores based on pore sizes. Pores are generally classified into micropores (<2 nm), mesopores (2–50 nm), and macropores (>50 nm). Micropores are further divided into ultramicropores (<0.7 nm) and supermicropores (0.7–2 nm). Since the porosity of carbons is responsible for their adsorption properties, the analysis of the porosity, especially the microporosity, is very important to foresee the behavior of these porous solids in CO_2 capture applications. Pores in carbon materials could be identified by several techniques depending on their type and size, such as gas adsorption (opened pores < 100 nm), mercury porosimetry (opened macropores), small angle scattering (SAS) (latent pores, including closed pores), transmission and scanning electron microscopy (TEM and SEM) (extrinsic pores on the surface of carbon materials), *etc.*

Physical adsorption of various gases is, undoubtedly, the most widely used technique. N_2 is the most widely used adsorptive. However, N_2 adsorption at 77 K is not suitable for the characterization of narrow microporosity (size <

0.7 nm), because diffusion kinetics of N_2 molecule is very slow in ultramicropores at a liquid N_2 temperature of 77 K. To overcome this problem, the use of CO_2 sorption at 273 K has been proposed. The higher temperature of CO_2 adsorption results in a larger kinetic energy of the molecules, which are able to enter the narrow pores. However, CO_2 adsorption analysis of the whole porosity should be performed at high pressures due to its high saturated vapor pressure (P^0) at 273 K (34.7 bar). In practice, the combination of N_2 and CO_2 adsorption allows us to obtain comprehensive information about the porosity.

In order to evaluate the specific surface area and PSD, various methods have been applied to gas isotherms, including, but not limited to, the BET (Brunauer–Emmett–Teller) method, D–R (Dubinin–Radushkevich) plot, BJH (Barret–Joyner–Halender) method, *t* plot, HK (Horvath–Kawazoe) method, and the DFT (density functional theory) method. Porous carbon materials, especially activated carbons, typically possess wide PSDs ranging from ultramicropores, supermicropores and mesopores to macropores. The NLDFT (non-linear density functional theory) method has been widely applied and featured in a standard by *ISO-15901-3*.[230] The standard NLDFT model assumes an idealized flat graphitic-like surface. Due to the heterogeneity in the chemistry and structure of the carbon materials, this model may not describe all the subtleties of a real carbon surface. The QSDFT (quenched solid-state density functional theory) model takes into account carbon surface roughness features. This model provides a better representation of the surface properties, including surface roughness features,[231,232] and has been recommended in recent years.[106,233] However, QSDFT models are not yet available for some of the measurement instruments (*e.g.*, micromeritics instruments) and sorption conditions (*e.g.*, CO_2, 0 °C). The NLDFT method is still widely used to evaluate porosity from N_2 sorption measurements.

Micropores, especially, feature in interactions between gas adsorbate molecules and adsorbents, and there are still no definitive methods to determine the PSD in the micropore region. For the analysis of microporosity, Dubinin and Radushkevitch have developed an equation:

$$V = V_0 \exp\left\{-k\left(RT/\beta\right)^2\left[2.303\lg\left(P_0/P\right)\right]^2\right\} \tag{1.14}$$

where V ($cm^3\ g^{-1}$) is the volume filled at a temperature T and a relative pressure (P/P_0), V_0 ($cm^3\ g^{-1}$) is the micropore volume, and k and β are constants for the pore structure and the affinity coefficient (*e.g.*, $\beta = 0.35$ for CO_2). This equation, based on the postulation that the mechanism for adsorption in micropores is that of pore filling rather than layer-by-layer surface coverage, generally applies well to adsorption systems involving only van der Waals forces and is especially useful to describe adsorption on activated carbon.[109,234] This equation has been used to calculate micropore volumes and to estimate the mean pore size widths (L) for both N_2 and CO_2 adsorption on different samples.[88,89,235,236] The mean pore width could been estimated

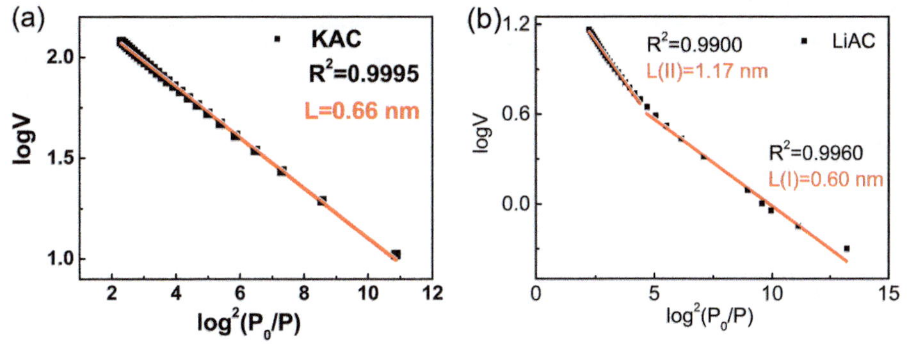

Figure 1.21 Evaluating the mean micropore size from the slope of D–R plots of: (a) KAC (b) LiAC. Adapted from ref. 89, with permission from John Wiley and Sons, Copyright 2016 Wiley-VCH Verlag GmbH & Co. KGaA, Weinheim.

using the equation of Stoeckli *et al.* (eqn (1.15))[237] when the characteristic energy (E_0) is between 42 and 20 kJ mol^{-1}:

$$L = 10.8/(E_0 - 11.4) \tag{1.15}$$

Its validity has been tested by direct determination of the distribution in the range 0.35–1.3 nm. A well-defined linear D–R plot over the entire $\log^2(P_0/P)$ range indicates a uniform micro-PSD, while the non-linear one usually indicates a wide microporosity with various micropore sizes (Figure 1.21).[238]

1.6.1.2 Micropore Filling Mechanism of CO_2 Adsorption

In order to reveal the adsorption mechanism, isotherms are simulated using two typical adsorption models including Langmuir and Dubinin–Astakhov (D–A) models (Figure 1.22).

The well-known Langmuir isotherm model can be expressed by eqn (1.16).

$$\frac{Q}{Q_0} = \frac{bP}{1 + bP} \tag{1.16}$$

where Q (cm^3 g^{-1}) denotes the amount adsorbed, Q_0 (cm^3 g^{-1}) denotes the saturated amount adsorbed, P (kPa) denotes the equilibrium pressure, and b/kP^{-1} denotes the adsorption affinity.

A linear expression for the Langmuir equation is

$$\frac{P}{Q} = \frac{1}{bQ_0} + \frac{1}{Q_0}P \tag{1.17}$$

Figure 1.22 shows the correlation of adsorption data with the Langmuir equation for a PANI-based activated carbon.[105] It was found that the CO_2 adsorption isotherms could not be fitted well by the Langmuir model.

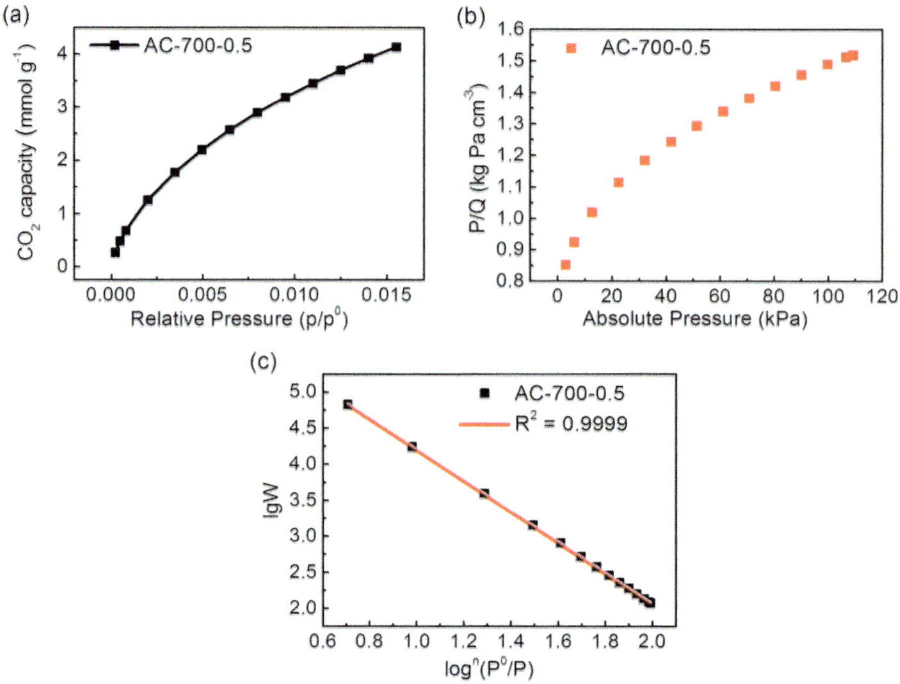

Figure 1.22 (a) CO_2 sorption isotherm of a PANI-based porous carbon at 25 °C, the correlation of CO_2 adsorption data with (b) the D–A model and (c) the Langmuir equation.

The Dubinin–Astakhov (D–A) model is applied to the adsorption equilibrium of CO_2 onto the activated carbons. The D–A equation is proposed to describe a wide microporosity as a general form of the D–R equation, which can be written as

$$\ln W = \ln W_0 - \left(\frac{RT}{E}\right)^n \ln^n\left(\frac{P_0}{P}\right) \tag{1.18}$$

or

$$\lg W = \lg W_0 - \left[2.303^{n-1}\left(\frac{RT}{E}\right)^n\right]\lg^n\left(\frac{P_0}{P}\right) \tag{1.19}$$

where W denotes the volume adsorbed, W_0 denotes the limiting micropore volume, E denotes the characteristic energy of the system, n denotes the heterogeneity parameter, R is the gas constant, T is the equilibrium temperature, and P_0 is the saturated pressure. Apparently, the experimental data can be fitted well by the D–A equation (Figure 1.22).

This simulation shows evidence that CO_2 adsorption on porous carbons can be depicted well by a micropore filling mechanism, not by monolayer adsorption. In other words, the CO_2 sorption capacity should depend on the pore volume of a certain pore size range but not the specific surface area.

1.6.1.3 Pore Size Effect at Different Sorption Pressures

The main contribution of micropores, especially ultramicropores, to CO_2 uptake, has been widely recognized. Therefore, it is necessary to finely explore the relationship between microporosity and CO_2 capture capacity, which is significant for designing high-performance carbon adsorbents for CO_2 capture. A pioneering study of the relationship between porosity and CO_2 capacity was conducted by Gogotsi and his co-workers.[106] They prepared a large number of different carbide-derived carbons (CDCs) derived from nano- and micrometer-sized precursors, with and without activation, and investigated their CO_2 sorption performances. Various porosity parameters, including specific surface area, pore size, and pore volume, were correlated with the resulting sorption properties. They found that there was no clear dependency between the total pore volume or the volume of pores larger than 1 nm and CO_2 capacities at 1 bar (Figure 1.23). Moreover, there is no clear trend correlating CO_2 uptake and average pore size. Hence, they concluded that the CO_2 sorption behavior at ambient conditions may be largely governed by the volume of micropores, similar to the result with H_2 sorption.

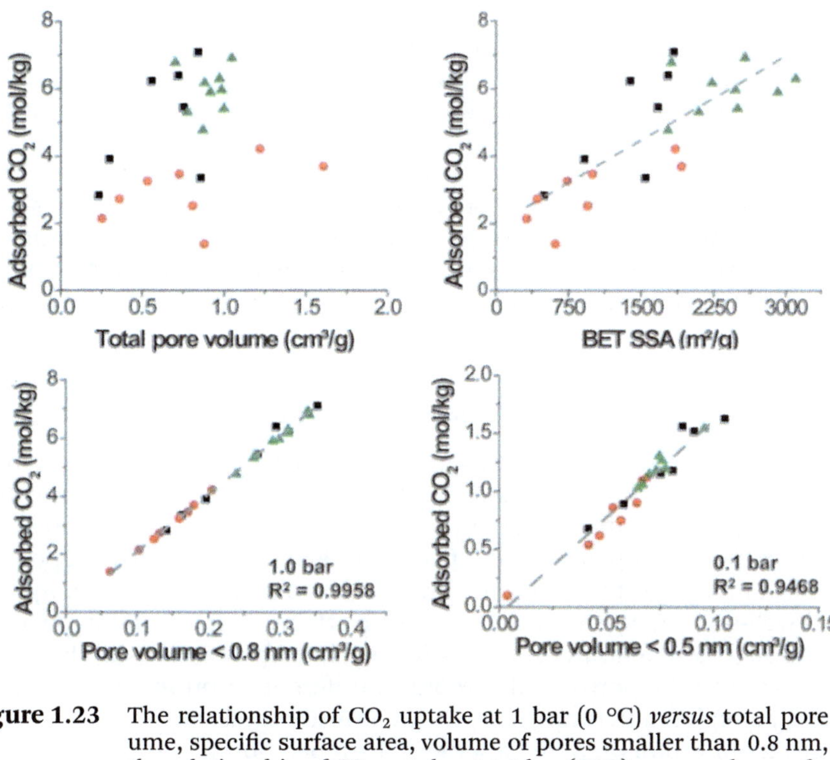

Figure 1.23 The relationship of CO_2 uptake at 1 bar (0 °C) *versus* total pore volume, specific surface area, volume of pores smaller than 0.8 nm, and the relationship of CO_2 uptake at 0.1 bar (0 °C) *versus* volume of pores smaller than 0.5 nm. Adapted from ref. 106 with permission from the Royal Society of Chemistry.

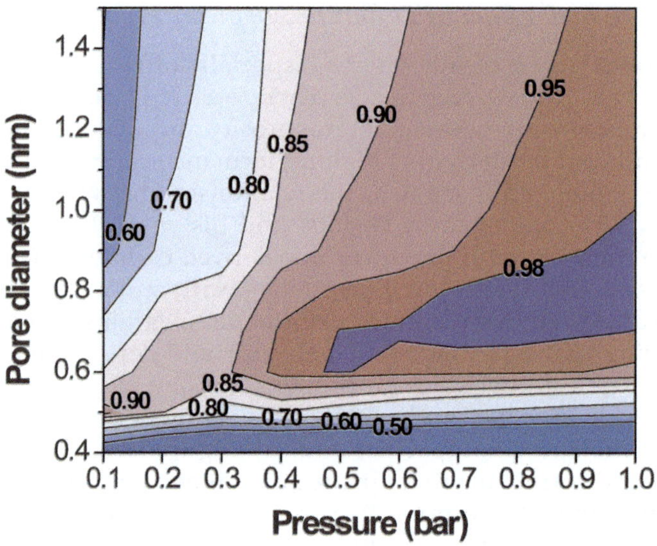

Figure 1.24 Coefficient of determination between CO_2 uptake from 0.1 to 1.0 bar and the volume of pores smaller than critical size. Reproduced from ref. 106 with permission from the Royal Society of Chemistry.

They further correlated the micropore volumes of various pore sizes with CO_2 uptake at different sorption pressures. At 1 bar, pores smaller than 0.8 nm were found to contribute the most to CO_2 uptake and at 0.1 bar, pores smaller or equal to 0.5 nm were preferred (Figure 1.23). This result is regardless of carbon particle size and activation/non-activation. Coefficients of determination between CO_2 uptake at a certain pressure (0.1–1.0 bar) and the cumulative pore volume of pores smaller or equal to a certain pore size (0.4–1.5 nm) were also obtained (Figure 1.24). With a lower total pressure, smaller pores contributed more to the measured CO_2 uptakes.

1.6.1.4 Pore Size Effect at Different Adsorption Temperatures

Almost all of the studies on CO_2 capture were carried out at low temperatures such as 0 °C and room temperature (Tables 1.1–1.3). However, industrial flue gas, the biggest contributor of CO_2, is at a relatively high temperature (>70 °C). The sorption temperature significantly influences gas uptake on porous sorbents, and a high temperature always leads to a low gas uptake. Apparently, a well-designed porosity for carbon adsorbents is required for CO_2 capture at a high temperature. An investigation into the relationship between CO_2 uptake and the porosity of carbons at different sorption temperatures, especially a high temperature (*e.g.*, 75 °C), is also necessary.

In 2012, our group prepared a series of N-doped porous carbon materials by using PANI as a carbon precursor *via* a pre-carbonization and post-KOH activation method.[105] The micropore size distribution of the prepared carbons varied with the activation conditions (KOH–carbon ratio and activation temperature). These carbons showed very high CO_2 uptakes of up to 1.86 and 1.39 mmol g^{-1} under 1 bar, 75 °C and 0.15 bar, 25 °C, respectively, which are amongst the highest of the known carbon materials for CO_2 capture.

It is well known that the smaller the micropore, the stronger the adsorption potential. The smaller micropores will be the preferred spaces for adsorption of CO_2 molecules, and could be filled to a higher degree due to their stronger adsorption potentials. Based on this view, the CO_2 sorption capacity per pore volume is used as a factor reflecting the adsorption capability of porous carbons. The correlations of CO_2 sorption capacities per pore volume with the volume fractions of micropores smaller than a critical size under different sorption conditions were investigated (Figure 1.25). At low temperatures (0 °C), the CO_2 sorption capacity per pore volume is strictly linear with the volume fraction of the pores smaller than 0.80 nm in total pore volume (high correlation coefficient of 0.97, Figure 1.25), agreeing well with the results of Gogosti.[106] Similar results are obtained at 25 °C and 75 °C, while this critical pore size (0.70 nm and 0.54 nm for 25 °C and 75 °C, respectively) decreases as the sorption temperature increases. Furthermore, the simulated lines plotted in Figure 1.25 are found to have an intercept of near zero, indicating that the small micropore is the most important (even exclusive) factor for CO_2 uptake. In a word, the CO_2 adsorption capacity of porous carbons is associated with the pores below a temperature-dependent size.

The correlation of CO_2 sorption capacity per pore volume at 0.15 bar, 25 °C, with the N contents of the carbons or the volume fractions of small micropores (<0.54 nm) in the total pores are also presented (Figure 1.26). Interestingly, the CO_2 sorption capacity per pore volume shows a very good linear correlation with the volume percentage of extremely small micropores (<0.54 nm). By contrast, the correlation of CO_2 sorption capacity per pore volume at 0.15 bar, 25 °C, with the N contents of the carbons obviously deviates from the linear. This fact further confirmed that the CO_2 capture capabilities of the investigated carbons are mainly contributed by the small micropores (<0.54 nm for 0.15 bar at 25 °C) but not the N-doping.

1.6.2 Surface Chemistry Effect on CO_2 Adsorption

To enhance the interactions of porous carbons with CO_2 molecules, many authors have tried to introduce polarity (*e.g.*, N-doping, S-doping, or cation-introducing, *etc.*) into the carbon frameworks because the CO_2 is highly quadrupolar and weakly acidic. However, the effect of heteroatom-doping, especially N-doping, on CO_2 capture is still unclear and is often controversial. Recently, there have been some experimental and simulation studies aimed at clarifying the effects of surface chemistry on the CO_2 uptake of carbons. Herein, we briefly summarize the recent relevant progress.

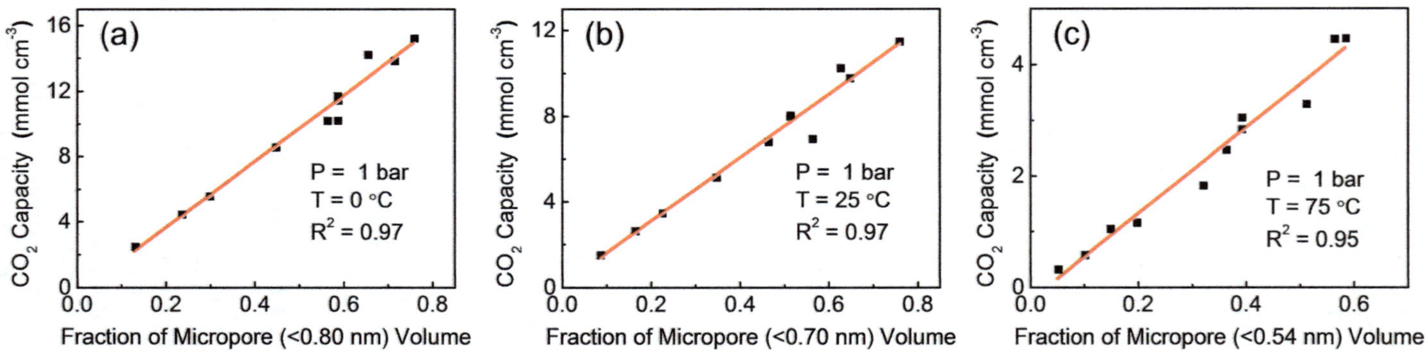

Figure 1.25 Linear fitting of the experimental data. A plot of CO_2 sorption capacity per pore volume *vs.* the volume percentage of small micropores in total pores at adsorption temperatures of 0, 25 and 75 °C. Adapted from ref. 105 with permission from the PCCP Owner Societies.

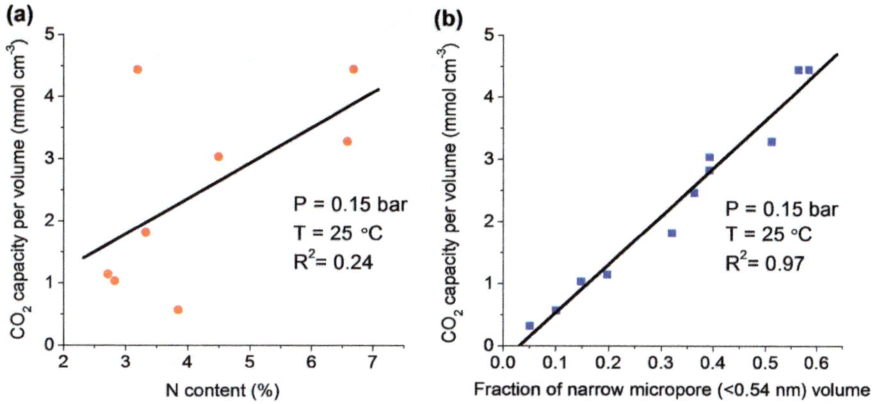

Figure 1.26 The correlation of CO_2 sorption capacity per pore volume at 0.15 bar, 25 °C, with (a) N contents of the carbons and (b) the volume fractions of small micropores (<0.54 nm) in the total pores. Adapted from ref. 105 with permission from the PCCP Owner Societies.

1.6.2.1 *Effect of Nitrogen Doping*

N-doping is widely used. Many authors have suggested that N-doping enhances the CO_2 capture performance of porous carbon adsorbents. In these works, the enhancement effect of N-doping is generally assumed to be chemical sorption based on a one-on-one base–acid interaction. Based on this view, our group prepared N-doped microporous carbons by a two-step casting process using zeolite NaY as a hard template, furfuryl alcohol as a carbon precursor and acetonitrile as the CVD (chemical vapor decomposition) substrate, in which the use of acetonitrile is to introduce nitrogen functionalities to the microporous carbons.[239] Based on the analysis results of X-ray photoelectron spectra, it was found that the type of nitrogen group strongly depends on the treatment temperature of the carbon samples (Figure 1.27). The nitrogen groups on the carbons prepared at 500 and 600 °C are mainly Lewis-basic species of amides/pyrrolic (400 eV) and pyridinic (398.9 eV) groups, while most of these basic groups decomposed at a higher temperature (>600 °C) and were converted into aromatic nitrogen species built into the carbon matrix (mainly quaternary N with typical Lewis acidity, ~401 eV). Although the carbon prepared at 500 °C possessed a very low CO_2 uptake (57 mg g^{-1}), this carbon showed the highest heat of adsorption (35–40 kJ mol^{-1}), much higher than the typical heat of physical adsorption (~20 kJ mol^{-1}), indicating a strong affinity between the surface of this carbon with CO_2 molecules. As the carbon temperature increased, the heat of adsorption significantly decreased due to a decrease in the interactions between CO_2 and carbon materials resulting from the loss of unstable basic nitrogen groups at high carbonization temperatures. These results demonstrated that the basic N groups (amides/pyrrolic and pyridinic) had a positive effect on CO_2 capture *via* a mechanism of base–acid

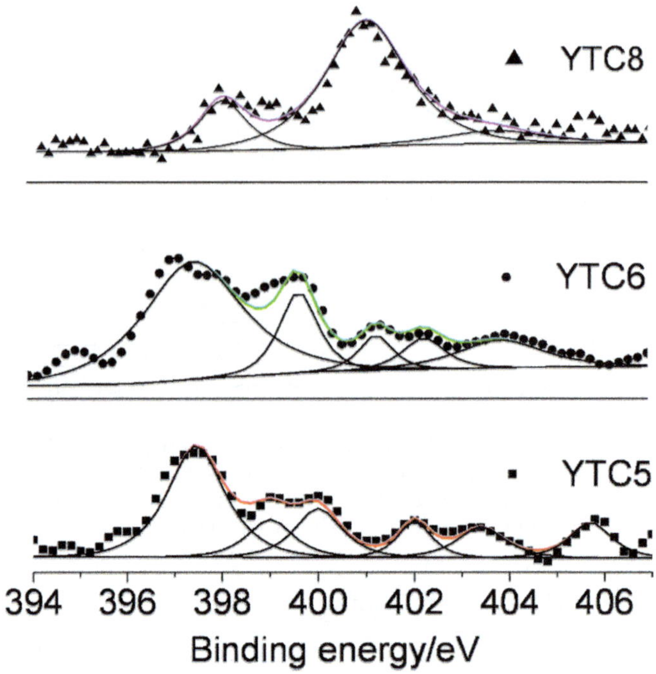

Figure 1.27 N1s spectra of the prepared carbons, the type of nitrogen group, reflected by the peaks of the N1s spectra, depends on the treatment temperature of acetonitrile. Reproduced from ref. 239 with permission from the Royal Society of Chemistry.

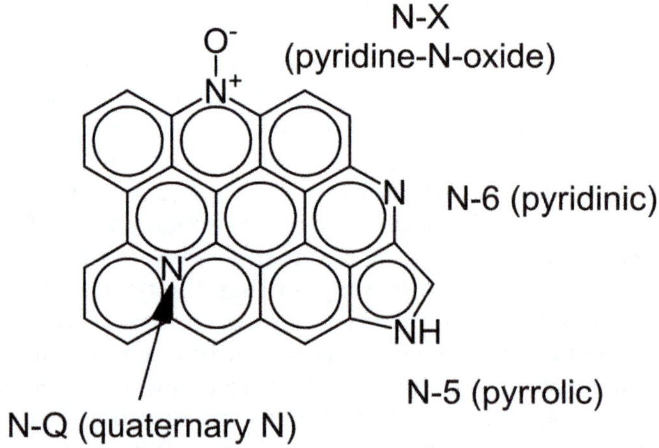

Figure 1.28 Nitrogen-containing groups appearing frequently in the carbon framework.

interactions, especially for the N-doped carbons prepared at a low carbon temperature.[239]

However, the preparation temperature of most carbon adsorbents is usually higher than 600 °C with the purpose of developing high microporosity. The nitrogen species in these N-doped carbons frequently appear in four weakly alkaline or Lewis-acidic types: pyridinic N, pyrrolic N, quaternary N, and pyridine-N-oxide (Figure 1.28). The one-on-one base–acid interactions could only be related to the lone pair of electrons in pyridinic-N and pyrrolic-N. Further, considering the relatively low N-doping level of the carbon adsorbents reported (<5 at%), the N-doping effect of acid–base interaction may be marginal.

In 2012, our group suggested a hydrogen-bonding interaction mechanism on the N-doped activated carbons through quantum chemical calculations.[240] A theoretical N-doped carbon surface model (NCSM) containing five different typical N-containing functional groups was developed. For comparison, a pure carbon surface model (CSM) was also devised, as is shown in Figure 1.29. Density functional theory B3LYP was employed to study the interactions between these models and CO_2 at 12 different positions, corresponding to 12 different H atoms in the NCSM model, and all the configurations were optimized with the 6-31+G* basis set for all atoms using the Gaussian-03 suite package. The optimized results at the 6-31+G* level showed that there were eight NCSM–CO_2 complexes and seven CSM–CO_2 complexes. In these complexes, hydrogen bonds between CO_2 and NCSM/CSM were formed due to the high electronegativity of the oxygen atom in the CO_2 molecule. This type of weak hydrogen bond has been widely studied in recent years, although the interaction of C–H/O was weaker than that of typical hydrogen bonds such as O–H/O and N–H/O. Computational results indicated that binding energies for such hydrogen bonds were different at various positions. The larger the binding energy ΔE (kJ mol^{-1}), the stronger the adsorption affinity. The average binding energy of eight NCSM–CO_2 complexes was 7.84 kJ mol^{-1}, and that of CSM–CO_2 complexes was only 1.26 kJ mol^{-1}, suggesting that the hydrogen bonds in the NCSM–CO_2 complexes were much stronger than those in CSM–CO_2 complexes. This demonstrated that the introduction of N into a carbon surface facilitated the hydrogen-bonding interactions between the carbon surface and CO_2 molecules. The existence of this weak hydrogen-bonding interaction was further confirmed by the FTIR analyses of SK-0.3-700 carbon (Figure 1.30), which showed that the peak due to a C–H anti-symmetric vibration was broadened and red-shifted to a low wavenumber under a CO_2 atmosphere compared with that measured under a N_2 atmosphere.

In the previous section, we discussed a temperature-dependent pore size effect on CO_2 uptake. In that work, the nitrogen content of the PANI-based carbons was also correlated with CO_2 sorption capacities at different sorption temperatures (0–75 °C). What was surprising was that there was no positive correlation between the CO_2 sorption capacity and the N content, concluded from the considerably low coefficient of determination (Figure 1.31, $R^2 \approx$

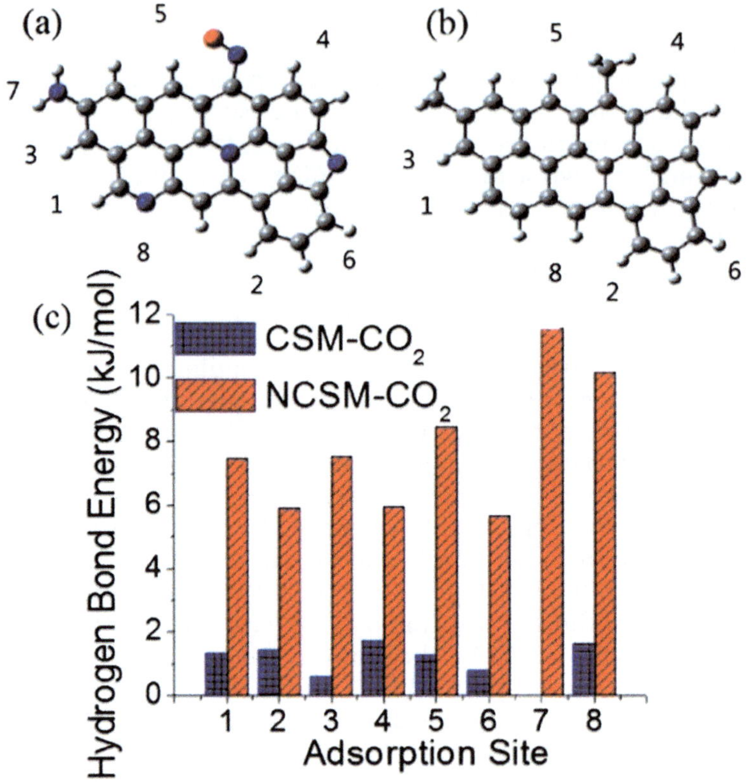

Figure 1.29 Theoretical models for (a) an N-doped carbon surface and (b) a pure carbon surface (red ball: oxygen atom; blue ball: nitrogen; grey ball: carbon; small grey ball: hydrogen). (c) Hydrogen bond energies at different adsorption sites. Reproduced from ref. 240 with permission from the Royal Society of Chemistry.

0.20). According to the D–A theory, the CO_2 molecules fill the micropores in a liquid-like state. We evaluated the CO_2 sorption capacities at 25 °C by multiplying the density of CO_2 at 25 °C by micropore volume for pore sizes smaller than 0.7 nm. These calculated values were very close to the measured CO_2 sorption capacities although the investigated carbons possess very different N contents (from 2.8 wt% to 6.7 wt%). Based on the these facts, we suggested an innovative view that the small micropore plays a critical role in high CO_2 uptake, and the N-doping doesn't enhance CO_2 capture capacities for porous carbons.

Sevilla *et al.* also investigated the role of micropore size and N-doping in CO_2 capture, and confirmed our view.[235] They prepared two types of activated carbons with analogous pore textures and different surface chemistry (N-free and N-doped), and comprehensively compared the CO_2 capture performances of these carbons. The CO_2 capture capacities of N-free and N-doped carbons were analogous, and no significant differences could be

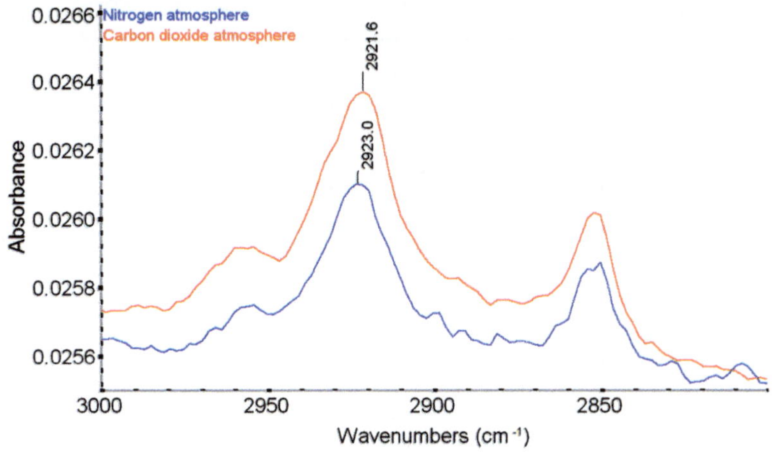

Figure 1.30 FT-IR spectra of the activated carbon measured under different atmospheres. Reproduced from ref. 240 with permission from the Royal Society of Chemistry.

observed between the isotherms of the N-free and N-doped samples over the whole pressure range at three sorption temperatures: −15, 0, and 25 °C. This observation further showed that the nitrogen functionalities present in these N-doped samples do not influence CO_2 adsorption capacities.

Kumar used molecular simulations to study the effects of quaternary N-doping on CO_2 uptake and CO_2–N_2 selectivity in representative carbon pore architectures (slit and disordered carbon structures) at 298 K.[241] The simulation results showed that N-doping can only deliver a marginal enhancement on CO_2 uptake, but it can improve CO_2–N_2 selectivity in smaller micropores. Therefore, they suggested that CO_2 uptake and CO_2–N_2 selectivity were predominantly controlled by the pore architecture as well as ultramicropores. They further found that the tendency of linear CO_2 molecules to lie flat on the carbon surface favored CO_2 uptake in slit pore architectures rather than disordered carbon pore structures. It should be noted that the N-doping effect demonstrated here may be difficult to exemplify experimentally if the material has a disordered pore architecture and complex surface chemistry (such as the presence of other functional groups).[241]

There are some other studies on the effect of N-doping. For instance, Sánchez-Sánchez *et al.* suggested that nitrogen functionalities acted as basic sites and oxygen and phosphorus groups as acidic ones toward the adsorption of CO_2 molecules, and among the N-containing groups, pyrrolic groups exhibited the highest influence, while the positive effect of pyridinic and quaternary functionalities was much smaller.[134] Their results agreed well with our experimental results.[239] On the contrary, a DFT simulation by Lim *et al.* indicated that pyridine groups were most effective at enhancing affinity with CO_2 *via* Lewis base–acid interactions, and pyrrolic groups generated weaker

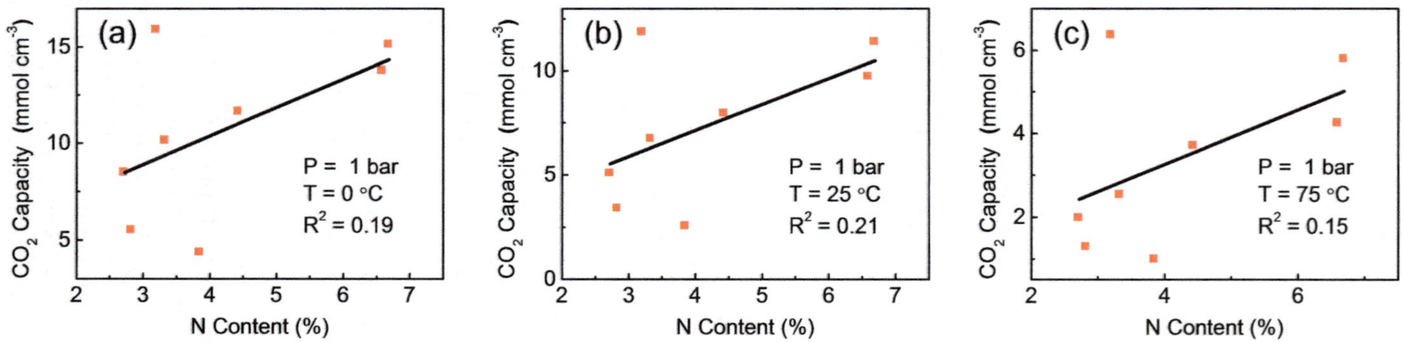

Figure 1.31 Linear fitting of the experimental data. A plot of CO_2 sorption capacity per pore volume *vs.* N-doping level at adsorption temperatures of 0, 25 and 75 °C. Adapted from ref. 105 with permission from the PCCP Owner Societies.

interactions with CO_2.[242] Babu *et al.* carried out experimental research on the CO_2 capture of mesoporous carbon nanotubes, and suggested that the presence of nitrogen functionalities has a beneficial influence on the CO_2 adsorption characteristics over a wide range of pressure (0–36 bar).[243] In a word, the influence of nitrogen incorporation on CO_2 capture is still a controversial research topic with conflicting conclusions, therefore, we can't draw firm conclusions here.

1.6.2.2 Effect of Other Heteroatom-doping

Besides N-doping, other types of heteroatom-doping, such as S-doping,[37,121,244–246] O-doping,[134,247–250] and P-doping,[251] *etc.* have also been researched.

There are several reports on the effects of S-doping on CO_2 capture. For instance, Xia *et al.* found that S-doped microporous carbon materials templated by EMC-2 possessed a homogeneous thiophenic S-doping, and exhibited a large isosteric heat of CO_2 adsorption. They suggested that the thiophenic groups strongly interact with CO_2 molecules *via* strong base–acid interactions and strong pole–pole interactions between the large quadrupole moment of CO_2 and the polar carbon surface associated with S-doping.[121] Seema and his co-workers suggested that the oxidized S content played a more significant role in increased CO_2 adsorption, while the plot of total S content (the sum of oxidized S and thiophenic S) *vs.* CO_2 uptake showed a negative correlation.[37] DFT calculations indicated that the oxidized S groups

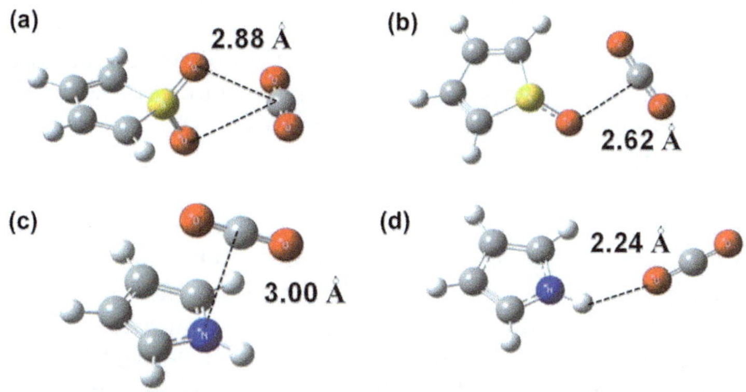

Figure 1.32 Molecular interactions of CO_2 with (a) the di-oxidized S of thiophene, (b) mono-oxidized S of thiophene, (c) pyrrolic N and (d) pyrrolic hydrogen. The binding energies are 4.3, 6.0, 3.4 and 3.1 kcal mol^{-1}, respectively. The dotted distances are denoted in Angstroms. All calculations were performed using the DFT (M062X/6-31+G*) method. Reprinted from H. Seema, K. C. Kemp, N. H. Le, S.-W. Park, V. Chandra, J. W. Lee and K. S. Kim, Highly selective CO_2 capture by S-doped microporous carbon materials, *Carbon*, **66**, 320–326, Copyright 2013, with permission from Elsevier.

attracted CO_2 molecules more strongly than pyrrolic N groups (Figure 1.32). The attraction energy of oxidized S groups with CO_2 was mainly due to the attraction energy between the negative O atom of thiophene (−0.94) and the positive C atom of the CO_2 molecule (+1.07). The enhanced effect of oxidized S (sulfone group) on CO_2 uptake was also reported for MOF-based CO_2 adsorbents.[252] Li *et al.* reported that sulfur, nitrogen, and phosphorus doping were all effective to enhance the gas adsorption capacity of carbon materials; among them, sulfur doping was the most positive.[251] However, Bandosz *et al.* recently reported that the S-containing groups were not stable in a CO_2 atmosphere; the thiophenes were oxidized by CO_2 molecules to sulfones and sulfonic acids, which were thermodynamically unstable and further decomposed forming SO/SO_2 and water, providing additional electrons for CO_2 reduction.[245,246,253]

Oxygen-containing functional groups inherently exist in the carbon frameworks. To some extent, O-containing groups, such as carbonyls, alcohols and ethers, can also induce Lewis base–acid interactions with CO_2 due to the weak electron-donating effect of the oxygen atom. CO_2 adsorption measurements showed that the oxidation process led to an increase in CO_2 adsorption capacity for the porous carbons.[247,248] Liu and Wilcox carried out a simulation investigation on the effect of oxygen-containing surface functionalities on the CO_2 adsorption of microporous carbons.[250] Bader charge analysis indicated that the oxygen atom was highly electronegative and thus can serve as a basic adsorption site to attract more CO_2 molecules. GCMC simulations predicted that oxygen-containing functional groups enhanced CO_2 adsorption in microporous carbon materials in the absence of water vapor. The surface heterogeneity leads to a more condensed packing pattern of CO_2 molecules in the micropores, allowing for maximum utilization of the limited pore space (Figure 1.33). They further observed that the induced polarity of the surface functionalization could enhance the selectivity of CO_2 over CH_4 and N_2, especially in a low-pressure regime.[249]

Metal ion-doping was also proved to enhance the CO_2 capture performance. For example, the K-doped porous carbons reported by Han *et al.* exhibited a high CO_2 adsorption capacity of 1.62 mmol g^{-1} at 0.1 bar and 25 °C and a high kinetic CO_2–N_2 selectivity of 44.[254] The calculation suggested that the highly ionic K–O bonds led to a strong polarization and charge separation within the carbon cluster and thus significant improvement of CO_2 adsorption. Zhu *et al.* reported a carbon adsorbent derived from pine-cone biomass, and suggested that the well-dispersed nitrogen and calcium dopants (Ca^{2+}) were positive to CO_2 capture.[255]

Undoubtedly, surface chemistry influences the CO_2 capture of carbon-based adsorbents, including CO_2 capacity, heat of adsorption, and adsorption selectivity of CO_2 over other gases. Many experimental and simulation investigations have proved this conclusion. Based on the results obtained so far, the interactions between the surface doping and CO_2 molecules could be generally summarized as Lewis acid–base interactions, hydrogen bonding, and electrostatic interactions. However, the exact role of heteroatom-doping, particularly N-doping, remains controversial and more studies are needed. Furthermore, the effect of heteroatom-doping should be evaluated in a practical (or simulated) condition with flue gas because the polar surface sites will enhance the affinity for the H_2O that saturates the flue gas, which will significantly degrade the CO_2 capture.

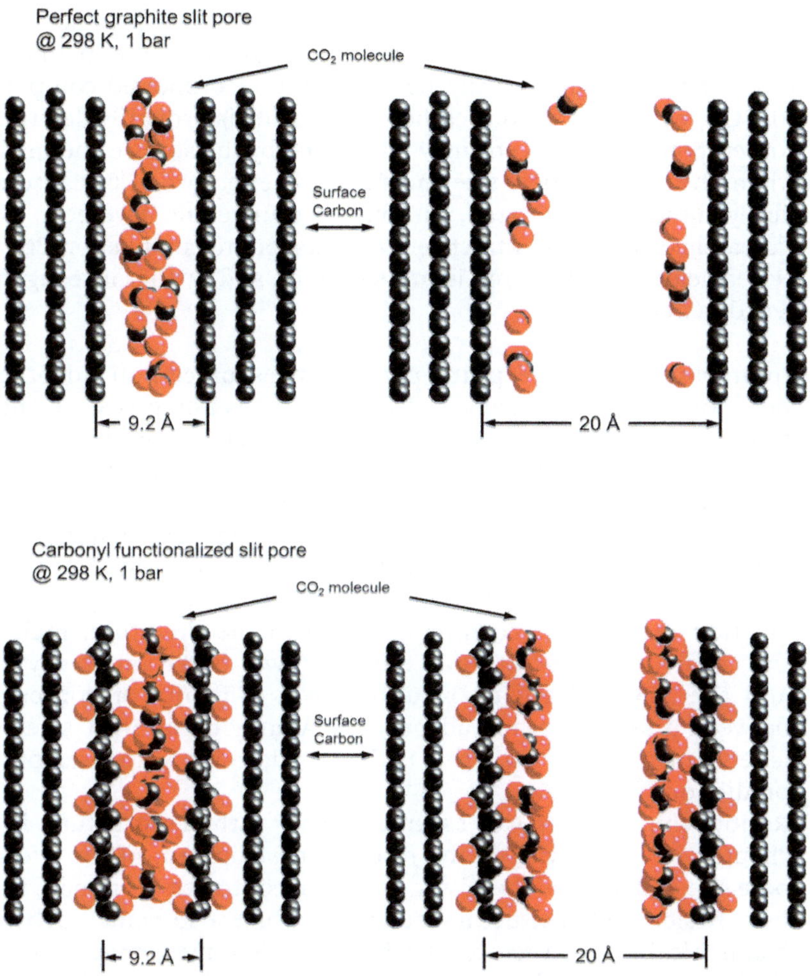

Figure 1.33 Comparisons of CO_2 adsorbed in functionalized micropores with that in the perfect graphite slit pore: left, side views of adsorbed CO_2 in various functionalized graphite slit pores with a pore width of 9.2 Å; right, side views of adsorbed CO_2 in various functionalized graphite slit pores with a pore width of 20 Å. Adapted with permission from Y. Liu and J. Wilcox, *Environ. Sci. Technol.*, 2012, **46**, 1940, Copyright 2012 American Chemical Society.

1.7 Summary and Outlook

In this chapter, the recent research progress in the field of carbon-based materials for post-combustion CO_2 capture has been reviewed. The carbon-based CO_2 adsorbents touched on here are classified into porous carbons, graphene-based carbons, carbon nanotubes, and carbon-based hybrid adsorbents. In the section on porous carbon materials, activated carbons, templated carbons, hierarchical porous carbons, and MOFs (ZIFs, COFs)-based carbons, were respectively discussed. The carbon-based hybrid adsorbents were divided into carbon–organic hybrid materials and carbon–inorganic hybrid materials according to the nature of the incorporated component. Since the CO_2 capture of porous carbon adsorbents is primarily determined by the microporosity, and is strongly influenced by the surface chemistry of the carbon, the effects of pore size and surface chemistry on CO_2 capture are specifically summarized. Overall, carbon-based materials have been extensively researched, and show attractive prospects for pre-combustion CO_2 capture. Nevertheless, this research is in its infancy, and further investigations are needed in many areas.

(1) Improving the CO_2 capture performances of carbon adsorbents: post-combustion CO_2 capture from flue gases possesses the characteristic of a low partial pressure of CO_2 (about 10–15% in flue gases) and selective separation of CO_2 and N_2. The capacity and selectivity of carbon materials are still low for this application. Tuning the pore size, inducing polarity by heteroatom-doping, and making hybrid composites, are efficient strategies for enhancing the CO_2 capture performances of carbons.[256] On the subject of selectivity, three calculation methods are commonly used in the literature, including the capacity ratio of CO_2–N_2, the selectivity based on Henry's law, and the IAST selectivity, resulting in a difficulty of comparison between different carbon adsorbents. Among them, IAST selectivity is recommended because this is the standard in mixture adsorption predictions.[257]

(2) Relationship between CO_2 capture and the carbon's structure: in this chapter, we have reviewed the research development of the effects of pore size and surface chemistry. However, controversies remain in this field. Besides, other structure factors of carbon adsorbents, such as the hierarchical pore structure, the ratio of micro/meso/macro porosity, the length of pores, the size and shape of the carbon particle, *etc.*, also influence CO_2 capture (*e.g.*, adsorption diffusion dynamics of CO_2), and need to be studied.

(3) Evaluation of CO_2 separation under real (or simulated) adsorption conditions: the current works mostly focus on the static adsorption of a single-component gas. However, real CO_2 separations are performed as dynamic adsorption processes. Furthermore, flue gas in the field of post-combustion capture is a mixture of mostly N_2, CO_2, and some

moisture. The co-existing gas components can cause a decline in the CO_2 adsorption capacity. Therefore, evaluation of CO_2 capture should be preferentially conducted on real (or simulated) flue gas under dynamic conditions.

(4) Development of low-cost and sustainable carbon adsorbents: the cost and sustainability of CO_2 sorbents is crucial in practical applications due to the huge demand for them. In view of this regard, utilization of renewable precursors (typically various biomass sources), reduction/elimination of the activating agents (*e.g.*, single-ion activation, self-template method), and lowering the energy consumption of carbonization (*e.g.*, low carbonization temperature, microwave heating, *etc.*), can be considered in the manufacture of low-cost carbon adsorbents.

Acknowledgements

This work was financially supported by the Natural Science Foundation of China (NSFC21576158, 51302156, 21476264), Natural Science Foundation of Shandong Province (ZR2017JL014) and Distinguished Young Scientist Foundation of Shandong Province (JQ201215).

References

1. B. Metz, O. Davidson, H. De Coninck, M. Loos and L. Meyer, *IPCC Special Report on Carbon Dioxide Capture and Storage*, Intergovernmental Panel on Climate Change, Geneva, Switzerland, Working Group III, 2005.
2. Y. Wang, Y. Qin, X. Wang and F. Wang, *ACS Catal.*, 2014, **5**, 393.
3. M. Cheng, D. R. Moore, J. J. Reczek, B. M. Chamberlain, E. B. Lobkovsky and G. W. Coates, *J. Am. Chem. Soc.*, 2001, **123**, 8738.
4. Y. Kou, Y. Nabetani, D. Masui, T. Shimada, S. Takagi, H. Tachibana and H. Inoue, *J. Am. Chem. Soc.*, 2014, **136**, 6021.
5. W. Tu, Y. Zhou and Z. Zou, *Adv. Mater.*, 2014, **26**, 4607.
6. M. S. Xie, B. Y. Xia, Y. Li, Y. Yan, Y. Yang, Q. Sun, S. H. Chan, A. Fisher and X. Wang, *Energy Environ. Sci.*, 2016, **9**, 1687.
7. D. D. Zhu, J. L. Liu and S. Z. Qiao, *Adv. Mater.*, 2016, **28**, 3423.
8. N. P. Wickramaratne and M. Jaroniec, *J. Mater. Chem. A*, 2013, **1**, 112.
9. J. Wang, I. Senkovska, M. Oschatz, M. R. Lohe, L. Borchardt, A. Heerwig, Q. Liu and S. Kaskel, *ACS Appl. Mater. Interfaces*, 2013, **5**, 3160.
10. J.-H. Kim, C.-H. Lee, W.-S. Kim, J.-S. Lee, J.-T. Kim, J.-K. Suh and J.-M. Lee, *J. Chem. Eng. Data*, 2003, **48**, 137.
11. A. Phan, C. J. Doonan, F. J. Uribe-Romo, C. B. Knobler, M. O'keeffe and O. M. Yaghi, *Acc. Chem. Res.*, 2010, **43**, 58.
12. S. Nandi, S. Collins, D. Chakraborty, D. Banerjee, P. K. Thallapally, T. K. Woo and R. Vaidhyanathan, *J. Am. Chem. Soc.*, 2017, **139**, 1734.

13. Y. Zeng, R. Zou and Y. Zhao, *Adv. Mater.*, 2016, **28**, 2855.
14. L. An, H. Wang, F. Xu, X.-L. Wang, F. Wang and J. Li, *Inorg. Chem.*, 2016, **55**, 12923.
15. J. Ludwinowicz and M. Jaroniec, *Carbon*, 2015, **94**, 673.
16. J. Wang and S. Kaskel, *J. Mater. Chem.*, 2012, **22**, 23710.
17. Y. Zhang, B. Li, K. Williams, W. Y. Gao and S. Ma, *Chem. Commun.*, 2013, **49**, 10269.
18. M. Yang, L. Guo, G. Hu, X. Hu, J. Chen, S. Shen, W. Dai and M. Fan, *Ind. Eng. Chem. Res.*, 2016, **55**, 757.
19. S.-Y. Lee, H.-M. Yoo, S. W. Park, S. H. Park, Y. S. Oh, K. Y. Rhee and S.-J. Park, *J. Solid State Chem.*, 2014, **215**, 201.
20. M. Sevilla, P. Valle-Vigón and A. B. Fuertes, *Adv. Funct. Mater.*, 2011, **21**, 2781.
21. J. He, J. W. F. To, P. C. Psarras, H. Yan, T. Atkinson, R. T. Holmes, D. Nordlund, Z. Bao and J. Wilcox, *Adv. Energy Mater.*, 2016, **6**, 1502491.
22. W. Shen, S. Zhang, Y. He, J. Li and W. Fan, *J. Mater. Chem.*, 2011, **21**, 14036.
23. A. Wahby, J. M. Ramos-Fernandez, M. Martinez-Escandell, A. Sepulveda-Escribano, J. Silvestre-Albero and F. Rodriguez-Reinoso, *ChemSuschem*, 2010, **3**, 974.
24. J. Silvestre-Albero, A. Wahby, A. Sepulveda-Escribano, M. Martinez-Escandell, K. Kaneko and F. Rodriguez-Reinoso, *Chem. Commun.*, 2011, **47**, 6840.
25. X. Hu, M. Radosz, K. A. Cychosz and M. Thommes, *Environ. Sci. Technol.*, 2011, **45**, 7068.
26. A. Arami-Niya, T. E. Rufford and Z. Zhu, *Carbon*, 2016, **103**, 115.
27. N. P. Wickramaratne and M. Jaroniec, *J. Mater. Chem. A*, 2013, **1**, 112.
28. M. Wei, Q. Yu, T. Mu, L. Hou, Z. Zuo and J. Peng, *Adsorption*, 2016, **22**, 385.
29. J. Cai, J. Qi, C. Yang and X. Zhao, *ACS Appl. Mater. Interfaces*, 2014, **6**, 3703.
30. J. Choma, L. Osuchowski, M. Marszewski and M. Jaroniec, *RSC Adv.*, 2014, **4**, 14795.
31. V. Chandra, S. U. Yu, S. H. Kim, Y. S. Yoon, D. Y. Kim, A. H. Kwon, M. Meyyappan and K. S. Kim, *Chem. Commun.*, 2012, **48**, 735.
32. C. Ge, J. Song, Z. Qin, J. Wang and W. Fan, *ACS Appl. Mater. Interfaces*, 2016, **8**, 18849.
33. Z. Liu, Z. Du, W. Xing and Z. Yan, *Mater. Lett.*, 2014, **117**, 273.
34. G. Sethia and A. Sayari, *Carbon*, 2015, **93**, 68.
35. J. Wang, I. Senkovska, M. Oschatz, M. R. Lohe, L. Borchardt, A. Heerwig, Q. Liu and S. Kaskel, *J. Mater. Chem. A*, 2013, **1**, 10951.
36. J. Wang and Q. Liu, *Nanoscale*, 2014, **6**, 4148.
37. H. Seema, K. C. Kemp, N. H. Le, S.-W. Park, V. Chandra, J. W. Lee and K. S. Kim, *Carbon*, 2014, **66**, 320.
38. H. Cong, M. Zhang, Y. Chen, K. Chen, Y. Hao, Y. Zhao and L. Feng, *Carbon*, 2015, **92**, 297.

39. G. P. Hao, W. C. Li, D. Qian and A. H. Lu, *Adv. Mater.*, 2010, **22**, 853.
40. G. P. Hao, W. C. Li, D. Qian, G. H. Wang, W. P. Zhang, T. Zhang, A. Q. Wang, F. Schuth, H. J. Bongard and A. H. Lu, *J. Am. Chem. Soc.*, 2011, **133**, 11378.
41. R. Banerjee, H. Furukawa, D. Britt, C. Knobler, M. O'Keeffe and O. M. Yaghi, *J. Am. Chem. Soc.*, 2009, **131**, 3875.
42. J. An, S. J. Geib and N. L. Rosi, *J. Am. Chem. Soc.*, 2009, **132**, 38.
43. J. W. Lee, B. Hawkins, D. M. Day and D. C. Reicosky, *Energy Environ. Sci.*, 2010, **3**, 1695.
44. D. L. Sivadas, S. Vijayan, R. Rajeev, K. N. Ninan and K. Prabhakaran, *Carbon*, 2016, **109**, 7.
45. Y. Hu, X. Tong, H. Zhuo, L. Zhong, X. Peng, S. Wang and R. Sun, *RSC Adv.*, 2016, **6**, 15788.
46. X. Fan, L. Zhang, G. Zhang, Z. Shu and J. Shi, *Carbon*, 2013, **61**, 423.
47. A. Alabadi, S. Razzaque, Y. Yang, S. Chen and B. Tan, *Chem. Eng. J.*, 2015, **281**, 606.
48. M. Sevilla and A. B. Fuertes, *Energy Environ. Sci.*, 2011, **4**, 1765.
49. J. Chen, J. Yang, G. Hu, X. Hu, Z. Li, S. Shen, M. Radosz and M. Fan, *ACS Sustainable Chem. Eng.*, 2016, **4**, 1439.
50. S. Deng, B. Hu, T. Chen, B. Wang, J. Huang, Y. Wang and G. Yu, *Adsorption*, 2015, **21**, 125.
51. G. Dobele, T. Dizhbite, M. V. Gil, A. Volperts and T. A. Centeno, *Biomass Bioenergy*, 2012, **46**, 145.
52. H. M. Coromina, D. A. Walsh and R. Mokaya, *J. Mater. Chem. A*, 2016, **4**, 280.
53. Z. Zhang, K. Wang, J. D. Atkinson, X. Yan, X. Li, M. J. Rood and Z. Yan, *J. Hazard. Mater.*, 2012, **229**, 183.
54. A. E. Creamer, B. Gao and M. Zhang, *Chem. Eng. J.*, 2014, **249**, 174.
55. H. Wei, S. Deng, B. Hu, Z. Chen, B. Wang, J. Huang and G. Yu, *ChemSuschem*, 2012, **5**, 2354.
56. R. Wang, P. Wang, X. Yan, J. Lang, C. Peng and Q. Xue, *ACS Appl. Mater. Interfaces*, 2012, **4**, 5800.
57. J. Wang, A. Heerwig, M. R. Lohe, M. Oschatz, L. Borchardt and S. Kaskel, *J. Mater. Chem.*, 2012, **22**, 13911.
58. W. Shen, Y. He, S. Zhang, J. Li and W. Fan, *ChemSusChem*, 2012, **5**, 1274.
59. S.-M. Hong, E. Jang, A. D. Dysart, V. G. Pol and K. B. Lee, *Sci. Rep.*, 2016, **6**, 34590.
60. J. S. Tan and F. N. Ani, *Sep. Purif. Technol.*, 2004, **35**, 47.
61. A. S. Ello, L. K. C. de Souza, A. Trokourey and M. Jaroniec, *J. CO_2 Util.*, 2013, **2**, 35.
62. L. Guo, J. Yang, G. Hu, X. Hu, L. Wang, Y. Dong, H. DaCosta and M. Fan, *ACS Sustainable Chem. Eng.*, 2016, **4**, 2806.
63. S. Shahkarami, A. K. Dalai, J. Soltan, Y. Hu and D. Wang, *Energy Fuels*, 2015, **29**, 7433.
64. H. Luo, C. C. Zhu, Z. C. Tan, L. W. Bao, J. J. Wang, G. Miao, L. Z. Kong and Y. H. Sun, *RSC Adv.*, 2016, **6**, 38724.

65. X. Ma, Y. Li, M. Cao and C. Hu, *J. Mater. Chem. A*, 2014, **2**, 4819.
66. A. A. Alhwaige, H. Ishida and S. Qutubuddin, *ACS Sustainable Chem. Eng.*, 2016, **4**, 1286.
67. Q. Wu, W. Li, S. Liu and C. Jin, *Appl. Surf. Sci.*, 2016, **369**, 101.
68. W. Travis, S. Gadipelli and Z. Guo, *RSC Adv.*, 2015, **5**, 29558.
69. Z. Zhao, Y. Wang, M. Li and R. Yang, *RSC Adv.*, 2015, **5**, 34803.
70. K. Li, S. Tian, J. Jiang, J. Wang, X. Chen and F. Yan, *J. Mater. Chem. A*, 2016, **4**, 5223.
71. T. Liang, C. Chen, X. Li and J. Zhang, *Langmuir*, 2016, **32**, 8042.
72. T. Wei, Q. Zhang, X. Wei, Y. Gao and H. Li, *Sci. Rep.*, 2016, **6**, 22646.
73. E. Haffner-Staton, N. Balahmar and R. Mokaya, *J. Mater. Chem. A*, 2016, **4**, 13324.
74. R. H. Hesas, A. Arami-Niya, W. M. A. W. Daud and J. N. Sahu, *J. Ind. Eng. Chem.*, 2015, **24**, 196.
75. C. Zhang, W. Song, Q. Ma, L. Xie, X. Zhang and H. Guo, *Energy Fuels*, 2016, **30**, 4181.
76. B. Huang, H. Shao, N. Liu, Z. J. Xu and Y. Huang, *RSC Adv.*, 2015, **5**, 88171.
77. A. Primo, A. Forneli, A. Corma and H. Garcia, *ChemSusChem*, 2012, **5**, 2207.
78. Z. Geng, Q. Xiao, H. Lv, B. Li, H. Wu, Y. Lu and C. Zhang, *Sci. Rep.*, 2016, **6**, 30049.
79. M. S. Shafeeyan, W. M. A. W. Daud, A. Houshmand and A. Arami-Niya, *Appl. Surf. Sci.*, 2011, **257**, 3936.
80. A. Arami-Niya, T. E. Rufford and Z. Zhu, *Energy Fuels*, 2016, **30**, 7298.
81. A. S. Ello, L. K. C. de Souza, A. Trokourey and M. Jaroniec, *Microporous Mesoporous Mater.*, 2013, **180**, 280.
82. A. S. Gonzalez, M. G. Plaza, F. Rubiera and C. Pevida, *Chem. Eng. J.*, 2013, **230**, 456.
83. W. Hao, E. Björkman, M. Lilliestråle and N. Hedin, *Appl. Energy*, 2013, **112**, 526.
84. Z. Xiong, Z. Shihong, Y. Haiping, S. Tao, C. Yingquan and C. Hanping, *BioEnergy Res.*, 2013, **6**, 1147.
85. L.-Y. Meng and S.-J. Park, *Mater. Chem. Phys.*, 2014, **143**, 1158.
86. A. Silvestre-Albero, J. Silvestre-Albero, M. Martinez-Escandell and F. Rodriguez-Reinoso, *Ind. Eng. Chem. Res.*, 2014, **53**, 15398.
87. J. Ludwinowicz and M. Jaroniec, *Carbon*, 2015, **82**, 297.
88. J. Zhou, Z. H. Li, W. Xing, T. T. Zhu, H. L. Shen and S. P. Zhuo, *Chem. Commun.*, 2015, **51**, 4591.
89. J. Zhou, Z. H. Li, W. Xing, H. L. Shen, X. Bi, T. T. Zhu, Z. P. Qiu and S. P. Zhuo, *Adv. Funct. Mater.*, 2016, **26**, 7955.
90. J. Wang, R. Krishna, X. Wu, Y. Sun and S. Deng, *Langmuir*, 2015, **31**, 9845.
91. L.-Y. Meng and S.-J. Park, *J. Solid State Chem.*, 2014, **218**, 90.
92. A. Heidari, H. Younesi, A. Rashidi and A. A. Ghoreyshi, *Chem. Eng. J.*, 2014, **254**, 503.

93. S. Khalili, B. Khoshandam and M. Jahanshahi, *RSC Adv.*, 2015, **5**, 94115.
94. K. Huang, S.-H. Chai, R. T. Mayes, S. Tan, C. W. Jones and S. Dai, *Microporous Mesoporous Mater.*, 2016, **230**, 100–108.
95. H. S. Kim, M. S. Kang and W. C. Yoo, *J. Phys. Chem. C*, 2015, **119**, 28512.
96. F. Caturla, M. Molina-Sabio and F. Rodriguez-Reinoso, *Carbon*, 1991, **29**, 999.
97. V. Subramanian, C. Luo, A. Stephan, K. Nahm, S. Thomas and B. Wei, *J. Phys. Chem. C*, 2007, **111**, 7527.
98. M. Zhi, F. Yang, F. Meng, M. Li, A. Manivannan and N. Wu, *ACS Sustainable Chem. Eng.*, 2014, **2**, 1592.
99. M. G. Plaza, F. Rubiera, J. J. Pis and C. Pevida, *Appl. Surf. Sci.*, 2010, **256**, 6843.
100. H. Teng and S.-C. Wang, *Carbon*, 2000, **38**, 817.
101. N. P. Wickramaratne and M. Jaroniec, *ACS Appl. Mater. Interfaces*, 2013, **5**, 1849.
102. N. P. Wickramaratne, J. Xu, M. Wang, L. Zhu, L. Dai and M. Jaroniec, *Chem. Mater.*, 2014, **26**, 2820.
103. R.-L. Tseng, F.-C. Wu and R.-S. Juang, *Sep. Purif. Technol.*, 2015, **140**, 53.
104. Y.-J. Heo and S.-J. Park, *Energy*, 2015, **91**, 142.
105. Z. Zhang, J. Zhou, W. Xing, Q. Xue, Z. Yan, S. Zhuo and S. Z. Qiao, *Phys. Chem. Chem. Phys.*, 2013, **15**, 2523.
106. V. Presser, J. McDonough, S.-H. Yeon and Y. Gogotsi, *Energy Environ. Sci.*, 2011, **4**, 3059.
107. M. Sevilla, J. B. Parra and A. B. Fuertes, *ACS Appl. Mater. Interfaces*, 2013, **5**, 6360.
108. J. Zhou, H. Shen, Z. Li, S. Zhang, Y. Zhao, X. Bi, Y. Wang, H. Cui and S. Zhuo, *Electrochim. Acta*, 2016, **209**, 557.
109. M. Dubinin, *Carbon*, 1989, **27**, 457.
110. X. Wang, J. Zhou, W. Xing, B. Liu, J. Zhang, H. Lin, H. Cui and S. Zhuo, *J. Energy Chem.*, 2017, **26**, 1007.
111. J. H. Knox, B. Kaur and G. Millward, *J. Chromatogr. A*, 1986, **352**, 3.
112. M. E. Davis, *Nature*, 2002, **417**, 813.
113. C. Liang, K. Hong, G. A. Guiochon, J. W. Mays and S. Dai, *Angew. Chem., Int. Ed.*, 2004, **43**, 5785.
114. S. Jun, S. H. Joo, R. Ryoo, M. Kruk, M. Jaroniec, Z. Liu, T. Ohsuna and O. Terasaki, *J. Am. Chem. Soc.*, 2000, **122**, 10712.
115. C. Liang, Z. Li and S. Dai, *Angew. Chem., Int. Ed.*, 2008, **47**, 3696.
116. Y. Meng, D. Gu, F. Zhang, Y. Shi, H. Yang, Z. Li, C. Yu, B. Tu and D. Zhao, *Angew. Chem.*, 2005, **117**, 7215.
117. D. K. Singh, K. S. Krishna, S. Harish, S. Sampath and M. Eswaramoorthy, *Angew. Chem., Int. Ed.*, 2016, **55**, 2032.
118. K. S. Lakhi, W. S. Cha, S. Joseph, B. J. Wood, S. S. Aldeyab, G. Lawrence, J.-H. Choy and A. Vinu, *Catal. Today*, 2015, **243**, 209.
119. C. Goel, H. Bhunia and P. K. Bajpai, *J. Environ. Sci.*, 2015, **32**, 238.
120. X. Peng, D. Can and W. Wang, *Chem. Eng. Sci.*, 2011, **66**, 2266.
121. Y. Xia, Y. Zhu and Y. Tang, *Carbon*, 2012, **50**, 5543.

122. P. Pachfule, B. P. Biswal and R. Banerjee, *Chem. - Eur. J.*, 2012, **18**, 11399.
123. L.-Y. Meng and S.-J. Park, *J. Colloid Interface Sci.*, 2012, **366**, 125.
124. K. Xia, Y. Yu, Y. Li, S. Li, Y. Wang, G. Wang, H. Zhang and A. Chen, *RSC Adv.*, 2016, **6**, 91557.
125. K. M. Nelson, S. M. Mahurin, R. T. Mayes, B. Williamson, C. M. Teague, A. J. Binder, L. Baggetto, G. M. Veith and S. Dai, *Microporous Mesoporous Mater.*, 2016, **222**, 94.
126. H.-M. Yoo, S.-Y. Lee and S.-J. Park, *J. Solid State Chem.*, 2013, **197**, 361.
127. Z. Zhang, B. Wang, C. Zhu, P. Gao, Z. Tang, N. Sun, W. Wei and Y. Sun, *J. Mater. Chem. A*, 2015, **3**, 23990–23999.
128. Z. Wu, P. A. Webley and D. Zhao, *J. Mater. Chem.*, 2012, **22**, 11379.
129. L. Liu, Q.-F. Deng, T.-Y. Ma, X.-Z. Lin, X.-X. Hou, Y.-P. Liu and Z.-Y. Yuan, *J. Mater. Chem.*, 2011, **21**, 16001.
130. R.-L. Liu, W.-J. Ji, T. He, Z.-Q. Zhang, J. Zhang and F.-Q. Dang, *Carbon*, 2014, **76**, 84.
131. J. Wei, D. Zhou, Z. Sun, Y. Deng, Y. Xia and D. Zhao, *Adv. Funct. Mater.*, 2013, **23**, 2322.
132. K. S. Lakhi, W. S. Cha, J.-H. Choy, M. Al-Ejji, A. M. Abdullah, A. M. Al-Enizi and A. Vinu, *Microporous Mesoporous Mater.*, 2016, **233**, 44.
133. Y. Wang, X. Bai, F. Wang, H. Qin, C. Yin, S. Kang, X. Li, Y. Zuo and L. Cui, *Sci. Rep.*, 2016, **6**, 26673.
134. A. Sanchez-Sanchez, F. Suarez-Garcia, A. Martinez-Alonso and J. M. D. Tascon, *ACS Appl. Mater. Interfaces*, 2014, **6**, 21237–21247.
135. D. Tiwari, C. Goel, H. Bhunia and P. K. Bajpai, *RSC Adv.*, 2016, **6**, 97728.
136. Y. Zhao, L. Zhao, K. X. Yao, Y. Yang, Q. Zhang and Y. Han, *J. Mater. Chem.*, 2012, **22**, 19726.
137. C. Ducrot-Boisgontier, J. Parmentier, A. Faour, J. Patarin and G. D. Pringruber, *Energy Fuels*, 2010, **24**, 3595.
138. L. Wang and R. T. Yang, *J. Phys. Chem. C*, 2012, **116**, 1099.
139. Y. Xia, R. Mokaya, G. S. Walker and Y. Zhu, *Adv. Energy Mater.*, 2011, **1**, 678.
140. S. Gadipelli and Z. X. Guo, *ChemSusChem*, 2015, **8**, 2123.
141. Z. Bian, X. Zhu, T. Jin, J. Gao, J. Hu and H. Liu, *Microporous Mesoporous Mater.*, 2014, **200**, 159.
142. J.-K. Sun and Q. Xu, *Energy Environ. Sci.*, 2014, **7**, 2071.
143. B. Liu, H. Shioyama, T. Akita and Q. Xu, *J. Am. Chem. Soc.*, 2008, **130**, 5390.
144. B. Liu, H. Shioyama, H. Jiang, X. Zhang and Q. Xu, *Carbon*, 2010, **48**, 456.
145. J. Wang, Y. Lin, Q. Yue, K. Tao, C. Kong and L. Chen, *RSC Adv.*, 2016, **6**, 53017.
146. Y. Zhang, B. Li, K. Williams, W.-Y. Gao and S. Ma, *Chem. Commun.*, 2013, **49**, 10269.
147. G. Srinivas, V. Krungleviciute, Z.-X. Guo and T. Yildirim, *Energy Environ. Sci.*, 2014, **7**, 335.

148. S. Ding, Q. Dong, J. Hu, W. Xiao, X. Liu, L. Liao and N. Zhang, *Chem. Commun.*, 2016, **52**, 9757.
149. X. Ma, L. Li, S. Wang, M. Lu, H. Li, W. Ma and T. C. Keener, *Appl. Surf. Sci.*, 2016, **369**, 390.
150. W. Zhao, S. Han, X. Zhuang, F. Zhang, Y. Mai and X. Feng, *J. Mater. Chem. A*, 2015, **3**, 23352.
151. A. Nikitin and Y. Gogotsi, *J. Nanosci. Nanotechnol.*, 2004, **7**, 553.
152. G. N. Yushin, E. N. Hoffman, A. Nikitin, H. Ye, M. W. Barsoum and Y. Gogotsi, *Carbon*, 2005, **43**, 2075.
153. V. Presser, M. Heon and Y. Gogotsi, *Adv. Funct. Mater.*, 2011, **21**, 810.
154. J. Chmiola, G. Yushin, Y. Gogotsi, C. Portet, P. Simon and P. L. Taberna, *Science*, 2006, **313**, 1760.
155. C. Largeot, C. Portet, J. Chmiola, P.-L. Taberna, Y. Gogotsi and P. Simon, *J. Am. Chem. Soc.*, 2008, **130**, 2730.
156. G. Yushin, R. Dash, J. Jagiello, J. E. Fischer and Y. Gogotsi, *Adv. Funct. Mater.*, 2006, **16**, 2288.
157. Y. Gogotsi, R. K. Dash, G. Yushin, T. Yildirim, G. Laudisio and J. E. Fischer, *J. Am. Chem. Soc.*, 2005, **127**, 16006.
158. H. S. Kim, J. P. Singer, Y. Gogotsi and J. E. Fischer, *Microporous Mesoporous Mater.*, 2009, **120**, 267.
159. M. R. Bonilla, J.-S. Bae, T. X. Nguyen and S. K. Bhatia, *J. Phys. Chem. C*, 2010, **114**, 16562.
160. W. Xing, C. Liu, Z. Zhou, J. Zhou, G. Wang, S. Zhuo, Q. Xue, L. Song and Z. Yan, *Nanoscale Res. Lett.*, 2014, **9**, 189.
161. S. K. Bhatia and T. X. Nguyen, *Ind. Eng. Chem. Res.*, 2011, **50**, 10380.
162. A. Silvestre-Albero, S. Rico-Francés, F. Rodríguez-Reinoso, A. M. Kern, M. Klumpp, B. J. M. Etzold and J. Silvestre-Albero, *Carbon*, 2013, **59**, 221.
163. M. Kormann and N. Popovska, *Microporous Mesoporous Mater.*, 2010, **130**, 167.
164. G. S. Chai, I. Shin and J. S. Yu, *Adv. Mater.*, 2004, **16**, 2057.
165. Y. Deng, C. Liu, T. Yu, F. Liu, F. Zhang, Y. Wan, L. Zhang, C. Wang, B. Tu and P. A. Webley, *Chem. Mater.*, 2007, **19**, 3271.
166. Q. Li, R. Jiang, Y. Dou, Z. Wu, T. Huang, D. Feng, J. Yang, A. Yu and D. Zhao, *Carbon*, 2011, **49**, 1248.
167. L. Estevez, R. Dua, N. Bhandari, A. Ramanujapuram, P. Wang and E. P. Giannelis, *Energy Environ. Sci.*, 2013, **6**, 1785.
168. A. Wilke and J. Weber, *J. Mater. Chem.*, 2011, **21**, 5226.
169. D. Li, Y. Chen, M. Zheng, H. Zhao, Y. Zhao and Z. Sun, *ACS Sustainable Chem. Eng.*, 2016, **4**, 298.
170. G.-P. Hao, W.-C. Li, D. Qian, G.-H. Wang, W.-P. Zhang, T. Zhang, A.-Q. Wang, F. Schueth, H.-J. Bongard and A.-H. Lu, *J. Am. Chem. Soc.*, 2011, **133**, 11378.
171. F. Liu, K. Huang, S. Ding and S. Dai, *J. Mater. Chem. A*, 2016, **4**, 14567.
172. Z. Tian, Y. Qiu, J. Zhou, X. Zhao and J. Cai, *Mater. Lett.*, 2016, **180**, 162.
173. M. Freitag, *Nat. Nanotechnol.*, 2008, **3**, 455.

174. Y. Zhu, S. Murali, M. D. Stoller, K. J. Ganesh, W. Cai, P. J. Ferreira, A. Pirkle, R. M. Wallace, K. A. Cychosz, M. Thommes, D. Su, E. A. Stach and R. S. Ruoff, *Science*, 2011, **332**, 1537.
175. X. Zhou, J. Qiao, L. Yang and J. Zhang, *Adv. Energy Mater.*, 2014, **4**, 1301523.
176. E. Yoo, J. Kim, E. Hosono, H.-S. Zhou, T. Kudo and I. Honma, *Nano Lett.*, 2008, **8**, 2277.
177. Z. Yin, J. Zhu, Q. He, X. Cao, C. Tan, H. Chen, Q. Yan and H. Zhang, *Adv. Energy Mater.*, 2014, **4**, 1300574.
178. H. Huang, Z. Song, N. Wei, L. Shi, Y. Mao, Y. Ying, L. Sun, Z. Xu and X. Peng, *Nat. Commun.*, 2013, **4**, 2979.
179. G. Srinivas, J. Burress and T. Yildirim, *Energy Environ. Sci.*, 2012, **5**, 6453.
180. M. Shan, Q. Xue, N. Jing, C. Ling, T. Zhang, Z. Yan and J. Zheng, *Nanoscale*, 2012, **4**, 5477.
181. W. Li, X. Geng, Y. Guo, J. Rong, Y. Gong, L. Wu, X. Zhang, P. Li, J. Xu and G. Cheng, *ACS Nano*, 2011, **5**, 6955–6961.
182. S. Gadipelli and Z. X. Guo, *Prog. Mater. Sci.*, 2015, **69**, 1.
183. R. Balasubramanian and S. Chowdhury, *J. Mater. Chem. A*, 2015, **3**, 21968.
184. G. Srinivas, J. Burress and T. Yildirim, *Energy Environ. Sci.*, 2012, **5**, 6453.
185. Z.-Y. Sui, Q.-H. Meng, J.-T. Li, J.-H. Zhu, Y. Cui and B.-H. Han, *J. Mater. Chem. A*, 2014, **2**, 9891.
186. S. Chowdhury and R. Balasubramanian, *Ind. Eng. Chem. Res.*, 2016, **55**, 7906.
187. K. S. Xia, X. L. Tian, S. X. Fei and K. You, *Int. J. Hydrogen Energy*, 2014, **39**, 11047.
188. A. K. Mishra and S. Ramaprabhu, *AIP Adv.*, 2011, **1**, 032152.
189. R. Kumar, V. M. Suresh, T. K. Maji and C. Rao, *Chem. Commun.*, 2014, **50**, 2015.
190. S. Chowdhury and R. Balasubramanian, *J. CO_2 Util.*, 2016, **13**, 50.
191. G.-P. Hao, Z.-Y. Jin, Q. Sun, X.-Q. Zhang, J.-T. Zhang and A.-H. Lu, *Energy Environ. Sci.*, 2013, **6**, 3740.
192. K. C. Kemp, V. Chandra, M. Saleh and K. S. Kim, *Nanotechnology*, 2013, **24**, 235703.
193. M. Saleh, V. Chandra, K. C. Kemp and K. S. Kim, *Nanotechnology*, 2013, **24**, 255702.
194. S. Yang, L. Zhan, X. Xu, Y. Wang, L. Ling and X. Feng, *Adv. Mater.*, 2013, **25**, 2130.
195. A. K. Mishra and S. Ramaprabhu, *J. Mater. Chem.*, 2012, **22**, 3708.
196. F.-Q. Liu, W. Li, J. Zhao, W.-H. Li, D.-M. Chen, L.-S. Sun, L. Wang and R.-X. Li, *J. Mater. Chem. A*, 2015, **3**, 12252.
197. A. I. Skoulidas, D. M. Ackerman, J. K. Johnson and D. S. Sholl, *Phys. Rev. Lett.*, 2002, **89**, 185901.
198. V. Sokhan, D. Nicholson and N. Quirke, *J. Chem. Phys.*, 2004, **120**, 3855.

199. C. Lu, H. Bai, B. Wu, F. Su and J. Fen-Hwang, *Energy Fuels*, 2008, **22**, 3050.
200. I. D. Mackie and G. A. DiLabio, *Phys. Chem. Chem. Phys.*, 2011, **13**, 2780.
201. D. Quinonero, A. Frontera and P. M. Deya, *J. Phys. Chem. C*, 2012, **116**, 21083.
202. D. G. A. Smith and K. Patkowski, *J. Phys. Chem. C*, 2015, **119**, 4934.
203. M. Rahimi, J. K. Singh and F. Mueller-Plathe, *Phys. Chem. Chem. Phys.*, 2016, **18**, 4112.
204. Z. Nickmand, S. F. Aghamiri, M. R. T. Khozanie and H. Sabzyan, *Sep. Sci. Technol.*, 2014, **49**, 499.
205. L. Liu and S. K. Bhatia, *J. Phys. Chem. C*, 2013, **117**, 13479.
206. M. Rahimi, J. K. Singh and F. Mueller-Plathe, *J. Phys. Chem. C*, 2015, **119**, 15232.
207. Y. Jin, S. C. Hawkins, C. P. Huynh and S. Su, *Energy Environ. Sci.*, 2013, **6**, 2591.
208. A. L. Ahmad, Z. A. Jawad, S. C. Low and S. H. S. Zein, *J. Membr. Sci.*, 2014, **451**, 55.
209. A. Houshmand, W. M. A. Wan Daud and M. S. Shafeeyan, *Sep. Sci. Technol.*, 2011, **46**, 1098.
210. C. Chen, J. Kim and W.-S. Ahn, *Korean J. Chem. Eng.*, 2014, **31**, 1919.
211. Z. Chen, S. Deng, H. Wei, B. Wang, J. Huang and G. Yu, *Front. Environ. Sci. Eng.*, 2013, **7**, 326.
212. J. Wang, H. Chen, H. Zhou, X. Liu, W. Qiao, D. Long and L. Ling, *J. Environ. Sci.*, 2013, **25**, 124.
213. Z. Tang, Z. Han, G. Yang and J. Yang, *Appl. Surf. Sci.*, 2013, **277**, 47.
214. M. Wang, L. Yao, J. Wang, Z. Zhang, W. Qiao, D. Long and L. Ling, *Appl. Energy*, 2016, **168**, 282.
215. F. Su, C. Lu and H.-S. Chen, *Langmuir*, 2011, **27**, 8090.
216. A. K. Mishra and S. Ramaprabhu, *RSC Adv.*, 2012, **2**, 1746.
217. P. Tamilarasan and S. Ramaprabhu, *J. Mater. Chem. A*, 2015, **3**, 101.
218. J. Wang, R. Krishna, J. Yang and S. Deng, *Environ. Sci. Technol.*, 2015, **49**, 9364.
219. W.-J. Liu, H. Jiang, K. Tian, Y.-W. Ding and H.-Q. Yu, *Environ. Sci. Technol.*, 2013, **47**, 9397.
220. L. K. G. Bhatta, S. Subramanyam, M. D. Chengala, U. M. Bhatta, P. Guha, R. P. H. Dinakar and K. Venkatesh, *Energy Fuels*, 2016, **30**, 4244.
221. D. Zhou, Q. Liu, Q. Y. Cheng, Y. C. Zhao, Y. Cui, T. Wang and B. H. Han, *Chin. Sci. Bull.*, 2012, **57**, 3059.
222. D. Qian, C. Lei, G.-P. Hao, W.-C. Li and A.-H. Lu, *ACS Appl. Mater. Interfaces*, 2012, **4**, 6125.
223. X.-W. Liu, T.-J. Sun, J.-L. Hu and S.-D. Wang, *J. Mater. Chem. A*, 2016, **4**, 3584.
224. M. Mohamedali, D. Nath, H. Ibrahim and A. Henni, *Greenhouse Gases*, InTech, 2016.
225. C. Cazorla, S. Shevlin and Z. Guo, *J. Phys. Chem. C*, 2011, **115**, 10990.

226. Y. Y. Li, M. M. Wan, X. D. Sun, J. Zhou, Y. Wang and J. H. Zhu, *J. Mater. Chem. A*, 2015, **3**, 18535.
227. N. N. A. H. Meis, A. M. Frey, J. H. Bitter and K. P. de Jong, *Ind. Eng. Chem. Res.*, 2013, **52**, 12812.
228. A. Garcia-Gallastegui, D. Iruretagoyena, V. Gouvea, M. Mokhtar, A. M. Asiri, S. N. Basahel, S. A. Al-Thabaiti, A. O. Alyoubi, D. Chadwick and M. S. P. Shaffer, *Chem. Mater.*, 2012, **24**, 4531.
229. J. Wang, L. Huang, R. Yang, Z. Zhang, J. Wu, Y. Gao, Q. Wang, D. O'Hare and Z. Zhong, *Energy Environ. Sci.*, 2014, **7**, 3478.
230. B. Standard and B. ISO, BS ISO, 2005, 15901.
231. A. V. Neimark, Y. Lin, P. I. Ravikovitch and M. Thommes, *Carbon*, 2009, **47**, 1617.
232. G. Y. Gor, M. Thommes, K. A. Cychosz and A. V. Neimark, *Carbon*, 2012, **50**, 1583.
233. X. Wu, W. Xing, J. Florek, J. Zhou, G. Wang, S. Zhuo, Q. Xue, Z. Yan and F. Kleitz, *J. Mater. Chem. A*, 2014, **2**, 18998.
234. M. Dubinin, *Pure Appl. Chem.*, 1989, **61**, 1841.
235. M. Sevilla, J. B. Parra and A. B. Fuertes, *ACS Appl. Mater. Interfaces*, 2013, **5**, 6360.
236. D. Fairén-Jiménez, F. Carrasco-Marín and C. Moreno-Castilla, *Carbon*, 2006, **44**, 2301.
237. F. Stoeckli and L. Ballerini, *Fuel*, 1991, **70**, 557.
238. H. Stoeckli, *J. Colloid Interface Sci.*, 1977, **59**, 184.
239. J. Zhou, W. Li, Z. Zhang, W. Xing and S. Zhuo, *RSC Adv.*, 2012, **2**, 161.
240. W. Xing, C. Liu, Z. Zhou, L. Zhang, J. Zhou, S. Zhuo, Z. Yan, H. Gao, G. Wang and S. Z. Qiao, *Energy Environ. Sci.*, 2012, **5**, 7323.
241. K. V. Kumar, K. Preuss, L. Lu, Z. X. Guo and M. M. Titiricit, *J. Phys. Chem. C*, 2015, **119**, 22310.
242. G. Lim, K. B. Lee and H. C. Ham, *J. Phys. Chem. C*, 2016, **120**, 8087.
243. D. J. Babu, M. Bruns, R. Schneider, D. Gerthsen and J. J. Schneider, *J. Phys. Chem. C*, 2017, **121**, 616.
244. M. Seredych, J. Jagiello and T. J. Bandosz, *Carbon*, 2014, **74**, 207.
245. M. Seredych, E. Rodriguez-Castellon and T. J. Bandosz, *Carbon*, 2016, **107**, 501.
246. M. Kwiatkowski, A. Policicchio, M. Seredych and T. J. Bandosz, *Carbon*, 2016, **98**, 250.
247. W. Xing, C. Liu, Z. Zhou, J. Zhou, G. Wang, S. Zhuo, Q. Xue, L. Song and Z. Yan, *Nanoscale Res. Lett.*, 2014, **9**, 189.
248. M. G. Plaza, K. J. Thurecht, C. Pevida, F. Rubiera, J. J. Pis, C. E. Snape and T. C. Drage, *Fuel Process. Technol.*, 2013, **110**, 53.
249. Y. Liu and J. Wilcox, *Environ. Sci. Technol.*, 2013, **47**, 95.
250. Y. Liu and J. Wilcox, *Environ. Sci. Technol.*, 2012, **46**, 1940.
251. D. Li, W. B. Li, J. S. Shi and F. W. Xin, *RSC Adv.*, 2016, **6**, 50138.
252. P. Xydias, I. Spanopoulos, E. Klontzas, G. E. Froudakis and P. N. Trikalitis, *Inorg. Chem.*, 2014, **53**, 679.

253. T. J. Bandosz, M. Seredych, E. Rodriguez-Castellon, Y. Cheng, L. L. Daemen and A. J. Ramirez-Cuesta, *Carbon*, 2016, **96**, 856.
254. Y. Zhao, X. Liu, K. X. Yao, L. Zhao and Y. Han, *Chem. Mater.*, 2012, **24**, 4725.
255. B. Zhu, C. Shang and Z. Guo, *ACS Sustainable Chem. Eng.*, 2016, **4**, 1050.
256. X.-Q. Zhang, W.-C. Li and A.-H. Lu, *New Carbon Mater.*, 2015, **30**, 481.
257. K. S. Walton and D. S. Sholl, *AIChE J.*, 2015, **61**, 2757.

CHAPTER 2

Zeolite and Silica-based CO_2 Adsorbents

GUILLAUME LAUGEL*[a], ROGERIA BINGRE[b] AND BENOÎT LOUIS*[b]

[a]Sorbonne Université, CNRS, Laboratoire de Réactivité de Surface, LRS, F-75005 Paris, France; [b]ICPEES – Institut de Chimie et Procédés pour l'Énergie, l'Environnement et la Santé, Energy and Fuels for a Sustainable Environment Team UMR 7515 CNRS, Université de Strasbourg, ECPM, 25 rue Becquerel, 67087 Strasbourg Cedex 2, France
*E-mail: guillaume.laugel@upmc.fr, blouis@unistra.fr

2.1 Introduction

Humankind has never faced such a considerable challenge. Indeed, there are seven billion people on Earth with the capacity to communicate and integrate, with a life-span never seen before. However, if this should enable us to become leading players in our future, expecting a better one, consumption patterns and property needs are a great part of our society. Unfortunately, this generates an economically and socially segregated world. Even worse, the population growth and the rise of capital goods warn us of the availability of energy and natural resources on our beloved planet. By staying on the track of an unsustainable system, *both socially and environmentally*, the economic structure and current production are endangered. It is likely that if

Inorganic Materials Series No. 2
Post-combustion Carbon Dioxide Capture Materials
Edited by Qiang Wang

Published by the Royal Society of Chemistry, www.rsc.org

the decisions taken at the *COP 21* protocol in Paris are not respected,[1] our growth may inevitably continue its march towards the destruction of our own species. Fortunately, the societal relevance of this issue is highlighted by the numerous actions from governments and industries established over the past decades: Intergovernmental Panel on Climate Change (IPCC), United Nations Framework Commission on Climate Change, Global Climate Change Initiative, European Strategic Energy Plans (SET-Plan, ZEP).[2,3] Despite the fact that the goal of reducing emissions of greenhouse gases (GHG) is far from being achieved, awareness of the need to minimize the environmental impact has been made in several segments of our society. Unfortunately, less than 1% of CO_2 produced in an anthropogenic manner is recycled or mitigated.[4] The inclusion of cleaner energy matrices along with more severe emission controls have already started to minimize the environmental impact. However, it is naive to believe that an environmental equilibrium will be reached without drastic changes in the consumption patterns and means of production.

Regarding GHG, it is fair to mention that NOx, chlorofluorohydrocarbons (CFCs), and CH_4 also contribute in a non-negligible manner. Nitrous oxide gas (N_2O) has a lifetime in the atmosphere of 150 years. However, the total human contribution to radiative forcing, used as an indicator of the contribution to global warming,[5,6] for CO_2 is four and 10 times superior to CH_4 and N_2O, respectively.

After the industrial era, a continuous increase of CO_2 concentration has been observed; it is close to 400 ppm today, which means an increase of 39% compared to the pre-industrial era.[7] For instance, Oppenheimer made precise temperature measurements and proved that such a rise drastically impacted Antarctica and subtropical forests.[8] An average rise of 0.6 °C on Earth's surface has been observed during the past 70 years. Several scenarios relating global warming throughout the years to GHG concentration have been proposed.[9] The symptoms that our planet suffers could be partly related to an increase in GHG concentration in the atmosphere. However, giving GHG the sole blame for this might be dangerous since it would ignore the relevance of other sources of global warming (non-anthropogenic causes such as solar radiation[10] or geological processes). Global warming relies on political, economic, societal, environmental and technological issues. Continuous scientific and technological search remains, therefore, mandatory to determine the factors, anthropogenic or not, that may be related to global warming.

One striking example is the inter-connection of CO_2 emissions and the Earth's carbon cycle. Firstly, it is worth mentioning that human beings have been emitting a CO_2 influx representing nearly 8 Gtons carbon per year.[5] This carbon, spread over the planet, is partly absorbed by several mechanisms.[11] The remaining CO_2 (roughly 50% of total emissions) accumulates in the atmosphere. On the other hand, it is known that the CO_2 influx from intrinsic processes on the planet is around 200 Gtons carbon per year, much higher than the 8 Gtons carbon per year produced by mankind. This huge difference between CO_2 influxes (man/planet) and the lack of a comprehensive knowledge of CO_2 sequestration mechanisms further complicates the global energy picture and hinders the setting of projects to mitigate and recycle villainous CO_2.

The convergence of several factors renders CO_2 sequestration the main alternative in the mid-term future until a clean energy source, such as nuclear fusion or energy manipulation with antimatter, can be found. When global behavior concerning consumption patterns, new types of consumer goods, and the ceaseless thirst for energy are taken into account, the worst option of all would be to 'wait and see'.

The mitigation of CO_2 emissions through key processes, like thermoelectric power plants and refineries, should be rapidly implemented to speed up the reduction of CO_2 emission. According to the International Energy Agency, approximately 40% of global CO_2 emissions are related to energy and heat generation.[3] Manufacturing and industrial processes (paper, chemicals, cement, steel) account for an additional 20%. Finally, transportation also accounts for 20% of the total CO_2 emissions. So, if capture technologies could be located and applied to power generation plants and huge industries, a decent target for reducing current CO_2 emissions by roughly 60% might be expected.

Three capture technologies exist: oxyfuel combustion, pre-combustion and post-combustion, the later representing the focus of this chapter. Post-combustion CO_2 capture technology involves treating the flue gas produced after the fuel is burned (Figure 2.1). The hot combustion gases leaving the boiler mainly consist of nitrogen, oxygen, water and CO_2. On a volume basis, the latter's concentration can be between 4% (gas turbine) and 30% (steel production).

Presently, chemical absorption in aqueous solutions of amines such as mono- or diethanolamine is a mature technology and the most efficient method of capturing 85–90% of CO_2 from flue gases. In spite of major improvements through process optimization, clear environmental concerns related to amine regeneration render it unrealistic to decrease the cost below 60 $ per ton of CO_2. Alternative technologies, which are beginning to be implemented commercially, like adsorption, conversion, chemical looping and membrane separation, have therefore arisen. Recently, a counter-flow/sweep membrane configuration allowed the recovery of 90% CO_2 at a

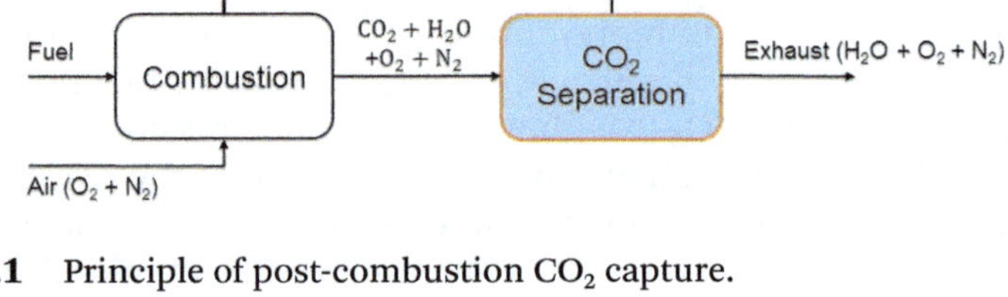

Figure 2.1 Principle of post-combustion CO_2 capture.

price below 20$ per ton of CO_2, highlighting the large potential of membrane technologies.[3]

The development of new solid-phase sorbents appears, therefore, to be a necessary research topic.[12] Solid adsorbents/sorbents can be classified into low-temperature (<200 °C), intermediate-temperature (200–400 °C), and high-temperature (>400 °C) CO_2 adsorbents/sorbents according to their working temperature range.[13,14]

High-temperature CO_2 sorbents are mainly used in the sorption enhanced steam methane reforming process operating between 500 and 800 °C. While properly using these sorbents, the conversion of methane to hydrogen can be improved, and concurrently, pure CO_2 can be captured.[15] To date, several types of high-temperature CO_2 sorbents have been developed, mainly CaO-based systems, or alkali dititanates, alkali zirconates, and finally alkali silicates.[16–19] Recently, alkali-metal-based silicate sorbents have attracted tremendous attention owing to their high CO_2 capture capacities and relatively low regeneration temperatures.[20] This chapter aims to discuss the nature of the main alkali-based silica sorbents, phyllosilicates and zeolites at low temperatures, their synthesis procedures as well as their performances in CO_2 capture efficiency, also in terms of potential operations during successive sorption cycles.

2.2 (Alkali) Silicates

2.2.1 Silicate Amine-based Adsorbents

2.2.1.1 CO_2 Absorption in Aqueous Alkanolamine Solutions

Carbon dioxide absorption in aqueous alkanolamine solutions is the commonest industrial technology for the capture of this greenhouse gas. This process is based on reactions between CO_2 and one or more basic absorbents, such as mono-, di-, and tri-ethanolamine. For industrial applications as post-combustion CO_2 absorbents, aqueous solutions of monoethanolamine (MEA), diethanolamine (DEA) and *N*-methyldiethanolamine (MDEA) are widely used. In the case of these examples, MEA is a primary amine (R_1NH_2), DEA a secondary amine ($R_1R_2NH_2$) and MDEA a tertiary amine ($R_1R_2R_3NH$), where R is an alkyl group. In the case of primary and secondary amines, the interaction with CO_2 in a water-free environment leads to the formation of carbamates (carbamation reaction) and protonated amines. Among the several reaction mechanisms proposed in the literature, the *zwitterion* mechanism is one of the most widely accepted to explain the interactions between the CO_2 and primary or secondary amines. Based on this mechanism, the global reactions can be described as follows:

$$\text{For a primary amine:} CO_2 + 2R_1NH_2 \leftrightarrow R_1NHCOO^- + R_1NH_3^+ \tag{2.1}$$

$$\text{For a secondary amine:} CO_2 + 2R_1R_2NH \leftrightarrow R_1R_2NCOO^- + R_1NH_3^+ \tag{2.2}$$

It is noteworthy that in these two cases, the reactions involve two amine groups per CO_2 molecule (*i.e.*, N–CO_2 = 2). In addition, the presence of water or a high partial pressure of CO_2, induces a low carbamate hydrolysis reaction. This hydrolysis leads to the formation of bicarbonate and free amine that can react with additional CO_2. The carbamate hydrolysis is therefore represented by the following reactions:

$$\text{For a primary amine:} R_1NHCOO^- + H_2O \leftrightarrow HCO_3^- + R_1NH_2 \quad (2.3)$$

$$\text{For a secondary amine:} R_1R_2NCOO^- + H_2O \leftrightarrow HCO_3^- + R_1R_2NH \quad (2.4)$$

It is important to note that at elevated pH (>10), bicarbonate ion (HCO_3^-) forms carbonate (CO_3^{2-}) according to the following reaction:

$$\text{For a primary amine:} HCO_3^- + R_1NH_2 \leftrightarrow R_1NH_3^+ + CO_3^{2-} \quad (2.5)$$

$$\text{For a secondary amine:} HCO_3^- + R_1R_2NH \leftrightarrow R_1R_2NH_2^+ + CO_3^{2-} \quad (2.6)$$

In the case of tertiary amines, with no hydrogen atom attached to the nitrogen, the carbamation reaction cannot occur, resulting in a low reactivity with respect to CO_2. Accordingly, the CO_2 absorption capacity of tertiary amines represents 1 mol of CO_2 per mol of amine. In fact, tertiary amines facilitate the formation of bicarbonate by the CO_2 hydrolysis reaction, thus reducing solvent regeneration costs since the heat released during bicarbonate formation remains lower than that of the carbamation reaction. To enhance the CO_2 absorption rates of tertiary amines, small amounts of primary or secondary amines can be added.

CO_2 capture *via* chemical absorption using alkanolamine solutions presents the advantage that amines are able to capture CO_2 from exhaust streams at low CO_2 partial pressures, and show high CO_2 capture selectivity with respect to other gases. Moreover, this chemical absorption can be reversed by sending the CO_2-rich absorbent to a stripper where the temperature is raised. A continuous recycling process is thus obtained because the regenerated absorbent is returned to the absorber. Nevertheless, the use of aqueous amines implies that the feed gas must be free of SO_2, oxygen, and particulates. Indeed, those components react with the amine, inducing its oxidation, degradation, and formation of heat-stable salts that ultimately result in absorbent loss and severe deterioration in capture efficiency. For example, it has been reported that for each ton of CO_2 captured, 1 to 4 kg of MEA need to be replaced[21] depending on the feed composition and amine concentration. In addition to the corrosive and toxic nature of the amine used, such a chemical absorption represents a high energy-consuming process.

It is then crucial to discuss post-combustion CO_2 capture technology that can be implemented on existing coal fired power plants for CO_2 emission mitigation by carbon capture and storage (CCS) in the energy production sector.

As mentioned, mature technologies mainly consist of aqueous amine scrubbing systems, which, however, significantly affect the overall efficiency of power plants after retrofitting, by at least 12%. Although new scrubbing systems with advanced solvents have led to decent improvements, it has been observed that the use of solid sorbent-based post-combustion capture (PCC) technologies is able to minimize this efficiency penalty. The next section will then focus on materials, such as sorption-based technologies and processes involving solid CO_2 adsorbents that are currently being scrutinized for large-scale applications.[22,23] The latter section will also summarize the recent performance assessment of various polyamine-supported materials and give some guidelines for possible research and development activities in the field. State-of-the-art amine-supported sorbent preparation is discussed in detail and related to CO_2 sorption performance as well as sorbent durability.

Accordingly, amine species were incorporated into the porous structures of various materials, such as porous carbon,[24,25] metal–organic frameworks (MOFs),[26,27] zeolites[28,29] and mesoporous silica.[30–32] These different solid adsorbents present lower CO_2 capture selectivities and poor performances under moist conditions, but usually require lower pressures for gas recovery and less energy for regeneration compared with liquid amine absorption technology.[22] Moreover, the accessibility of a supported porous structure has been identified as a key parameter impacting overall CO_2 capacity under cyclic conditions. In general, highly open pore structures favor easy amine functionalization methods and distribution of amine groups into the solid matrix, thus rendering it easier to transport CO_2 to the sorption sites. These sorbents are seen as immediate alternatives to solvent systems as opposed to ionic liquids or MOF materials, which are still in the early phase of R&D sorption technologies.

Keeping this in mind, mesoporous silica with its large surface area, large pore volume, adjustable uniform pore size and easy surface functionalization, appears as an extremely well adapted support to allow homogeneous distribution of amine species. Furthermore, the use of a wide range of surfactants and preparation methods gave rise to many kinds of mesoporous silica (M41S,[33,34] SBA,[35,36] AMS,[37] HMS,[38] MSU,[39] FDU,[40] KIT,[41,42] *etc.*). These materials offer a large variety of mesostructures with ordered hexagonal porous systems in two dimensions (P6mm symmetry group), in three dimensions ($P6_3/mmc$) or cubic in three dimensions (Pm3m, Pm3n, Fd3m, Fm3m, Im3m, *etc.*). The CO_2 adsorption capacities of various supports (carbon nanotubes, mesoporous silica) as well as their properties are presented in Table 2.1. According to this table, the CO_2 adsorption capacity of the different mesoporous silicas is relatively low compared to the results obtained with graphene and carbon nanotubes. In the case of silica, the best CO_2 adsorption result was obtained for MCM-41 silica in pure CO_2 at 25 °C. It should be emphasized that an increase in temperature up to 75 °C led to a drastic decrease in the adsorption capacity. At this temperature, a change in the gas composition, including a decrease in the CO_2 partial pressure and the presence of O_2, diminished the adsorption capacity. In addition, the results

Table 2.1 CO_2 adsorption capacities and operating conditions for mesoporous silica supports.

Support				Adsorption capacity	Operating conditions		
Type	S^a ($m^2\ g^{-1}$)	Vp^b ($cm^3\ g^{-1}$)	Dp^c (nm)	(mmol g^{-1})	Gas composition	T (°C)	Reference
MCM-41	1229	1.15	2.7	0.62	100% CO_2	25	48
MCM-41	1229	1.15	2.7	0.20	100% CO_2	75	48
MCM-41	1229	1.15	2.7	0.14	15% CO_2 (4.3% O_2) in N_2	75	48
SBA-15	725	1.12	9.3	0.04	100% CO_2	75	62
As-SBA-15	345	0.71	8.9	0.05	100% CO_2	75	62
KIT-6	895	1.22	6.0	0.02	100% CO_2	75	59

[a]Surface area.
[b]Pore volume.
[c]Pore size.

of CO_2 adsorption for silica with different morphologies such as MCM-41, SBA-15, as-SBA-15 (presence of template inside the pores) and KIT-6, did not show a direct correlation between their porous structures (specific surface, porous volume and pore size) and their adsorption capacities. The following paragraphs deal with the synthesis of amine–silica composites and their utilization as CO_2 adsorbents.

2.2.2 Synthesis of Amine–Silica Adsorbents

Several methods have been developed to incorporate functional groups, such as organic amines, into mesostructured silica and have recently been exhaustively described in a few reviews.[43–47] According to the classification found in the literature, the functionalization methods of mesoporous silica can generally be classified as: impregnation, post-synthesis grafting and direct condensation; this last functionalization method is the least used in the context of CO_2 capture with amine–silica based adsorbents.

2.2.2.1 Impregnation Method

As suggested by its name, the impregnation method generally involves the adsorption of organic molecules onto the surface of a material *via* hydrogen bonds, van der Waals forces, or electrostatic attractions. In a typical synthesis, the desired amount of amine is dissolved in a solvent (such as methanol), followed by the addition of the mesostructured silica. Finally, the solvent is evaporated to obtain the functional material. It is noteworthy that the total pore volume of the support and the density of the amine limit the amount of amine located inside the pores. The impregnated samples often present high

CO_2 capture capacities. Nevertheless, their performance and long-term viability are limited by the leaching of amines over multiple regeneration cycles.

Xu *et al.*[48,49] were the first to successfully incorporate a polymer with a high concentration of amine groups, such as polyethylenimine (PEI), into a mesostructured material. In this study, the PEI was dissolved in methanol, then MCM-41 mesoporous silica was added and finally the solvent was evaporated. The obtained material showed up to 50% polymer loading but the specific surface area was significantly reduced compared to MCM-41 pristine support (474 $m^2\ g^{-1}$ compared to 1229 $m^2\ g^{-1}$). This decrease suggests that the pore structure is extensively blocked. However, those materials were relatively efficient for CO_2 capture, absorbing up to 2.55 mmol of CO_2 per gram of adsorbent.

2.2.2.2 *Post-synthesis Grafting*

In contrast to the impregnation method, post-synthesis grafting requires a modification of the silica-based material surface by forming a covalent bond. Usually, organosilanes $(R'O)_3Si$-**R** (**R** representing the functional group of interest such as an amino group), or less frequently chlorosilanes ($ClSi\mathbf{R}_3$) and silazanes ($HNSi\mathbf{R}_3$) are used to react with silanol groups from the silica surface. In a typical synthesis, the mesostructured silica is dispersed in a solvent (typically toluene), followed by the addition of aminosilane and the mixture is heated under reflux. The solid obtained after evaporation has to be washed several times to remove unreacted aminosilane.[50] As-obtained materials usually retain a high degree of ordering since the inorganic mesostructure is already formed before the grafting. Moreover, the leaching problem in the case of the impregnation method can be relieved by covalently attaching the amines onto solid surfaces. However, the location of the functional group within the pores is usually accompanied by a decrease in pore volume and surface area, like in the case of the impregnation method.

Other original post-synthesis grafting procedures exist in the literature such as the one developed by Hicks *et al.*[51,52] In this study, the authors proposed a smart approach to control the density of the amine groups present on the surface of the silica material. The aim of their work was to obtain isolated amine sites, thus to limit the interactions between amines and residual surface silanols or with other amine groups. In this way, a two-step grafting method was developed. Initially, the amine group contained in the (3-aminopropyl)trimethoxysilane was protected by reaction with an aldehyde to form an imine. Silane was then grafted onto the surface of SBA-15, and finally deprotected in an acidic medium to produce the amine group. Protective groups with different sizes allowed control of the space between the amine groups. Fluorescence measurements have notably demonstrated that the amine groups were isolated and did not interact with each other.

2.2.2.3 Direct Condensation

Contrary to the previously described methods, which involve post-functionalization, direct condensation (also called co-condensation) allows 'one-pot' incorporation of organic groups during synthesis. A $(R'O)_3$Si-**R** precursor containing the function of interest, such as **R** = amino group, is added to a silica precursor (tetraethoxysilane, tetramethoxysilane or sodium silicate) during the synthesis, in the presence of an organic template. During the condensation of the silicic precursor, the functional group will be incorporated in the framework of the material. Since the functional groups are an integral part of the silica matrix, problems of pore obstruction are barely observed. It is generally recognized that the distribution of organic functions is more homogeneous in the whole material, compared with the materials obtained by grafting. Nevertheless, the modification of the synthesis parameters caused by the addition of the aminosilane usually requires numerous optimization steps, thus raising the complexity of such a protocol.

2.2.3 CO_2 Capture by Amine–Silica-based Adsorbents

2.2.3.1 Amine Impregnated on Silica

Table 2.2 shows the CO_2 adsorption capacities of amines impregnated on different mesoporous silicas such as MCM-41, SBA-15, KIT-6, *etc.* The operating conditions during the CO_2 adsorption as well as the characteristics of the pore structure of the silica are also reported. The first studies on this topic have focused on the use of PEI, linear or plugged, as an adsorbent group and most of the current studies are still focused on this attractive polymer. Linear polyethyleneimines contain all secondary amines, in contrast to branched PEIs, which contain primary, secondary and tertiary amino groups. Table 2.3 summarizes the different amines that have been impregnated on mesoporous silica in order to act as CO_2 adsorbents. Many publications have exhaustively detailed CO_2 capture *via* the use of amines impregnated or grafted onto different types of silica.[31,53] The following paragraphs will develop a few important points in the process of CO_2 adsorption, *i.e.*, operating conditions (temperature, partial pressure of CO_2, moisture content), effect of the percentage of amine impregnated on the support, and also the porous structure of the support.

2.2.3.1.1 Effect of Temperature. As demonstrated by Xu *et al.*,[48,54] a decrease in temperature from 75 °C to 25 °C causes a dramatic decrease from 2.55 to 0.75 mmol g^{-1} of CO_2 adsorption capacity for PEI (50 wt%) impregnated on MCM-41. This reduction may initially seem quite surprising because the adsorption of CO_2 is an exothermic process; a decrease in the temperature should therefore induce a positive influence on the adsorption. However, at low temperatures, the bulk-like formation of PEI within the pores of MCM-41 implies that CO_2 adsorption will be a diffusion-limited process.[55]

Table 2.2 CO_2 adsorption capacities and operating conditions for amine-impregnated mesoporous silica support.

Amine (wt%)	Support				Adsorption capacity	Operating conditions		Reference
	Type	S^a ($m^2\ g^{-1}$)	Vp^b ($cm^3\ g^{-1}$)	Dp^c (nm)	($mmol\ g^{-1}$)	Gas composition	T (°C)	
PEI (75)	MCM-41	1229	1.15	2.7	3.02	100% CO_2	75	48
PEI (50)	MCM-41	1229	1.15	2.7	2.55	100% CO_2	75	48
PEI (50)	MCM-41	1229	1.15	2.7	0.75	100% CO_2	25	48
PEI (50)	MCM-41	1229	1.15	2.7	2.03	15% CO_2 in N_2	75	54
PEI (50)	MCM-41	1229	1.15	2.7	2.84	13% CO_2 (10% H_2O) in N_2	75	54
PEI (50)	MCM-41	1042	0.85	2.8	2.52	100% CO_2	75	59
PEI (50)	MCM-48	1162	1.17	3.1	2.70	100% CO_2	75	59
PEI (50)	SBA-15	753	0.94	5.5	2.89	100% CO_2	75	59
PEI (50)	SBA-16	736	0.75	4.1	2.93	100% CO_2	75	59
PEI (50)	KIT-6	895	1.22	6.0	3.07	100% CO_2	75	59
PEI (60)	HMS	561	1.44	9.8	4.18	100% CO_2	75	60
PEI (50)	SBA-15	950	1.31	6.6	3.18	15% CO_2 in N_2	75	61
TEPA (70)	As-SBA-15	345	0.71	8.9	3.93	100% CO_2	100	62
TEPA (60)	As-MCM-41	76	0.36	10.5	5.39	100% CO_2	75	63
TEPA/DEA (35/15)	SBA-15	725	1.12	9.3	4.00	100% CO_2	75	64

[a]Surface area.
[b]Pore volume.
[c]Pore size.

Table 2.3 Structures of amines impregnated on mesoporous silica supports.

Name	Structure
Monoethanolamanine (MEA)	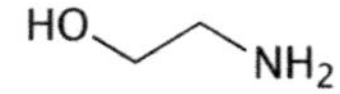
Diethanolamine (DEA)	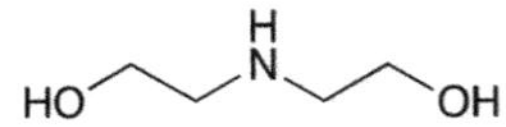
Triethanolamine (TEA)	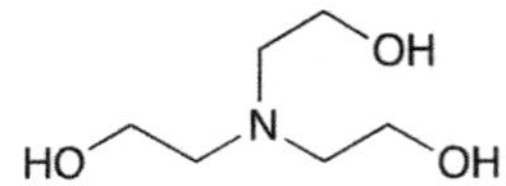
Polyethyleneimine (PEI)	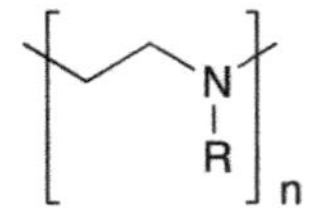 (R = H for linear, R = H or CH_x for branched)
Diethylenetriamine (DETA)	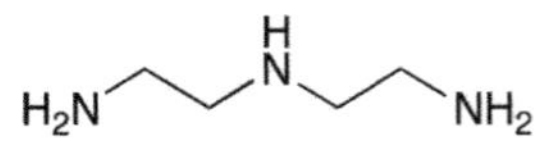
Tetraethylenepentamine (TEPA)	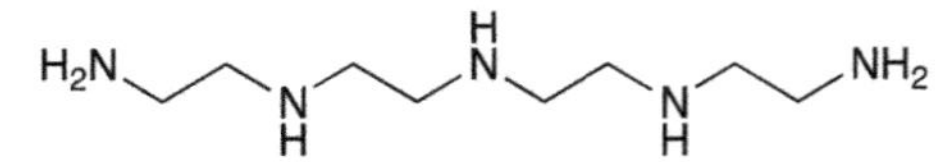
Tetraethylenepentamine-acrylonitrile (TEPAN)	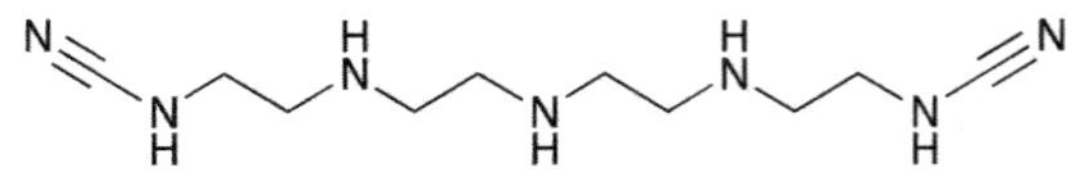
Pentaethylenehexamine (PEHA)	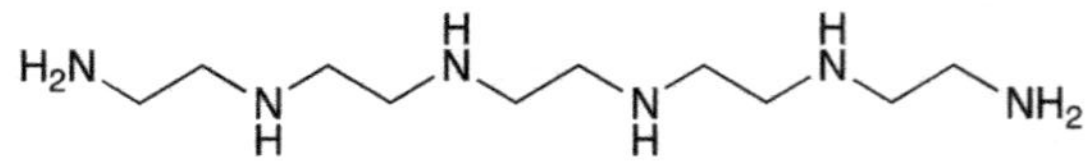
2-Amino-2-methyl-1,3-propanediol (AMPD)	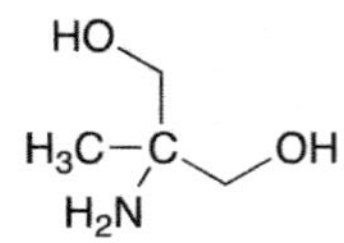
2-(2-Aminoethylamino)ethanol (AEAE)	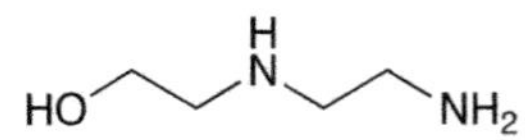

At higher temperatures, the diffusion rate of CO_2 in the bulk-like PEI is greater, which increases the reaction rate of CO_2 with the PEI and thus allows a better adsorption capacity. This phenomenon has also been highlighted in other studies.[56,57]

2.2.3.1.2 Effect of CO_2 Partial Pressure. CO_2 adsorption capacity is also influenced by the partial pressure of CO_2 in the gas feed. Regarding a PEI-type adsorbent (50 wt%) impregnated on MCM-41, the CO_2 adsorption capacity diminishes from 2.55 mmol g^{-1} to 2.03 mmol g^{-1} at the same time as the CO_2 partial pressure diminishes from 1 atm to 0.15 atm.[49,54]

2.2.3.1.3 Effect of Water. Xu *et al.* also studied the influence of water on CO_2 adsorption capacity using PEI (50 wt%)/MCM-41. The authors showed that the adsorption capacity increased from 2.03 mmol g^{-1}, in the case of a gaseous composition containing 15% of CO_2 diluted in nitrogen, to 2.84 mmol g^{-1}, when 10% of water was added to a mixture of 13% CO_2 in nitrogen.[54] This improvement in the CO_2 adsorption capacity can be explained by the adsorption mechanism, especially the formation of carbamate and bicarbonate, whether under anhydrous conditions or not.[58] It should be emphasized that the beneficial effect of water is only observed for molar concentrations of water lower than the molar concentration of CO_2.

2.2.3.1.4 Effect of Percentage of Impregnation. According to the work of Xu *et al.*,[48] the amount of amine (PEI) deposited on MCM-41 mesoporous silica has a great influence on the CO_2 adsorption capacity. Indeed, in pure CO_2 at 75 °C, the CO_2 adsorption capacity increases from 2.55 mmol g^{-1} to 3.02 mmol g^{-1} when the percentage of PEI impregnation on MCM-41 was enhanced from 50% to 75%. It should be highlighted that the increase in the adsorption capacity is not directly proportional to the percentage of amine deposited on the support. The reason for this difference relies on the formation of aggregates during amines impregnation on the support, which leads to a decrease in CO_2 accessibility at the sites.

2.2.3.1.5 Effect of the Support Structure. There is also a correlation between the CO_2 adsorption capacities of amines impregnated on different silicas and the porous structures of these silicas. Indeed, various studies carried out on adsorbents, such as PEI coated on mesoporous silica, have demonstrated that under the same operating conditions of temperature and CO_2 pressure, the CO_2 adsorption capacities of these materials vary as a function of the pore size of the silica.[59,60] In the case of mesoporous silica with hexagonal geometry, the order of adsorption capacity was as follows: HMS > KIT-6 > SBA-16 ≈ SBA-15 > MCM-48 > MCM-41. This classification follows the decreasing order of the pore diameter. This trend is also confirmed when CO_2 is diluted in nitrogen. Hence, in the case of PEI/MCM-41, the CO_2 adsorption capacity was 2.03 mmol g^{-1} whilst this adsorption increased to 3.18 mmol g^{-1} for PEI/SBA-15 at the same temperature and gaseous composition (75 °C, 15% CO_2 in N_2).[54,61] On the other hand, it should be noted that the influence of the pore size on mesoporous silica not impregnated with amine has not been demonstrated.

2.2.3.1.6 Effect of the Mesopores Structuring Agent. CO_2 capture was tested with tetraethylenepentamine (TEPA) impregnated on MCM-41 and SBA-15 with and without removal of the structuring agent, allowing the synthesis of these two mesoporous silicas.[62,63] According to Table 2.2, the CO_2 adsorption capacity was 3.93 mmol g^{-1} when 70% in weight of TEPA was grafted on as-SBA-15 while the adsorption increased to 5.39 mmol g^{-1} in the case of TEPA impregnated on as-MCM-41. In this study, the authors also showed an approximate 10% improvement in the adsorption capacity of as-synthesized supports compared to the corresponding calcined supports, which seems to indicate that the elimination of the structuring agent is not necessarily required. Such an increase can be explained by the fact that the presence of the template in the pores of the silica prevents the formation of TEPA aggregates. The same authors then studied a mixture of two amines (TEPA and diethanolamine (DEA)) deposited on as-SBA-15 silica, still with its structuring agent inside its mesopores.[64] In this case, the best CO_2 adsorption capacity of 4.00 mmol g^{-1} was obtained for a mixture of impregnated amine consisting of 35 wt% TEPA and 15 wt% DEA under an anhydrous CO_2 atmosphere. This adsorption capacity is slightly higher than the one previously found for TEPA (70 wt%) on as-SBA-15. Finally, the authors demonstrated the high stability of the CO_2 adsorption capacity of the mixture of TEPA and DEA impregnated on as-SBA-15. Indeed, the adsorption of CO_2 slightly diminished from 3.77 to 3.61 mmol g^{-1} after six cycles of adsorption and desorption under an anhydrous flow of CO_2 diluted in N_2.

2.2.3.2 Amine Grafted on Silica

Despite their attractive CO_2 adsorption capacities, amines impregnated on silica possess very low thermal stabilities, especially during the desorption phenomenon.[65] In order to overcome this drawback, aminosilanes can be covalently grafted by silylation reaction onto mesopores' silica surface. The main aminosilanes encountered in the literature in the context of CO_2 capture are presented in Table 2.4. As in the case of amines impregnated on silica, some reviews have already exhaustively presented work on amines grafted onto silica as CO_2 absorbents.[31,53] The following paragraphs will therefore be dedicated to the influence of important parameters on the CO_2 adsorption capacities of these materials (Table 2.5).

2.2.3.2.1 Effect of CO_2 Partial Pressure. In the case of APTES (3-aminopropyltriethoxysilane)-type aminosilanes for grafting on MCM-48-type silica, Huang *et al.*[66] showed that at room temperature, reducing the CO_2 partial pressure from 1 atm to 0.05 atm dramatically affected the CO_2 adsorption capacity of the grafted material, which diminished from 2.05 mmol g^{-1} to 1.14 mmol g^{-1}. A similar effect on amines impregnated on silica was described earlier.

Table 2.4 Structures of aminosilanes grafted on mesoporous silica.

Name	Structure
3-Aminopropyltrimethoxysilane (APTMS)	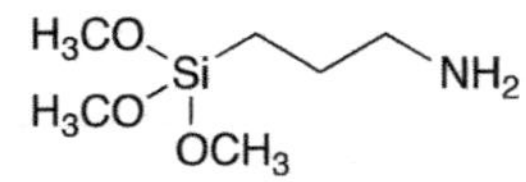
3-Aminopropyltriethoxysilane (APTS)	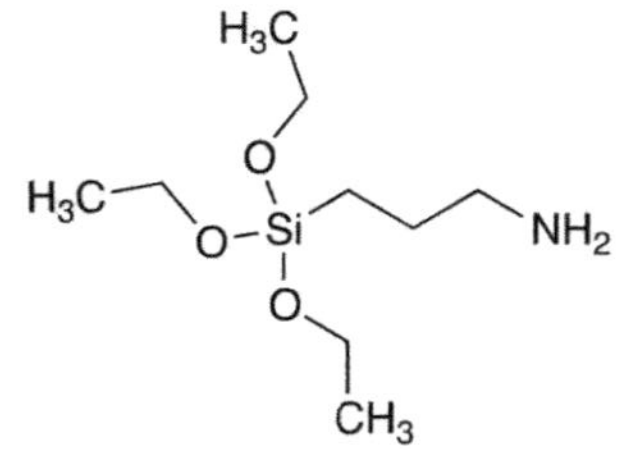
N-[3-(Trimethoxysilyl)propyl]-ethylenediamine (2N-APMS)	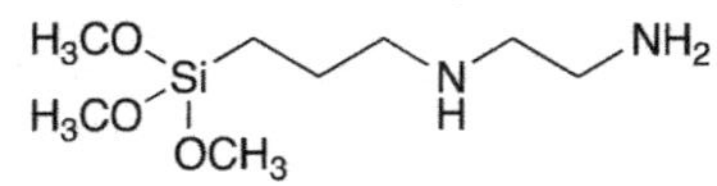
N-[3-(Trimethoxysilyl)propyl]-diethylenetriamine (3N-APMS)	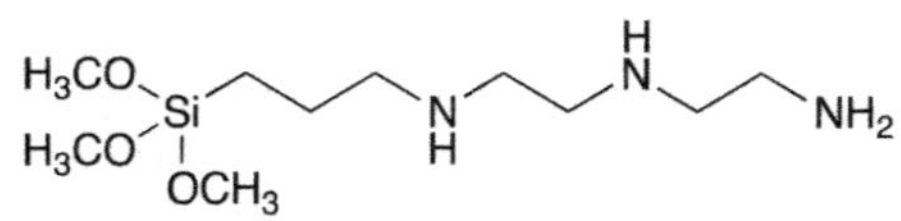
Ethylhydroxyl-aminopropyl-trimethoxysilane (EHAPTS)	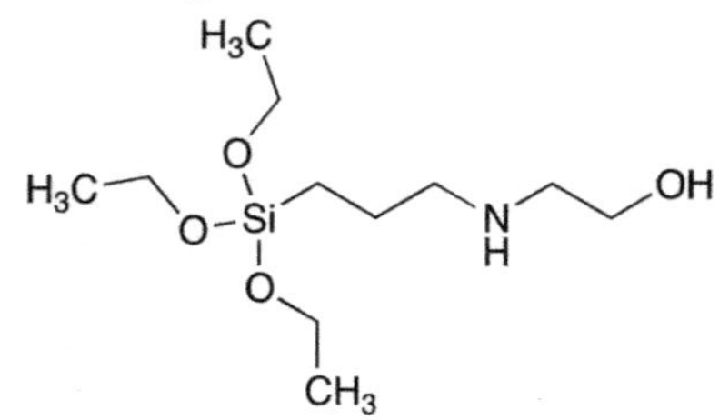
Diethylhydroxyl-aminopropyl-trimethoxysilane (DEHAPTS)	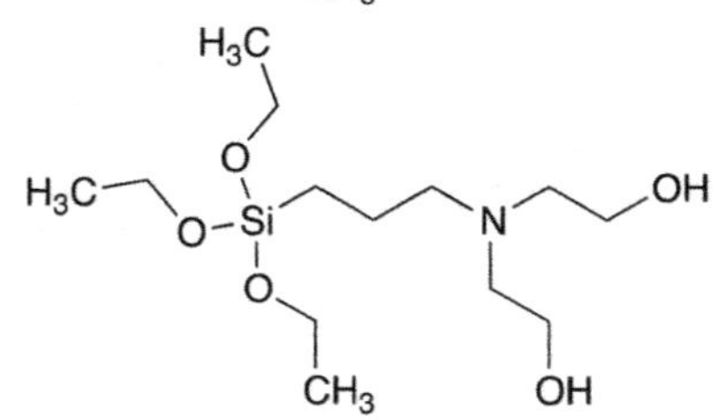

Table 2.5 CO_2 adsorption capacities and operating conditions for amine-grafted mesoporous silica supports.

	Support				Operating conditions		
Amine	Type	S^a ($m^2\ g^{-1}$)	Adsorption capacity (mmol g^{-1})	N content (mmol g^{-1})	Gas composition	T (°C)	Reference
APTES	MCM-48	1389	2.05	2.3	100% CO_2	RT	50
APTES	MCM-48	1389	1.14	2.3	5% CO_2 in He	RT	50
APTES	Silica gel	340	0.89	1.27	100% CO_2 (humid)	27	67
APTES	Silica gel	340	0.41	1.27	100% CO_2	27	67
APTES	SBA-15	204	0.41	—	4% CO_2 (humid) in He	25	68
APTES	SBA-15	204	0.22	—	4% CO_2 in He	25	68
3N-APMS	PE-MCM-41	1230	2.51	7.90	5% CO_2 (74% RH) in N_2	25	70
3N-APMS	PE-MCM-41	1230	2.05	7.90	5% CO_2 in N_2	25	70
APTES	SBA-15	782	1.02	1.89	15% CO_2 (78% RH) in N_2	60	69
APTES	SBA-15	782	1.06	1.89	15% CO_2 in N_2	60	69
2N-APMS	SBA-15	782	2.09	3.25	15% CO_2 (78% RH) in N_2	60	69
2N-APMS	SBA-15	782	1.73	3.25	15% CO_2 in N_2	60	69
3N-APMS	SBA-15	782	3.06	3.68	17% CO_2 (78% RH) in N_2	40	69
3N-APMS	SBA-15	782	2.74	3.68	17% CO_2 in N_2	40	69
APTES	HMS	1125	1.59	2.29	90% CO_2 (humid) in Ar	20	71
APTES	SiO_2 gel	567	0.68	1.14	90% CO_2 (humid) in Ar	20	71
APTES	Silica xerogel	816	1.15	1.7	100% CO_2	RT	50
2N-APMS	SBA-16	479	0.73	3.06	15% CO_2 in N_2	60	72
3N-APMS	HMS	1268	1.34	4.57	90% CO_2 (humid) in Ar	20	73
3N-APMS	SiO_2 gel	567	1.05	2.64	90% CO_2 (humid) in Ar	20	73
APTES	MCM-41	864	0.39	2.02	5% CO_2 in N_2	30	69
APTES	SBA-15	782	0.45	1.89	5% CO_2 in N_2	30	69
APTES	SBA-15	820	0.66	2.61	15% CO_2 (12% H_2O) in He	60	74
2N-APMS	SBA-15	820	1.51	4.61	15% CO_2 (12% H_2O) in He	60	74
3N-APMS	SBA-15	820	1.80	5.80	15% CO_2 (12% H_2O) in He	60	74

[a]Surface area; RH (relative humidity).

2.2.3.2.2 Effect of Water. Similar to the results obtained in the case of amines impregnated on silica, the presence of water during CO_2 adsorption significantly improved the adsorption capacities of the adsorbents with grafted amines. Many works highlight this promoting effect of water.[67–70] Among the numerous studies, it should be noted that the first work on the use of grafted amines on silica as adsorbents was published by Leal *et al.*[67] in 1995. For this purpose, the authors used an APTES-type aminosilane and silica gel, and proposed that CO_2 adsorption occurred through the formation of ammonium carbamate in dry conditions, and formation of ammonium bicarbonate in the presence of water. This APTES-grafted silica gel exhibits a CO_2 adsorption capacity of 0.41 mmol g^{-1} under anhydrous conditions and approximately double under moist conditions (0.89 mmol g^{-1}). In addition, Serna-Guerrero *et al.*[70] used MCM-41 silica with expanded pores (pore-expanded MCM-41, also called PE-MCM-41). The grafted aminosilane is 3N-APMS and a CO_2 adsorption capacity of 2.05 mmol g^{-1} was achieved, with a dry gas consisting of 5% of CO_2 diluted in N_2 at 25 °C and 1 atm. They have also shown that under dry conditions, the adsorbent became less and less effective because urea was formed during adsorption–desorption cycles, especially below 20 °C. The adsorption capacity was altered after 700 cycles. The authors also found that with the same stream composition (5% CO_2 in N_2) but with a relative humidity of 74%, the CO_2 adsorption capacity of the material was enhanced from 2.05 mmol g^{-1} to 2.51 mmol g^{-1}. In addition, this promoting effect of water in CO_2 adsorption was demonstrated by using different amines grafted on a variety of silica substrates: APTES-grafted SBA-15,[68,69] 2N-APMS-grafted SBA-15[69] and 3N-APMS-grafted SBA-15.[69]

2.2.3.2.3 Effect of the Support. As shown in Table 2.5, the important influence of the support on CO_2 adsorption capacities can also be demonstrated. Indeed, Knowles *et al.*[71] showed that under the same operating conditions: *i.e.*, 90% CO_2 (moisture) in Ar at 20 °C, with APTES as the aminosilane, a change of silica caused a modification in the CO_2 sorption capacity. The shift from a HMS-type silica with a high specific surface area (1125 m^2 g^{-1}) to a silica gel with a lower specific surface area (567 m^2 g^{-1}) led to a strong decrease in the adsorption capacity from 1.59 mmol g^{-1} to 0.68 mmol g^{-1}. The decrease in the surface area of the support also led to a decrease in the N content in the sample, from 2.29 mmol g^{-1} to 1.14 mmol g^{-1}. Another study carried out on APTES-grafted MCM-48 and silica xerogel also revealed that a decrease in the silica surface area from 1389 m^2 g^{-1} to 816 m^2 g^{-1} induced a clear decrease in the CO_2 adsorption capacity, from 2.05 mmol g^{-1} to 1.15 mmol g^{-1}.[50] In addition, a decrease in CO_2 adsorption capacity has been demonstrated in the case of longer chain aminosilane used with a SBA-15-type silica and another silica of a different porosity, SBA-16, which have specific surface areas of 782 m^2 g^{-1} and 479 m^2 g^{-1}, respectively.[69,72] Nevertheless, it should be emphasized that the decrease in the adsorption capacity is not linearly correlated with the decrease in the specific surface area of the silica. Under the same operating conditions and with the same aminosilane (3N-APMS), despite a decrease of more than half of the specific surface area of the silica gel (567 $m^2 g^{-1}$) compared to a silica-type HMS (1268 $m^2 g^{-1}$), the adsorption capacity of these two absorbents did not vary much – 1.05 mmol g^{-1} and 1.34 mmol g^{-1}, respectively.[73] Chang *et al.*[69] have similarly shown that a direct correlation between the specific surface area of silica and CO_2 adsorption capacity did not necessarily exist. In this study, APTES as the aminosilane was grafted on two types of supports, such as MCM-41 and SBA-15, with moderate specific surface areas of 864 $m^2 g^{-1}$ and 782 $m^2 g^{-1}$, respectively. The authors demonstrated that the best CO_2 adsorption capacity (0.45 mmol g^{-1}) was obtained with the silica possessing the lowest specific surface area (SBA-15).[69]

2.2.3.2.4 Effect of Organosilane Type. Hiyoshi *et al.*[74] have shown that the nature of the aminosilanes used in the synthesis of amine-grafted absorbents plays an important role in the absorption capacities of these materials. In this study, the synthesis of the absorbent is performed from three different aminosilanes (APTES, 2N-APMS and 3N-APMS) and a mesoporous silica-type SBA-15, with a specific surface area of 820 $m^2 g^{-1}$. In an atmosphere composed of 15% CO_2 and 12% H_2O in He, at 60 °C, the authors demonstrated that the CO_2 adsorption capacity was enhanced from 0.66 mmol g^{-1} for APTES-grafted-SBA-15, to 1.51 mmol g^{-1} for 2N-APMS-grafted-SBA-15 and finally to 1.80 mmol g^{-1} for 3N-APMS-grafted-SBA-15. It seems therefore that an increase in the carbon chain of the organosilane as well as the presence of a secondary amine group allowed a great enhancement of the adsorption capacity of the material.

2.3 Alkali Silicate-based Sorbents

As already mentioned, oxides can react with CO_2 to form carbonates, which after heating can regenerate both the oxide and CO_2 in a pure manner. The choice of potentially interesting oxides is based on several criteria such as adsorption capacity, thermal stability, regeneration temperature and, of course, the cost. In addition, a screening methodology based on thermodynamic data allowed a more efficient selection of metal oxides in which there is a real interest for CO_2 capture.[75–77] Based on this study, the two most promising alkaline earth metal oxide candidates are CaO and MgO.[78] Nevertheless, recent studies have shown that calcium-, sodium- and lithium-based silicates may also have great potential for CO_2 capture. In the next paragraphs, we will try to provide an overview of the latest advances in the use of these silicates as CO_2 absorbents.

2.3.1 Calcium Silicate ($CaSiO_3$)

The interest in calcium silicates is mainly due to the fact that these materials have a lower CO_2 adsorption temperature compared to that encountered in the case of CaO. Wang *et al.*[79] studied the adsorption–desorption properties of a $CaSiO_3$ sample, synthesized by a chemical reaction between $CaCO_3$ and SiO_2 at a high temperature (800 °C). It appears that the adsorption of CO_2 on these materials starts at 400 °C. At this temperature, the adsorption capacity of $CaSiO_3$ is 3.41 mmol g^{-1} in the case of a gas stream with 15% of CO_2 diluted in N_2. The regeneration temperature of these materials is about 800 °C. Nevertheless, their adsorption capacity diminished drastically after repeated adsorption–desorption cycles. This decrease was correlated with the decrease in the surface area due to the sintering of the materials. Other studies have shown that the CO_2 adsorption capacity of $CaSiO_3$ reached 6.33 mmol g^{-1} at 700 °C under 1 atm and that water could play a promoting role in the adsorption.[80,81] Despite these relatively good adsorption capacities, the interest in $CaSiO_3$ as an adsorbent appears to be largely limited by its low stability over time.

2.3.2 Sodium Metasilicate (Na_2SiO_3)

So far, metasilicate sodium-type materials have not yet been extensively examined as CO_2 absorbers. The first study in this field was published by Rodriguez *et al.*[82] In this study, Na_2SiO_3 samples were prepared using two different methods: solid-state and precipitation. The authors carried out a complete kinetic study of the adsorption process of CO_2 on these materials and were able to demonstrate that the adsorption proceeded *via* a two-step mechanism: (1) superficial chemical sorption and (2) sodium diffusion. The calculated values of activation energies in these two steps were different, being 17.5 kJ mol^{-1} for the surface reaction step and 24.0 kJ mol^{-1} for the sodium diffusion, respectively. This last step was evaluated by the authors as the kinetically determining step in the carbonation process. In addition, it has been shown that a decrease in the size of the Na_2SiO_3 particles positively influenced the adsorption capacity. This promoting effect was attributed to an increase in the specific surface area due to the decrease in particle size. Another study conducted by the same group on the use of Na_2SiO_3 as a CO_2 adsorbent has shown the importance of water in the adsorption reaction.[83] Two synthesis methods were performed to obtain Na_2SiO_3: (1) a solid-state reaction (Na_2CO_3 + SiO_2) and (2) a combustion method (NaOH, SiO_2, and urea). The authors highlighted that the CO_2 adsorption capacity of Na_2SiO_3 reached 16.39 mmol g^{-1} under humid conditions, while this adsorption decreased by a factor of 2 under anhydrous conditions. It should be emphasized that this study was carried out *at very low adsorption temperatures (30 to 60 °C)*. Such a promoting effect of water, already described in the previous paragraphs, was explained by the authors as the superficial reaction between Na_2SiO_3 and water, leading to the formation of Si–OH and Na–OH surface species. These species can be easily combined with CO_2 to form $NaHCO_3$. In a parallel study, the authors showed that the method of preparation influenced the adsorption capacities of the materials. In the case of Na_2SiO_3 prepared by the solid-state reaction, the specific surface was greater than the one obtained by the combustion method (1.6 m^2 g^{-1} and 0.5 m^2 g^{-1}). The solid-state Na_2SiO_3 thus exhibited a CO_2 adsorption capacity of 8.6 mmol g^{-1}

under anhydrous conditions, which is more than double that obtained with the other sample. These results showed that *Na_2SiO_3 can be considered as a potential candidate for CO_2 capture at low temperatures.*

2.3.3 Lithium Orthosilicate (Li_4SiO_4) and Other Lithium Silicates

2.3.3.1 Li_4SiO_4

The attention being paid to lithium silicates in the field of CO_2 capture at high temperatures has increased considerably recently.[78,84] In particular, lithium orthosilicate (Li_4SiO_4) appears to be a good candidate for CO_2 capture between 500 and 850 °C. The main advantages of these materials, compared to other high temperature CO_2 adsorbents such as CaO, are that they possess a high CO_2 adsorption capacity and a low regeneration temperature (<750 °C). In addition, lithium silicates exhibit excellent thermal stabilities, which enable them to maintain excellent adsorption capacities, despite a rather significant number of sorption cycles.

The adsorption mechanism of CO_2 with Li_4SiO_4 proceeds according to the double-shell theory.[85] This theory is based on the shrinking model principle, which is used to describe phenomena when a solid particle is being consumed by dissolution or reaction with the external surface. The model is called the shrinking core model, as the inside core of unreacted material reduces in size or shrinks. When Li_4SiO_4 reacts with CO_2, the core dissolves to ionic species and forms a Li_2SiO_3 shell that covers the unreacted core (eqn (2.7)) and (Figure 2.2).[86] CO_2 diffuses to the surface and reacts with Li^+ and O^{2-} to leave a lithium carbonate (Li_2CO_3) shell on top of the lithium metasilicate (Li_2SiO_3) solid crust, hence it is a double shell (eqn (2.8)).

$$Li_4SiO_4 = Li_2SiO_3 + 2Li^+ + O^{2-} \tag{2.7}$$

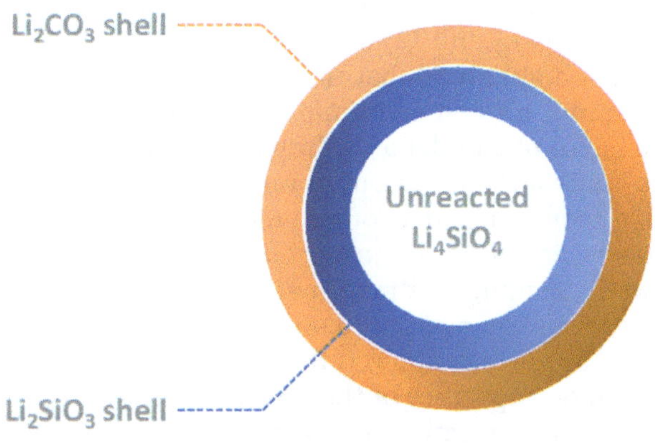

Figure 2.2 Illustration of a Li_4SiO_4 particle for the kinetic modeling of double-shell theory.

$$CO_2 + 2Li^+ + O^{2-} = Li_2CO_3 \tag{2.8}$$

A molten carbonate is formed, which improves diffusion through the shell.[87] Then, the CO_2 capture process can be described as a two-step process. The first step is a superficial reaction with the external surface that quickly forms a carbonate shell and a lithium silicate. The second step is a slow diffusion of CO_2 through the shell to the inner core that continues the chemisorption. This step is also the rate determining step that dictates the kinetics and the uptake capacity.[88]

2.3.3.1.1 Influence of the Operating Conditions. Theoretically, Li_4SiO_4 can adsorb up to 0.367 g of CO_2 per gram of Li_4SiO_4 (~8.34 mmol CO_2 per g Li_4SiO_4). The carbonation reaction with Li_4SiO_4 is given in the following equation:

$$Li_4SiO_4 + CO_2 = Li_2CO_3 + Li_2SiO_3 \quad \Delta_f H° = -143.5\,kJ\,mol^{-1} \tag{2.9}$$

Experimentally, a CO_2 adsorption capacity close to 8.0 mmol g^{-1} can be approached at a temperature of 700 °C. However, the adsorption properties of these materials depend very much on the operating conditions, such as the adsorption temperature, and the presence of other gases. Thus, Essaki *et al.*[89] have demonstrated the influence of the CO_2 adsorption temperature on Li_4SiO_4 pellets. According to this study, the optimal adsorption temperature is between 550 and 600 °C in a flow containing 10% CO_2. At a temperature above 600 °C, the amount of CO_2 adsorption drastically drops. Other experimental parameters have to be considered during CO_2 capture using Li_4SiO_4; among them, the presence of water in the feed allowed a significant rise in the amount of CO_2 adsorbed by Li_4SiO_4, as in the cases of other adsorbents discussed in this chapter.[90,91] Quinn *et al.*[91] reached a CO_2 adsorption capacity of 5 to 6 mmol g^{-1} in a gas mixture containing 15% of CO_2 and 10% of H_2O in N_2, while the adsorption properties drastically diminished in an anhydrous flow of 15% of CO_2 diluted in N_2 (0.34 to 0.77 mmol g^{-1}). On the other hand, it has been shown that the presence of SO_2 has a significant inhibitory effect on the adsorption by Li_4SiO_4. Pacciani *et al.*[92] demonstrated that SO_2 reacts with Li_4SiO_4 at 550 °C, leading to the formation of Li_2SO_4. This reaction is irreversible, thus leading to a drastic decrease in the adsorption capacity. In addition, such poisoning of the adsorbent is visible even at extremely low SO_2 concentrations (0.002%). Industrial implementation of Li_4SiO_4 as a CO_2 adsorbent thus requires an upstream desulfurization step.

2.3.3.1.2 Improvement of the Absorption Performance of Li_4SiO_4. The improvement of the CO_2 adsorption performance of Li_4SiO_4 requires modifications of the synthesis routes and doping.

Improvement *via* Li_4SiO_4 Synthesis. Several methods for synthesizing Li_4SiO_4 have been investigated such as solid-state reactions,[85,93,94] precipitation methods,[95] urea combustion[96] and sol–gel methods.[97–99] With all methods, one of the key points remains the choice of an appropriate silica source. In most cases, quartz, amorphous silica, fumed silica or tetraethyl orthosilicate (TEOS) are used as a silicon source, and LiOH or $LiNO_3$ are commonly used as precursors for the formation of Li_4SiO_4. Despite its low cost, quartz is not very attractive as it is inert under most conditions and does not react easily with LiOH or $LiNO_3$. In a solid-state reaction, Li_2CO_3 was mixed homogeneously with amorphous silica at different ratios (0.5, 1, 2 and 4) then calcined at 700, 800 and 900 °C.[94] XRD analysis highlighted that at 700 °C, with ratios of 0.5, 1 and 2, all samples contained amorphous silica and Li_2SiO_3 as the main lithium silicate phase. At Li/Si = 4 and at 700 °C, close to complete phase purity of Li_4SiO_4 could be achieved. However, at a higher temperature (900 °C), the decomposition of Li_4SiO_4 started. An average temperature of 800 °C is necessary to get a pure Li_4SiO_4 phase. Solid-state reaction synthesis can become challenging, as high temperatures are required, leading to volatilization and lack of composition and microstructure control. The sol–gel method is preferred since a synthesis at low temperatures is possible and there is a better control of both microstructure and morphology. Alkoxide precursors or TEOS and $LiOH{\cdot}H_2O$-based sol–gel synthesis are able to produce Li_4SiO_4 powders but require an organic medium.[97] An alternate sol–gel method with citric acid as a chelating agent is based on aqueous production of Li_4SiO_4 powders, thanks to LiOH and aerosil SiO_2.[98,99] At a ratio of citrate-to-metal (C–M) equal to 1.4, a high pure Li_4SiO_4 phase can be produced. The method resulted in primary particle sizes of 100 nm once calcined at 675 °C. Moreover, the smaller particle size and high surface area led to powders with excellent sinterability. Numerous studies have shown that an improvement in the adsorption capacity of Li_4SiO_4 can be observed by a decrease in the sizes of the adsorbent particles.[93,100,101] Romero-Ibarra *et al.*[100] have shown that a 10-fold increase in the surface area of Li_4SiO_4 after ball milling favorably influenced the adsorption properties of the material as well as its durability.

Improvement *via* Doping. The CO_2 adsorption performance of Li_4SiO_4 can be improved mainly by (1) promotion with alkali cations, or (2) transition metals.

Alkali Promoter. The interest in doping Li_4SiO_4 with sodium or potassium arises from the fact that such a promoting effect has already been demonstrated in the case of Li_2ZrO_4. Indeed, alkaline solid solutions such as lithium-sodium zirconates ($Li_{2-x}Na_xZrO_3$) and lithium-potassium zirconates ($Li_{2-x}K_xZrO_3$) exhibit much better adsorption capacities for CO_2 than pure lithium or sodium zirconates.[18,102,103] Mejia-Trejo *et al.*[97] were interested in the synthesis of $Li_{4-x}Na_xSiO_4$ solid solutions and the performances of these materials in CO_2 capture. According to the authors, a maximum adsorption capacity of 4.4 mmol g^{-1} was reached for $Li_{3.85}Na_{0.15}SiO_4$ and this value remains higher than the one found under the same conditions for Li_4SiO_4

(3.9 mmol g^{-1}). In addition, it should be emphasized that the adsorption kinetics in the Na-doped sample was greatly improved. Doping with alkali metals also involved the preparation of eutectic mixing. Mixing dopant with precursor during the initial synthesis steps allows the formation of eutectic-doped adsorbents. This is different from simple doping where the dopant is added after the solid adsorbent has been produced. Many studies on doping CO_2-capture-based sorbents have dealt with alkali carbonates, mostly K_2CO_3 and Na_2CO_3. Seggiani *et al.*[104] showed that the CO_2 adsorption capacities of modified Li_4SiO_4 with different alkali carbonates (K_2CO_3, Na_2CO_3), binary ($K_2CO_3/LiCO_3$, $Na_2CO_3/LiCO_3$) or ternary ($K_2CO_3/Na_2CO_3/LiCO_3$) eutectic mixtures have been greatly improved. According to this study, the best adsorption performance, 5.2 mmol g^{-1} at 580 °C, was obtained when 30 wt% Na_2CO_3 or K_2CO_3 was added to Li_4SiO_4. In addition, the authors noted that in the case of the use of Na_2CO_3, the sorption capacities drastically diminished as a function of the number of sorption–desorption cycles, certainly due to a sintering effect. This is not the case with the addition of K_2CO_3, which allowed them to maintain a good capacity for adsorbing CO_2 even after 25 sorption–desorption cycles.

Transition Metal Doping. The promoting effect of transition metal doping comes directly from the mechanism (double shell theory) of CO_2 adsorption with Li_4SiO_4, already described in the previous paragraph. As a reminder, in this mechanism, Li_4SiO_4 decomposes at high temperatures into $LiSiO_3$ with the formation of Li^+ and O^{2-} (eqn (2.7)). These cations will then diffuse on the surface of the solid and react with CO_2 to form Li_2CO_3 (eqn (2.9)). It is therefore necessary to facilitate the diffusion of Li^+ and O^{2-} to the surface to theoretically improve the CO_2 adsorption performance of Li_4SiO_4. To allow a good diffusion, it is necessary to generate defects in the crystal structure of Li_4SiO_4 by appropriate doping with heteroelements such as Al and Fe. The original idea for this type of doping came from "know-how" gathered from solid-state ionic conductors where ionic mobility is an important factor.[105] In this context, studies were focused on the influence of the substitution of Si^{4+} by Al^{3+} and Li^+ by Al^{3+}.[105–107] In the former case, the substitution creates interstitial Li^+ ions whilst in the latter, vacancies can be generated by the substitution of Li^+ by Al^{3+}. In both cases, Li^+ mobility and conductivity are greatly improved with respect to non-modified Li_4SiO_4.[105,106] However, the diffusion of O^{2-} to the surface remains the limiting step of this mechanism, as the ionic radius of O^{2-} is higher than the ionic radius of Li^+. To improve the CO_2 adsorption performance of Li_4SiO_4, it seems more advisable to favor the appearance of vacancies in Li_4SiO_4, such as the substitution of Li^+ by Al^{3+}, to facilitate O^{2-} hopping. This hypothesis was confirmed by a study conducted by Gauer *et al.*[107] in which the effect of Li_4SiO_4 doping by heteroelements such as Al or Fe was investigated. The authors showed that the sample in which part of Li^+ was replaced by Al^{3+} ($Li_{3.7}Al_{0.1}SiO_4$) exhibited better CO_2 sorption capacity when compared to the $Li_{4.1}Al_{0.1}Si_{0.9}O_4$ sample.

This result tends to prove that it is beneficial to generate vacancies, rather than creating interstitial Li^+. In addition, the use of Fe in place of Al in $Li_{3.7}Fe_{0.1}SiO_4$ is more advantageous because CO_2 is more easily desorbed. Other studies have also highlighted the fact that the choice of the heteroelement in the substitution has a great importance in CO_2 adsorption capacity. Ortiz-Landeros *et al.*[108] demonstrated that the addition of Al and V in Li_4SiO_4 greatly influenced the CO_2 chemisorption of $Li_{4+x}(Si_{1-x}Al_x)O_4$ and $Li_{4-x}(Si_{1-x}V_x)O_4$ solid solutions. In the sample doped with Al, the adsorption capacity was much higher compared to $Li_{4-x}(Si_{1-x}V_x)O_4$. This difference can be explained by the composition of the product's external shell. Thus, in these two samples, Li_2CO_3 and Li_2SiO_3 were both found in the external shell and the difference in composition relies on the presence of Li_3VO_4 or $LiAlO_2$. The decrease in the CO_2 sorption of the V-doped sample was explained by the formation of Li_3VO_4 during adsorption, which led to a decrease in lithium diffusion since these species have a lower diffusion coefficient than Li_2CO_3, Li_2SiO_3 and $LiAlO_4$.

2.3.3.2 Other Lithium Silicates

In parallel to the work done on Li_4SiO_4, the use of other lithium silicates as CO_2 adsorbents at high temperatures has been reported in a few studies.[88,109] Over a wide temperature range, CO_2 adsorbents with a high lithium content silicate, such as Li_8SiO_6, have been described in the literature.[109] In the case of Li_8SiO_6, Wang *et al.* achieved a high CO_2 adsorption capacity of 11.8 mmol g^{-1}. They were also able to establish that the adsorption reaction with these adsorbents occurred *via* a two-step mechanism, dependent on the temperature. At first, Li_8SiO_6 chemisorbs 2 moles of CO_2 to produce Li_4SiO_4, which can then optionally capture a third mole of CO_2 to form Li_2SiO_3 and Li_2CO_3. It should be noted that the second stage of the process only occurs at temperatures below 550 °C. In addition, an *ab initio* thermodynamic and experimental data approach was developed by Duan *et al.*[88] for the study of lithium silicates with different Li_2O–SiO_2 ratios as CO_2 adsorbents. The authors demonstrated a better adsorption capacity, a higher heat of reaction and higher turnover temperatures when the Li_2O–SiO_2 ratio was increased (compounds ranging from $Li_2Si_3O_7$ to Li_8SiO_6). It should be noted that Li_8SiO_6, the lithium silicate with the highest Li_2O–SiO_2 ratio, required a high regeneration temperature and thus consumed greater energy during the CO_2 capture process. In addition, a high CO_2 adsorption capacity was found in the case of Li_2O-rich lithium silicates (Li_8SiO_6, Li_4SiO_4, $Li_6Si_2O_7$) when the sorption temperature was below the turnover temperature of Li_2SiO_3. At these temperatures, these materials adsorb CO_2 and form Li_2CO_3 and SiO_2. On the other hand, if the capture temperature was close to the turnover temperature of Li_2SiO_3, the products obtained after adsorption of CO_2 were Li_2CO_3 and Li_2SiO_3 with a low adsorption capacity. In the case of SiO_2-rich lithium silicates ($Li_2Si_2O_5$, $Li_2Si_3O_7$), these materials can thermodynamically adsorb CO_2 at a relatively low temperature, but with a low adsorption capacity. Also, the adsorption kinetics is very slow when the temperature remains too low.

2.4 Clays-based Adsorbents

2.4.1 Phyllosilicates

Clays are very fine-grained natural minerals that can be found in abundance particularly in soils or as sedimentary rocks. Usually, these geologic clay deposits consist of hydrated phyllosilicate minerals, quartz and organic fragments. These remarkable minerals are particularly of interest due to their unique properties, leading to plastic when moist, but hardening when they are dried and backed to make bricks, pottery or ceramics. An exhaustive review of clays and their applications is proposed in a detailed book by Mukherjee.[110] Two elementary units are important to build phyllosilicates: tetrahedral silicon species (SiO_4^{4-}) and divalent or trivalent metals octahedrally coordinated to hydroxyl groups ($M^{2+}/M^{3+}(OH)_6$). The silicon tetrahedron shares three basal oxygens with other silicon tetrahedra to build a hexagonal two-dimensional pattern, the tetrahedral sheet (T-sheet), and the remaining apical oxygens point toward the same direction to link an octahedral cation. The octahedra share edges to form a two-dimensional unit, the so-called octahedral sheet (O-sheet). If the octahedron is centered with M^{2+}, all octahedral sites are occupied and the sheet is called trioctahedral. If the cation is trivalent (M^{3+}), then only 2/3 of the octahedral sites are occupied and the resulting two-dimensional unit sheet is named dioctahedral.[110] In an ideal condition, the apical oxygen of the tetrahedral sheets can be linked to one octahedron, building the T–O structure (1 : 1 structure). On one side, the basal plane consists of the O^{2-} of the T-sheet, on the other side there is a plane of OH^-. Between them, an anionic plane consists of the apical oxygens of the T-sheets and the unshared OH^- of the octahedra.[110] On both sides, the octahedra possess hydroxyls to share, one octahedron can be sandwiched by two tetrahedral sheets, originating the T–O–T structure (2 : 1 structure). Depending on the structure and the chemical nature of the clay minerals, a classification of clays can be established (Figure 2.3).[110]

2.4.2 Clays for CO_2 Capture

The numerous advantages of clays, such as their low cost, availability, large-scale production, regeneration capability, and mechanical and thermal stability, have attracted attention in industry, leading to extensive uses as bricks and insulating materials. Clays were thoroughly investigated and some studies have demonstrated additional peculiar properties: a high surface area,[111] selective adsorption and surface modification.[112,113] Their weak-base character and ion-exchange properties lead to the formation of weak van der Waals

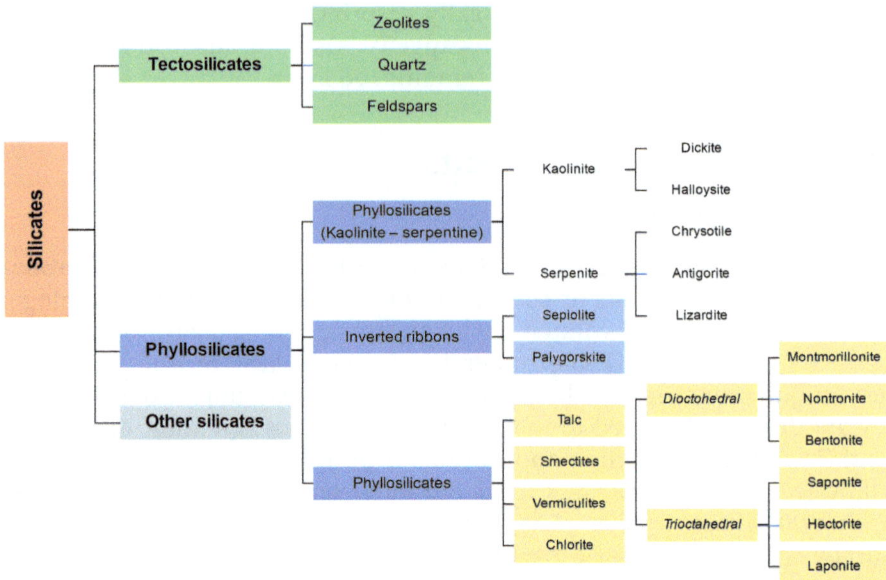

Figure 2.3 Classification of silicate minerals.[110]

forces, able to trap carbon dioxide molecules.[114–117] As a consequence, clays can potentially be used for CO_2 capture. An interesting review from Wang *et al.*[78] highlighted recent advances in solid sorbents for CO_2 capture, and a more recent review from Chan *et al.*[118] focused specifically on low cost adsorbents from clays. Most studies dealing with CO_2 adsorption have focused on dioctahedral smectite clays such as montmorillonite and bentonite.[118] Interesting studies were also carried out on trioctahedral smectites such as saponite (Mg-rich),[119] hectorite (Li-rich)[120] and laponite (Li-rich).[121,122] CO_2 capture was also studied on sepiolite.[112,123,124] Table 2.6 summarizes the CO_2 adsorption results obtained in this work.

2.4.2.1 Smectite

Smectite is the general name for a group of hydroxyl alumino-silicates, with trioctahedral or dioctahedral T–O–T structures, containing alkali and alkaline earth elements. This group includes several mineral species such as montmorillonite, bentonite, saponite and hectorite. The difference between these species from this smectite group relies on variations in chemical structure and composition, particularly the substitutions of Al by Si in tetrahedral cation sites and Al, Fe, Mg and Li in octahedral cationic sites. Isomorphous substitution generates an excess of negative charge in the layers, which needs to be compensated by corresponding cations. In 2008, smectite was selected by Pires *et al.*[113] as a porous clay mineral with potential for CO_2 capture. Treatment of smectite with cetyltrimethylammonium bromide (CTAB) was

Table 2.6 Summary of CO_2 adsorption data for selected clay-based sorbents presented in this section.

Materials	CO_2 adsorption capacity	Conditions	Reference
Smectite			
Modified with tetraethoxysilane (TEOS) and phenyltriethoxysilane (PhOS)	1.2 mmol g^{-1}	25 °C and 85 kPa	113
Montmorillonite			
With immobilized diamine	(1) 1.8 mmol g^{-1} (2) 2.4 mmol g^{-1}	(1) 15% CO_2 in N_2 (2) In pure CO_2 at 100 °C	128
Intercalated with three soya-oil-derived polyglycerol dendrimers	0.012–0.016 mmol g^{-1}		116
(1) Modified with aminopropyltrimethoxysilane (APTMS) (2) Modified with polyethylenimine (PEI)	(1) 1.7 mmol g^{-1} (2) 3.9 mmol g^{-1}	1 bar and at *ca.* 20 bar, in a temperature range of 75–85 °C	132
Modified with PEI	(1) 1.8 mmol g^{-1} (2) 3.6 mmol g^{-1}	(1) At room temperature (25–30 °C) (2) 17 vol% of moisture in flue gas conditions	117
Bentonite			
(1) Non-modified bentonite (2) PEI-modified bentonite	(1) 0.14 mmol g^{-1} (2) 1.1 mmol g^{-1}		115
Pre-treated with sulfuric acid and modified with TEPA (loading of 50 wt%)	(1) 2.95 mmol g^{-1} (= 130 mg CO_2 per g) (2) 4.32 mmol g^{-1} (= 190 mg CO_2 per g)	(1) At 75 °C under dry conditions (2) 18% moisture added to the flue gas	136
(1) Raw bentonite (2) Porous clay heterostructures (PCH) from bentonite (3) PCH grafted with 3-(triethoxysilyl)propylamine (APTES) (4) PCH impregnated with polyethyleneimine (PEI) (5) PCH impregnated with tetraethylenepentamine (TEPA)	(1) 0.112 mmol g^{-1} (2) 0.640 mmol g^{-1} (3) 1.023 mmol g^{-1} (4) 1.465 mmol g^{-1} (5) 1.644 mmol g^{-1}	1 bar and at 25 °C	137

(*continued*)

Table 2.6 (*continued*)

Materials	CO_2 adsorption capacity	Conditions	Reference
Saponite			
Alumina-pillared saponite			
(1) Calcined at 200 °C (2) Calcined at 350 °C (3) Calcined at 500 °C	(1) 1.30 mmol g^{-1} (2) 0.50 mmol g^{-1} (3) 0.25 mmol g^{-1}	At 0 °C and 35 kPa	119
Hectorite			
Modified with 1-butyl-3-methylimidazolium bromide (BMI), 1,3-didecyl-2-imidazolium chloride (DDMI) or 1-decyl-3-methylimidazolium chloride (DMI)	(1) 0.18–0.31 mmol g^{-1} (2) 0.26–0.42 mmol g^{-1} (3) 0.33–0.53 mmol g^{-1} (4) 0.38–0.60 mmol g^{-1} (5) 0.44–0.69 mmol g^{-1}	(1) 20 kPa, 30 °C (2) 40 kPa, 30 °C (3) 60 kPa, 30 °C (4) 80 kPa, 30 °C (5) 101.3 kPa, 30 °C	120
Sepiolite			
Immobilized tetraethylenepentamine (TEPA) onto acid-modified nanosepiolite (loading of 60 wt%)	3.8 mmol CO_2 per g-sorbent,	1 vol% CO_2 in N_2 along with 1 vol% H_2O at 60 °C	123
Polyethylenimine impregnated on acid-modified sepiolite	1.70 mmol g^{-1}	At 338 K and 1 bar	124

followed by a reaction with octylamine for 3 h. Their study showed that a high surface area (634 $m^2\,g^{-1}$) and a better micropore size distribution 0.45 mmol g^{-1} were exhibited by porous smectite with a mixture of tetraethoxysilane (TEOS) and phenyltriethoxysilane (PhOS). According to the Jaroniec–Gadkaree–Choma model, the pore width range was 0.58–0.63 nm.[125] However, the phenyl groups present on the smectite surface are disadvantageous for CO_2 capture, leading to a CO_2 adsorption capacity of 1.20 mmol g^{-1} at 25 °C and 85 kPa. According to Sirca's investigation,[126] interactions with CO_2 molecules are favored with –OH groups rather than phenyl groups, because of the strong quadrupole moment of 4.3×10^{-26} esu cm^2 for CO_2 molecules. Hence, it was deduced that the surface chemistry of the adsorbents has to be taken into consideration to improve adsorption capacity. This point was particularly detailed during studies on montmorillonites.

2.4.2.2 Montmorillonite

The term 'montmorillonite' is based on a large deposit of clay located in Montmorillon (France). It gained popularity as both an individual mineral name and a term for all smectite minerals. The term 'montmorillonite' has been restricted more recently to the dioctahedral mineral with a T–O–T structure, in which small amounts of octahedral aluminium are substituted by octahedral magnesium. This member of the smectite group is therefore a 2 : 1 clay, with a $(OH)_4Si_8(Al_{4-x}Mg_x)O_{20}$ chemical structure. It consists of two tetrahedral sheets of silica sandwiching a central octahedral sheet of alumina. Selective adsorption and catalyst support properties were intensively studied.[111] Montmorillonite can be used directly or after treatment with concentrated sodium hydroxide (2–5 M, 80 °C, 3–4 h) as synthesized by Azzouz *et al.* in 2003.[127] However, as mentioned for smectites, modification of montmorillonites is necessary to improve CO_2 capture. Immobilized diamine on montmorillonite was synthesized by Stevens *et al.*,[128] by a water-aided exfoliation and grafting route. CO_2 adsorption capacities of 1.8 mmol g^{-1} and 2.4 mmol g^{-1} were obtained in 15% CO_2 in N_2 at 95 °C and in pure CO_2 at 100 °C, respectively. Polyol-dendrimer-modified montmorillonites are also particularly attractive, especially for their acute compromise between the highest adsorption capacities and lowest desorption temperatures. These properties are remarkably advantageous in topical subjects such as greenhouse effect reduction. For instance, Azzouz *et al.*[114,116,129–131] highlighted the importance of polyol-montmorillonites as re-usable, eco-friendly natural materials. For this purpose, the research group synthesized montmorillonite intercalated with three soy-oil-derived polyglycerol dendrimers with molecular weights of 500, 1100 and 1700, and also polyol-intercalated montmorillonites. These modifications led to effective CO_2 capture, thanks to the weak base-like character of the main adsorption sites, such as OH groups grafted on the dendritic moiety. CO_2 adsorption capacities up to 0.012–0.016 mmol g^{-1} were reached, but with much lower desorption temperatures than amine-based adsorbents. Moreover, in the study on montmorillonites intercalated with polyol dendrimers, Azzouz *et al.*[130] demonstrated that dendrimer size and content are key parameters. As a consequence, the specific surface area and the number of accessible OH groups diminished, leading to lower CO_2 retention capacity. These authors also suggested that the retention of larger amounts of CO_2 than predicted by stoichiometry shows that more than one CO_2 molecule adsorbed on each OH group. These results confirm that in these cases, OH groups should interact with CO_2 only through physical interactions.[131] Roth *et al.*[132] also confirmed that CO_2 can potentially be captured in an efficient manner by an amine-modified montmorillonite nanoclay. This group noticed that untreated nanoclay led to low CO_2 adsorption. In contrast, once doped with amine by a reaction between nanoclay with aminopropyltrimethoxysilane (APTMS) and polyethylenimine (PEI), a significant increase of CO_2 capture was achieved. CO_2 adsorption capacities of 1.7 mmol g^{-1} and 3.9 mmol g^{-1} were respectively obtained at 1 bar and at *ca.* 20 bar, in a temperature range of 75–85 °C. It is interesting to note that regeneration of these modified nanoclays was possible using several pathways: thanks to sweep gases such as nitrogen at 100 °C or CO_2 at 155 °C, or under vacuum at 85 °C under a pressure swing operation mode. According to recent work on montmorillonite modifications done by Wang *et al.*,[117] it appears that the largest surface area was obtained with PEI-treated montmorillonite (183–253 m^2 g^{-1}) compared to polyglycerol, CTAB or ethylene glycol treatments (13–68 m^2 g^{-1}).[118] Porosity and surface area were both improved by acid and base treatment on montmorillonite. This led to an increase in CO_2 capture, with an impressive CO_2 adsorption capacity of 80 mg CO_2 per g adsorbent (1.8 mmol g^{-1}), at room temperature (25–30 °C). The interaction between CO_2 and the active site is indeed enhanced by the amine group present on the montmorillonite's surface. The same research group also noticed that the remarkable activity of PEI-treated montmorillonite in CO_2 adsorption could even be increased to 161 mg g^{-1} thanks to the 17 vol% of moisture in flue gas. Xu *et al.* described that CO_2 is captured as a carbamate, by reaction of the immobilized amine with CO_2 under dry conditions.[48,133] Under moist conditions, bicarbonates are readily formed. Indeed, an interaction between one amine group and CO_2 can form a bond in moisture.

2.4.2.3 Bentonite

The name bentonite comes from a large deposit of clay located in Fort Benton, Wyoming, US. It consists mostly of montmorillonite, but occasionally, other types of smectite may be present. Pioneering work from Leliveld *et al.* in 1999[134] related the sorption properties to the nature of the parent clay, taking into account the Si–Al ratio and textural properties. As mentioned by Gil *et al.*,[135] bentonite's surface is negatively charged, because of the isomorphous substitution of Al^{3+} for Si^{4+} within the layers. Bentonite in its natural state is not very efficient for CO_2 capture. As previously described for montmorillonites, it is necessary to treat bentonite to enhance CO_2 capture. Chen *et al.*[115] highlighted recently the notable increase in CO_2 adsorption capacity from 0.14 to 1.1 mmol g^{-1} using PEI-modified bentonite, compared to pristine bentonite. The advantages of this PEI-modified bentonite are, in particular, its excellent CO_2 adsorption selectivity over N_2 and stability during successive runs. Nevertheless, one drawback of this material is the smaller pore volume, leading to lower CO_2 capture by using PEI-modified bentonite compared to PEI-impregnated mesoporous silica. Sulfuric acid treatment was reported to solve this hindrance, by inducing significant improvements in the pore volume and surface area of bentonite.[136] A CO_2 breakthrough sorption of 2.95 mmol g^{-1} was obtained at 75 °C under dry conditions, with a TEPA loading of 50 wt%. Once 18% moisture was added to the flue gas, bicarbonate formation allowed the CO_2 sorption capacity to be further increased to 4.32 mmol g^{-1}. Finally, these compounds exhibited a great regeneration ability and thermal stability below 130 °C. Vilarrasa-García *et al.*[137] recently synthesized porous clay heterostructures (PCH) from bentonite, and functionalized them with amines by grafting using 3-(triethoxysilyl) propylamine (APTES) or by impregnation of amine-rich polymers using polyethylenimine (PEI) and tetraethylenepentamine (TEPA). These functionalized PCH present higher specific surface areas and pore volumes than PCH and raw bentonite, leading to higher CO_2 adsorption capacities. First of all, CO_2 adsorption increased from 0.11 mmol CO_2 g^{-1} from bare bentonite to 0.64 mmol CO_2 g^{-1} with PCH, at 1 bar and 25 °C. A CO_2 adsorption capacity of 1.02 mmol CO_2 g^{-1} was achieved over grafted PCH with APTES, whereas it reached 1.64 and 1.46 mmol CO_2 g^{-1} using PCH impregnated with TEPA and PEI, respectively. The authors explained this reactivity by the coexistence of physical and chemical interactions when grafted PCH with APTES was used, whilst chemical interactions between CO_2 and the amines located in the porous structure and/or on the surface of the adsorbent play a major role in CO_2 adsorption using PCH impregnated with TEPA or PEI. It is noteworthy to indicate that too high a polymer content should be avoided, as it creates stacking of the amine-rich polymer, leading to a decrease in CO_2 capture efficiency.

2.4.2.4 Saponite

Saponite is a trioctahedral mineral of the smectite group, considered as Mg-rich, with the chemical formula $Ca_{0.25}(Mg,Fe)_3((Si,Al)_4O_{10})\ (OH)_2{\cdot}n(H_2O)$. CO_2 capture on alumina pillared saponite was studied by Gil *et al.*[119] They found that an increase in calcination temperature in the modified saponite had a dramatic effect on CO_2 capture. Indeed, the CO_2 adsorption capacity was about 1.30 mmol g^{-1} at 0 °C and 35 kPa with alumina pillared saponite calcined at 200 °C, but it diminished to 0.50 and 0.25 mmol g^{-1} when the modified saponite was calcined at 350 and 500 °C, respectively. The authors explained this effect by the different amounts of micropore formation in the modified saponite, due to the redistribution of alumina pillars in saponite upon thermal treatment. The micropores were partially blocked and thus the related specific area was lost when the calcination temperature was enhanced.[138,139]

2.4.2.5 Hectorite

The name hectorite comes from a large deposit of clay located in Hector, California, US. Like saponite, it is also a trioctahedral mineral of the smectite group, but considered as Li-rich, with the chemical formula $Na_{0.3}(Mg,Li)_3$-$Si_4O_{10}(OH)_2$ (Figure 2.4). Synthetic hectorites with tuned porosities were prepared by hydrothermal crystallization by Pawar *et al.*[120] Lithium fluoride (LiF), magnesium oxide (MgO) and silica sol (SiO_2) were refluxed for 48 h. The different hectorites were modified with 1-butyl-3-methylimidazolium bromide (BMI), 1,3-didecyl-2-imidazolium chloride (DDMI) and 1-decyl-3-methylimidazolium chloride (DMI), respectively. Once calcined, these specific hectorites exhibited mainly mesopores, with specific surface areas between 335–355 $m^2\ g^{-1}$. Pawar's group demonstrated that elevated pressure enhanced CO_2 capture. Indeed, when the reaction with CO_2 was performed at 30 °C with increasing pressures (from 20 to 101.3 kPa), then CO_2 adsorption capacities were enhanced as a consequence from 0.18 to 0.69 mmol g^{-1}, respectively. As CO_2 interaction and adsorption on synthetic hectorites requires more energy, the elevation of pressure can be helpful to solve this drawback.

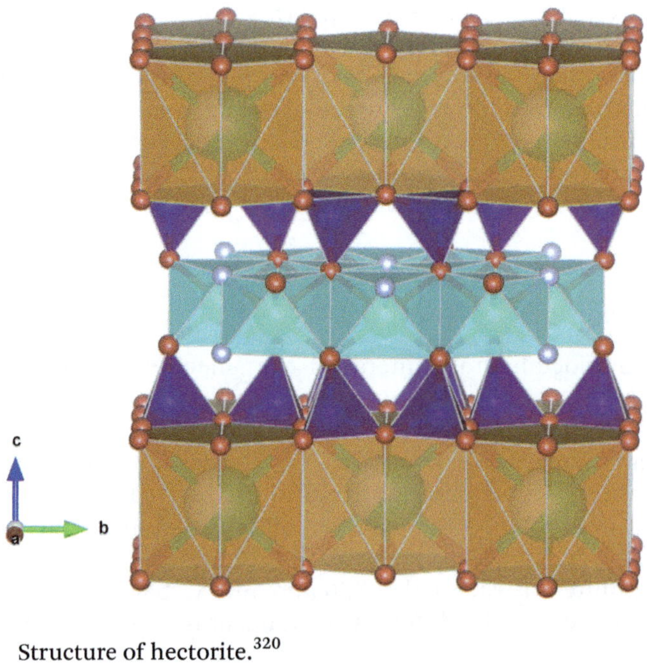

Figure 2.4 Structure of hectorite.[320]

2.4.2.6 Laponite

Laponite is a lithium sodium magnesium silicate clay with the chemical formula $Na_{0,55}(Mg_{5,45}Li_{0,55})(Si_8O_{20})(OH)_4{\cdot}nH_2O$. This clay has a trioctahedral T–O–T structure, and small amounts of Mg octahedra are substituted by Li octahedra. This synthesized phyllosilicate has a similar structure to hectorite, but has shorter sheets. Shah *et al.* recently presented their work[121,122] on modified laponite as a solid adsorbent with a highly selective capture capacity. Sericite and hydrotalcite clays were also investigated in this study. To enhance CO_2 adsorption capacity, modified laponite organoclays were prepared by immobilizing poly(amidoamine) (PAMAM) dendrimer through an ion-exchange process on laponite. These amine-terminated dendrimers are organic molecules with highly branched three-dimensional spherical shapes, involving a high number of terminal functional groups. In this study, the authors used fourth generation (G4) PAMAM dendrimers, which can load 64 primary amine functions. Thanks to the numerous binding sites on the PAMAM dendrimer, the CO_2 adsorption on modified laponite was more efficient than on pristine laponite, and also increased with the amount of loaded dendrimers. Moreover, it was noticed that the binding sites on the dendrimers prevent the desorption process. After reaching an adsorption–desorption equilibrium, CO_2 bonded to the dendrimer was conserved in the organoclay, whereas CO_2 adsorbed on the binding site of laponite was almost desorbed.

2.4.2.7 Sepiolite

Sepiolite is a hydrated magnesium-silicate clay from the palygorskite-sepiolite group, with the chemical formula $[(Si_{12})(Mg_9)O_{30}(OH)_6(OH_2)4H_2O]$ (Figure 2.5). This mineral is usually associated with other clays and non-clay minerals

Figure 2.5 Structure of sepiolite.[321]

such as phosphates, carbonates and quartz. Like other phyllosilicates, sepiolite possesses T-sheets, but the cationic octahedra are linked together to form ribbons or chains, instead of O-sheets like other phyllosilicates. It is thus composed of large bundles of crystalline fibers. Its specific chemical structure, with channels of molecular dimensions oriented along the longitudinal direction of the fibers, and non-microporous surface coming from the external surface of the fibers, is notably responsible for its adsorptive properties. As a consequence, sepiolite presents several interesting advantages for CO_2 capture. Also, the high concentration of Si–OH groups on the surface leads to a polar character, important for CO_2 adsorption. Finally, the high availability of sepiolite, particularly in Spain, and its low cost are also attractive. In 2007, Delgaldo *et al.* studied[112] the use of this mineral for the removal of carbon dioxide from a carbon dioxide–methane mixture by a pressure swing adsorption (PSA) process. They obtained a methane product with a purity over 97% from a feed mixture with 34% of carbon dioxide, highlighting that sepiolite can be a mineral of choice used in a PSA cycle for carbon dioxide removal from landfill gases. Irani *et al.* studied[123] tetraethylenepentamine (TEPA) immobilized onto acid-modified nano-sepiolite. The surface area was initially increased from 103 to 273 $m^2 g^{-1}$ thanks to the acid treatment of nano-sepiolite. A CO_2 adsorption capacity of 3.8 mmol CO_2 per g sorbent, for 1 vol% CO_2 in N_2 along with 1 vol% H_2O at 60 °C, was obtained with a TEPA loading of 60 wt% on modified nano-sepiolite. Villarrasa-García *et al.*[124] also performed an acid treatment on sepiolite to enhance the surface area by progressive depletion of Mg^{2+} and formation of an amorphous silica phase. Microwave radiation was used to speed up heating during the acid treatment, followed by impregnation of the sepiolite with polyethylenimine. A CO_2 adsorption capacity of 1.70 mmol g^{-1} at 338 K and 1 bar was obtained using this modified sepiolite.

2.5 Mineral Silicates for Carbonation

2.5.1 Mineral Carbonation

To reduce carbon dioxide emissions, one of the most promising approaches remains the capture and sequestration of this greenhouse gas in geological formations. Mineral carbonation can be defined as the fixation of CO_2 as a stable carbonate mineral, such as dolomite ($CaMg(CO_3)_2$), magnesite ($MgCO_3$), calcite ($CaCO_3$), or siderite ($FeCO_3$). In nature, it is known as a natural phenomenon called a silicate-weathering process.[140] This kind of reaction occurs with a very slow kinetics and takes place on a geological timescale. The main advantages of mineral carbonation are that the reaction is exothermic and thermodynamically possible under normal conditions, with the permanent storage of CO_2 as a stable carbonate.[141,142] Several research groups have investigated different mineral carbonation methods to find a suitable way to enhance the reaction.[140,142] Several detailed reviews present an exhaustive overview of these studies.[141–143]

2.5.2 Silicates as Natural Minerals for Carbonation

The materials generally used for mineral carbonation are magnesium (Mg) and calcium (Ca) silicates. Mineral availability and percentage of impurities and reactive components (Mg and Ca) are some of the main criteria for the selection of minerals for the carbon capture and sequestration (CCS) process. The presence of components that can inhibit the reaction and the possibility of forming a stable and environmentally safe by-product are also interesting points for mineral selection.[140] Suitable materials for mineral carbonation would be magnesium or calcium oxides and hydroxides, but they are not easily found in nature. The most frequently encountered structures in nature are magnesium silicates. These minerals have been exhaustively studied because of their relative abundance, high content of reactive component and their R_{CO_2} value (ratio of the mass of mineral–mass of fixed CO_2).[144–147] Indeed, the MgO concentration is higher in magnesium silicate (up to 50% in weight) than in calcium silicate (10% in weight). This leads to a better CO_2 capture capacity for magnesium silicates (0.55 kg CO_2 per kg rock for magnesium silicate compared to 0.08 kg CO_2 per kg rock for calcium silicate).[144] The most studied magnesium silicates for mineral carbonation are olivine, forsterite and serpentine. Some other suitable minerals for the CCS process exist, but in smaller amounts. In the case of calcium silicate, wollastonite ($CaSiO_3$) can be mentioned. The mineral **olivine** is a magnesium iron silicate with the chemical formula $(Mg,Fe)SiO_4$. The ratio of magnesium and iron varies between two end members: forsterite (Mg-endmember: Mg_2SiO_4) and fayalite (Fe-endmember: Fe_2SiO_4). The composition of olivine is thus usually written as molar percentages of forsterite and fayalite. The **serpentine** group describes a group of hydrous magnesium iron phyllosilicate minerals $(Mg,Fe)_3Si_2O_5(OH)_4$.[148] These minerals can contain traces of other elements such as manganese or nickel. There are three important mineral polymorphs of serpentine: chrysotile (asbestos) (Figure 2.6), lizardite and antigorite. The chemical formula is the same but the crystal structure is different. These

Figure 2.6 Structure of chrysotile (serpentine polymorph).[322]

three 1:1 phyllosilicates present a cyclic cylindrical structure,[149] a planar structure[150] or a wavy structure,[151] respectively: depending on the difference between tetrahedral layers (T) and octahedral layers (O).

2.5.3 Mineral Pre-treatments

Carbonation is in nature a very slow phenomenon. Several types of pre-treatment have been proposed to 'activate' the mineral and speed up the reaction rate, to improve the efficiency of the mineralization process. Commonly used pre-treatment techniques include: grinding, magnetic separation, chemical treatment and thermal activation.[152,153] Conventional **grinding** reduces grain size and significantly improves the mineral carbonation reaction rate, by enhancing the surface area and generating imperfections in the crystalline matrix, as demonstrated by Gerdemann *et al.* and O'Connor *et al.*[154,155] A specific type of grinding is attrition at high energy. This mechanical treatment leads to a change in the structure and increases the specific surface area, thus promoting the carbonation capacity of the mineral.[156–158] **Magnetic separation** is a technique widely used to separate magnetic elements, such as iron, for the purification of minerals and the concentration of magnetic compounds, but also to separate iron from food.[159] When mineral carbonation is carried out in an oxidizing atmosphere, an iron oxide layer is formed and this drawback slows down the diffusion of CO_2 towards the reactive species. Performing carbonation in a non-oxidizing atmosphere is quite expensive and difficult, so magnetic separation has been investigated for the removal of magnetic parts.[140,160] If magnetic separation is done prior to heat treatment or carbonation, mineral carbonation will be enhanced as the formation of a passive red blood cell layer is avoided and the reaction rate is enhanced, respectively.[140,152,160,161] **Chemical treatment** remains a method of choice for mineral activation by increasing surface areas. It enhances the dissolution kinetics of minerals. H_2SO_4 was found to be one of the most efficient acid treatments on serpentine,[141,142,153,161,162] but other compounds can be used for chemical treatments, such as HCl, H_3PO_4, KOH, $Mg(OH)_2$, NH_3 or NH_4Cl. Finally, **thermal treatment** can be a means to increase carbonation rate.[153] This process removes water from the structure, which allows the creation of a more open structure. The energy required for this thermal treatment is relatively high, as shown by many studies; *i.e.*, thermal treatment of serpentine has to be carried out at a temperature between 600–650 °C to be the most effective.[163–166]

2.5.4 Thermodynamics of Mineral Carbonation

CO_2 reacts with metal oxides to produce the corresponding carbonates while releasing heat. The chemical reaction corresponds to eqn (2.10), with metal oxides as MO, where M is a divalent metal (*i.e.*, calcium, magnesium, or iron). The amount of heat released depends on the metal content and the chemical composition of the metal oxides. Eqn (2.11) and (2.12) describe the reactions between magnesium oxides and calcium oxides and CO_2.[167] Eqn (2.13) highlights the general carbonation pathway starting from magnesium or calcium silicates. Eqn (2.14)–(2.16) illustrate the transformation of magnesium silicates, such as serpentine and forsterite (olivine), and the conversion of calcium silicates, such as wollastonite, respectively. These reactions occur in nature over a geological timeframe.

$$\mathrm{MO} + \mathrm{CO_2} \rightarrow \mathrm{MCO_3} + \mathrm{heat} \tag{2.10}$$

$$\mathrm{MgO} + \mathrm{CO_2} \rightarrow \mathrm{MgCO_3} (\Delta H_\mathrm{r} = -118\,\mathrm{kJ\,mol^{-1}}) \tag{2.11}$$

$$\mathrm{CaO} + \mathrm{CO_2} \rightarrow \mathrm{CaCO_3} (\Delta H_\mathrm{r} = -179\,\mathrm{kJ\,mol^{-1}}) \tag{2.12}$$

$$(\mathrm{Mg,Ca})_x\,\mathrm{Si}_y\mathrm{O}_{x+2y+z}\mathrm{H}_{2z} + x\mathrm{CO_2} \rightarrow x(\mathrm{Mg,Ca})\mathrm{CO_3} + y\mathrm{SiO_2} + z\mathrm{H_2O} \tag{2.13}$$

$$\mathrm{Mg_3Si_2O_5(OH)_4} + 3\mathrm{CO_2} \rightarrow 3\mathrm{MgCO_3} + 2\mathrm{SiO_2} + 2\mathrm{H_2O} (\Delta H_\mathrm{r} = -64\,\mathrm{kJ\,mol^{-1}}) \tag{2.14}$$

$$\mathrm{Mg_2SiO_4} + 2\mathrm{CO_2} \rightarrow 2\mathrm{MgCO_3} + \mathrm{SiO_2} (\Delta H_\mathrm{r} = -90\,\mathrm{kJ\,mol^{-1}}) \tag{2.15}$$

$$\mathrm{CaSiO_3} + \mathrm{CO_2} \rightarrow \mathrm{CaCO_3} + \mathrm{SiO_2} (\Delta H_\mathrm{r} = -90\,\mathrm{kJ\,mol^{-1}}) \tag{2.16}$$

Theoretically, the carbonation of olivine, serpentine and wollastonite is thermodynamically spontaneous at ambient conditions as enthalpy is negative. However, under these conditions, the mineral carbonation remains very slow. To speed up the process, an increase in temperature was investigated by Huijgen and Comans.[141] Nevertheless, they found that at high temperatures, chemical equilibrium favors CO_2 in gaseous form rather than binding CO_2 in solid form as carbonates. Thermodynamically favorable carbonation can be obtained by raising both pressure and temperature.[168]

2.5.5 Processes for Mineral Carbonation

Many processes for mineral carbonation have been investigated. Different criteria are involved in the selection of the appropriate carbonation process: the minimum energy required and the maximum efficiency, whether the chemical compounds are recovered, the kinetics of the reaction, and obtaining a stable and eco-compatible product. Carbonation can be classified into three types. It can be either ***in situ*** or ***ex situ***, and if it does not fall into one of these two categories, it will be classified as 'other mineral carbonation route'. *In situ* mineral carbonation can be carried out within a geological formation such as basalt where the rock reacts with injected CO_2 to form a carbonate. *Ex-situ* mineral carbonation takes place in a reactor, according to industrial processes.

2.5.5.1 *Mineral Carbonation* Ex Situ

Ex-situ mineral carbonation is divided into direct and indirect mineral carbonation. Direct carbonation takes place in a single step, carried out either in an aqueous medium or along the gas–solid reaction pathway. Indirect carbonation occurs when the reaction proceeds in two or more consecutive steps, carried out in an aqueous medium or according to the gas–solid reaction. Table 2.7 summarizes some aqueous and gas–solid mineral carbonation results. Nevertheless, *ex-situ* mineral carbonation is not easily set up industrially due to major bottlenecks. Indeed, the slow reaction rate, which requires an acceleration to be efficient, the high mass of rock needed to fix CO_2, and the high cost and technical supports required (*i.e.*, transport, heat...) are some of the limitations hindering this process.

2.5.5.1.1 Mineral Carbonation '*Ex-Situ*': Aqueous Carbonation. Mineral carbonation in an aqueous medium occurs according to three important mechanisms: eqn (2.17) shows the dissolution of CO_2 in water, eqn (2.18) presents the leaching of cations from the reactive metal during the dissolution of the silicate, and, finally, eqn (2.19) describes the carbonate precipitation. These three mechanisms can be realized in the same reactor (direct method) or in different successive assemblies (indirect method). The first studies on wollastonite carbonation with an aqueous method were carried out by Kojima *et al.*[169]

$$CO_2 + H_2O \rightarrow H_2CO_3 \rightarrow H^+ + HCO_3^- \tag{2.17}$$

$$(Ca/Mg)\text{-silicate} + 2H^+ \rightarrow (Ca/Mg)^{2+} + SiO_2 + H_2O \tag{2.18}$$

$$(Ca/Mg)^{2+} + HCO_3^- \rightarrow (Ca/Mg)CO_3 + H^+ \tag{2.19}$$

Since dissolution of the silicate remains the limiting step, numerous investigations have been carried out, both on the dissolution step and on various types of pre-treatments.[146,154,158,161,170–173] The use of acetic acid, in the indirect method, was developed by Kakizawa *et al.*,[174] but one of the drawbacks relied on the difficulties associated with recovering the acid. Mineral carbonation of mainly forsterite has been studied by Kwak *et al.*[175]

2.5.5.1.2 Mineral Carbonation '*Ex-Situ*': Gas–Solid Route. The gas–solid method is the easiest approach towards mineral carbonation. The magnesium or calcium mineral reacts directly with CO_2 to produce the corresponding carbonate.[142] The advantages of this method are its exothermic characteristics, which could theoretically balance the energy requirement for

Table 2.7 Summary of selected aqueous and gas–solid mineral carbonation results.

Materials	Efficacy	Conditions	Reference
Magnesium silicates			
Serpentine	80%	Aqueous direct carbonation, 0.64 M $NaHCO_3$, 1 M NaCl, 185 °C, 150 bar	155
Activated serpentine	60%	Aqueous direct carbonation, 0.64 M $NaHCO_3$, 1 M NaCl, 155 °C, 126 bar	153
Serpentine	25%	Dry, direct gas–solid carbonation Pure CO_2, 500 °C, 340 bar	173
Meta-chrysotile	70%	Direct gas–solid carbonation with 1 bar H_2O Pure CO_2, 130 °C, 31 bar	190
$Mg(OH)_2$ extracted from serpentine	>50%	Dry, indirect gas–solid carbonation Pure CO_2, 500 °C, 20 bar	177,314
Olivine	50%	Aqueous direct carbonation, 0.64 M $NaHCO_3$, 1 M NaCl, 185 °C, 150 bar	142
Olivine	57%	Dry, direct gas–solid carbonation Argon and pure CO_2, 400 °C, 1000 bar	315
Olivine	0.14 g CO_2/9 olivine	Dry and with H_2O (8.3 vol%) direct gas–solid carbonation 10 vol% CO_2 balance N_2, 200 °C	192
Calcium silicates			
Wollastonite	75%	Aqueous indirect carbonation, acetic acid, 60 °C, 30 bar	174
Wollastonite	70%	Aqueous direct carbonation, deionized water, 200 °C, 20 bar	316
Wollastonite	82%	Aqueous direct carbonation, distilled water, 100 °C, 40 bar	142

the total process.[176] Gas–solid carbonation can be carried out either in a single step (direct method) or in several steps (indirect method).[177] Eqn (2.20) illustrates the direct gas–solid reaction of solid Ca/Mg silicate, whilst eqn (2.21)–(2.24) present the conversion of a Mg-silicate, such as serpentine, into $Mg(OH)_2$ and a subsequent indirect carbonation. According to Zevenhoven *et al.*,[176] the reaction in eqn (2.21) is not favorable thermodynamically, so the alternative routes described by eqn (2.22)–(2.24) may be favored.

$$(\mathrm{Ca/Mg})\text{-silicate} + \mathrm{CO_2} \rightarrow (\mathrm{Ca/Mg})\mathrm{CO_3} + \mathrm{SiO_2} \tag{2.20}$$

$$\mathrm{Mg_3Si_2O_5(OH)_4} + \mathrm{H_2O} \rightarrow 3\mathrm{Mg(OH)_2} + 2\mathrm{SiO_2} \tag{2.21}$$

$$\mathrm{Mg_3Si_2O_5(OH)_4} \rightarrow 3\mathrm{MgO} + 2\mathrm{SiO_2} + 2\mathrm{H_2O} \tag{2.22}$$

$$\mathrm{MgO} + \mathrm{H_2O} \rightarrow \mathrm{Mg(OH)_2} \tag{2.23}$$

$$\mathrm{Mg(OH)_2} + \mathrm{CO_2} \rightarrow \mathrm{MgCO_3} + \mathrm{H_2O} \tag{2.24}$$

Lackner *et al.* developed the first trials on gas–solid carbonation with a hydrochloric acid extraction of hydroxides from silicate minerals.[146,173] However, this method has a very low reaction rate and is thermodynamically limited. The indirect gas–solid reaction, in several stages, with the production of magnesium hydroxides from serpentine and subsequently their carbonation in a separate step has therefore been investigated.[177–184] However, the low amount of carbonation obtained in a dry atmosphere led researchers to move towards the addition of small amounts of water because its presence appeared to enhance gas–solid carbonation.[177,185–193] The impact of steam on the carbonation of pure mineral phases has been investigated.[188,192] Other studies have focused on the reaction mechanisms and the increase in the carbonation rate, notably on serpentine minerals.[189,194]

2.5.5.2 *Mineral Carbonation* In Situ

In this *in-situ* process, CO_2 is injected directly into porous rocks, so this process avoids some of the drawbacks of an *ex-situ* process. Indeed, it is noteworthy to highlight that transports, pre-treatments of minerals and the use of additives are not required for *in-situ* mineral carbonation. One of the key points of this route is related to the porosity and permeability of the host rock. Easily dissolved metal cations are also necessary. As presented by Olijare,[143] the challenges of this method are first related to the availability of water for CO_2 injection. Indeed, the first step of *in-situ* carbonation consists of CO_2 dissolution. CO_2 was dissolved in water before its injection, but this method requires a huge amount of water: 27 and 341 tons of water were required to dissolve 1 t of CO_2 at 25 °C at 25 bar partial pressure and 2 bar partial pressure, respectively.[143] CO_2 can also be injected directly into the rock to have direct dissolution within the mineral, as suggested by Brady, Gislason and Goldberg[195–197] for oceanic basalts. The need of impermeable cap rocks is also considered as a challenge for *in-situ* carbonation. Indeed, to allow the carbonation to occur efficiently, the reactive species from the rock and the dissolved CO_2 have to be in contact for a sufficient time. An impermeable cap rock is thus necessary to keep them together, but this property is missing in usual rocks or else rocks can be altered. The possible mobilization of trace and toxic metals can also be considered as a drawback.[196,198] Basalt and peridotite, an ultramafic rock, are rich in divalent cations and poor in silica, and hence could be potentially attractive for *in-situ* mineral carbonation.[199–202] Oelkers *et al.* indeed presented a study called the CarbFix project, which consists of injecting water-dissolved CO_2 from a geothermal power plant into basalt.[203] Kelemen and Matter also proposed some studies on peridotite.[202] Some groups have worked on induced *in-situ* localized and accelerated carbonation, by injection of reactive mineral slurries such as fly ash.[204]

2.5.5.3 Other Mineral Carbonation Routes

If mineral carbonation cannot be classified into *ex-situ* or *in-situ* carbonation, it will be classified as 'other mineral carbonation routes'. Biomineralization falls into this category. This process does not employ mineral silicates as others processes described previously, but cyanobacteria and microalgae.[205] Biomineralization of CO_2 by calcium carbonate precipitation thanks to biotic pathways is indeed a natural way of global calcification in the oceans.

2.6 Zeolites and Related Materials

2.6.1 Foreword

Physical adsorption processes by solid sorbents have been employed for hydrogen purification by CO_2 removal although this has not yet reached a commercial stage for bulk CO_2 recovery from flue gases. The technical feasibility is dictated by the adsorption step because there must be a sufficiently strong affinity to selectively remove an undesired component from the gas mixture. It must be stated that *the stronger the affinity is, the higher the energy requirement for regeneration.*

The main advantages of physical adsorption on solid sorbents over chemical or physical absorption by liquids are their simple and energy efficient operation and regeneration, which can be achieved by pressure or temperature swing modes.

Zeolitic molecular sieves are under consideration for flue gas although both scale-up and the need to develop CO_2 specific adsorbents remain issues to be solved.[206] Indeed, the gas stream has to be cooled and dried, which limits the attractiveness of such PSA (pressure swing adsorption) processes, or other TSA (temperature swing adsorption) mainly for pre-combustion applications. In spite of higher hydrophobic properties,[207] activated carbons often lead to larger CO_2 capacities than zeolite at high pressure.[208,209] A drying step is therefore necessary in most cases to get rid of the water. Some metal–organic frameworks and organic–inorganic hybrids have shown quite promising performances for selective CO_2 adsorption, outperforming zeolites and activated carbons for high pressure applications such as those encountered in pre-combustion.[23] Unfortunately, several critical issues, such as material stability in the presence of air or steam, for which only limited information is available (so far) in the literature, seriously hinder the applicability of the MOF family.[210]

2.6.2 Peculiarities of Zeolites

Zeolites are crystalline aluminosilicates belonging to the group of tectosilicate molecular sieves. These porous solids possess well-ordered, interconnected microporous channels with diameters ranging from 0.3 to 2 nm (Figure 2.7). The three-dimensional structure arises from the corner-like assembly of $[SiO_4]^{4-}$ and $[AlO_4]^{5-}$ tetrahedra (TO_4, where T represents Si or Al). Each Si has its 4 + charge balanced by the four surrounding O (each with a 2− charge), so the silica tetrahedron in the lattice is electrically neutral. Since the trivalent Al is also bonded to four shared oxygen anions, the residual charge on each AlO_4 unit is 1−. Hence, each AlO_4 tetrahedron requires a 1+ charge from a cation (mono- or divalent) located in the channels or cavities to maintain the electrical neutrality throughout the structure. In zeolites, these cations are usually alkalis, ammonium, or protons, easily replaced by ion exchange. In addition, water molecules are present in the void volume of the cavities and channels. The unit cell formula of zeolites is usually written as:

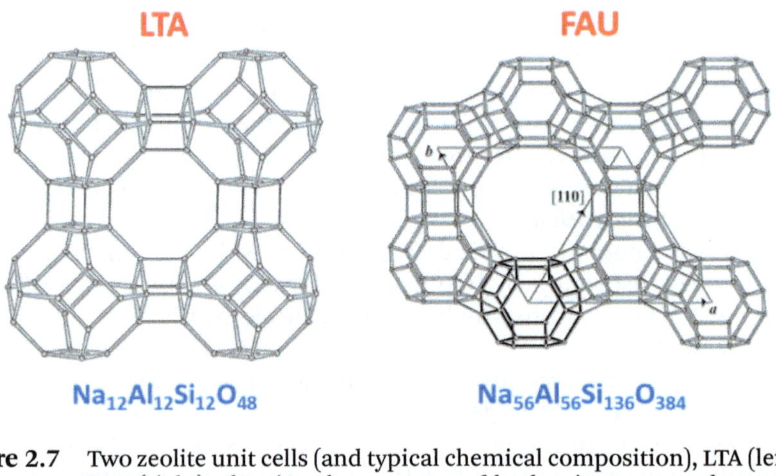

Figure 2.7 Two zeolite unit cells (and typical chemical composition), LTA (left) and FAU (right), showing the presence of both microporous channels and cavities.

$$M_{x/n}\left[(AlO_2)_x(SiO_2)_y\right]zH_2O$$

where M represents the cation (alkali or alkaline earth) counter-balancing the negative charge associated with the framework aluminium ions, n is the charge of the cation, x is the number of Al per unit cell, y represents the number of Si, and z is the number of water molecules entrapped in the channel system.

The zeolite properties can therefore be tuned by changing the Si–Al ratio and also by the nature of M. In addition, the sizes of the pores and their three-dimensional organization seriously impact the physical and chemical steps involved during the interaction between a molecule and the zeolite surface. These extremely porous materials are therefore of prime interest for catalytic, ion exchange and sorption applications.

2.6.3 CO_2 Sorption in Zeolites: Main Issues

2.6.3.1 Preamble

CO_2 usually undergoes physical adsorption (physisorption) or chemical adsorption (chemisorption) at the surface of zeolites. In these processes, several aspects influence their sorption efficiencies and they have been studied by several groups to better understand different adsorption mechanisms.[211–214]

Key factors strongly impacting the sorption potential have therefore been identified: **pore size and topology, polarizing power, nature and number of cations** compensating the framework negative charge, their **distribution, Si–Al ratio, the polarity and size of absorbed molecules, the presence of water** and other gases, the presence of carbonates at the surface and finally **pressure and temperature**.

CO_2 capture over solid sorbents occurs mainly by physisorption, which involves (relatively) little energy demand for the regeneration of the adsorbent, typically 40–50 kJ mol^{-1} of adsorbed species,[211] this being quite efficient when compared to amine scrubbing or cryogenic distillation. As materials of choice, low-cost high Al-content FAU and LTA zeolites (Figure 2.7) possess a good combination of high CO_2 uptake, high carbon dioxide over nitrogen selectivity, as well as high thermal and mechanical stability and were therefore thoroughly investigated.[215–218]

When considering zeolites' peculiar but attractive properties, the following question arises: *which properties render these microporous materials valuable sorbent materials?* The sorption properties of zeolites are strongly dependent on the cations' (alkali, protons, metal cations) nature and size, located at extra-framework positions within zeolite cages and channels.[219–221]

The carbon dioxide capture properties of acidic zeolites are well reported.[215,222,223] The continuous exclusion of nitrogen, dominant in flue gas, renders acidic zeolites promising materials for carbon dioxide separation. Indeed, the different affinity for the two gases stems to a large extent from the interaction between the zeolite's acid sites and the quadrupole moment of the adsorbate molecule, which for carbon dioxide is approximately four times

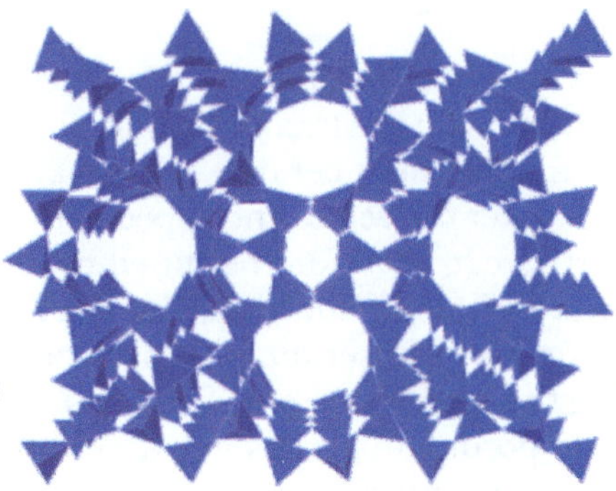

Figure 2.8 Research activities on zeolite-based carbon dioxide sorbents.

greater than for nitrogen.[224] In spite of the promising results in gas separation technology and applicability in the pressure swing adsorption (PSA) process achieved with acidic zeolites, their adsorption capacities rapidly declined at temperatures above 30 °C, even becoming negligible at 200 °C.[78]

As mentioned, the separation of gases by a zeolite depends on the following factors: the structure of the framework, its chemical composition, its micro/mesoporosity, crystallinity, the nature of the charge compensating cation but also on the operating conditions (CO_2 concentration, moisture, temperature,...).

Figure 2.8 aims to cover the current and main research fields related to zeolite-based sorbents.

Harlick *et al.*[225] tested 13 zeolites in CO_2 sorption and highlighted a linear adsorption isotherm. Importantly, they noticed that the lower the Si–Al ratio was, the higher the sorption capacity. This is directly connected to the cation content to compensate the framework's negative charge. In addition, they demonstrated that the sorption was extremely dependent on the cation nature, *i.e.*, *higher sorption with stronger electrostatic interaction between the cation and CO_2*. Indeed, CO_2 is an acidic gas and therefore sorption by a solid may be interpreted in terms of an acid–base interaction. Calleja *et al.* reported that the preference of the adsorbent for polar molecules (*e.g.*, CO_2) increases as the Si–Al ratio decreases due to the higher surface heterogeneity and the stronger electrostatic field generated inside the zeolite pores.[226] For instance, Siriwardane *et al.*[208] reported that the CO_2 adsorption capacities in high Al-containing FAU and LTA zeolites were 3.64 and 3.07 mmol g^{-1} (at 25 °C and 1 atm of CO_2), respectively.

For the same FAU topology (Figure 2.7), the CO_2 adsorption isotherms of zeolites displayed a wide range of adsorption capacities ranging from 1.2 mmol g^{-1} (HY-5) to 4.5 mmol g^{-1} (13X) at 25 °C and 1 bar.[225] In general, a high capacity can be achieved at a relatively low temperature, typically below 100 °C, but this diminishes dramatically as the temperature is increased above 200 °C. The main bottleneck in the separation processes using zeolites remains the competition between water and CO_2 adsorption. Water adsorption is indeed favored and as a consequence, CO_2 capacity drops dramatically when the separation process is carried out under humid conditions. Hence, operating CO_2 adsorption under relatively dry conditions is a pre-requisite but may impose additional costs for an upstream flue gas drying process in a power plant. However, Diaz reported the impact of an alkaline treatment of zeolite NaY with aqueous solutions of CsOH or NaOH, or with carbonate precursors, preparing the corresponding alkali-exchanged zeolites.[227] They studied the influence of H_2O on the CO_2 adsorption capacities of these modified zeolites by temperature programmed desorption, when the modified zeolites were saturated with H_2O at 298 K between the first and second adsorption steps with pure CO_2. The results showed that the CO_2 capacity of pristine NaY substantially diminished during the second cycle owing to the presence of pre-adsorbed water. In stark contrast, the capacities of Cs–Y and Na–Y zeolites were maintained or even increased during the second cycle, suggesting that doping with an alkali cation acts as an efficient tool to mitigate the detrimental effect of moisture on the CO_2 adsorption capacities.

2.6.3.2 *Key Parameters and Mechanistic Considerations*

In this section, the nature of the cation (H^+, Li^+, K^+, Na^+...) will be investigated in-depth with the aim of deciphering how its presence can influence the capacity of adsorption by zeolites.

In order to compensate the framework's negative charge, exchangeable cations are present near $[AlO_4]^-$, thus generating a strong electrostatic field within the pores. Hence, molecules possessing dipole moments, such as water or those with high quadrupole moments such as CO_2 (and N_2), interact with the electrostatic field, being potentially (strongly) adsorbed on the surface. Such an interaction can be estimated by eqn (2.25):

$$U_{E\Theta} = \frac{-\Theta ze(3\cos^2\theta - 1)}{\left[4r^3\left(4\pi\varepsilon_0\varepsilon_r\right)\right]} \tag{2.25}$$

This equation depends on e (elementary charge), ε_0 (permittivity of vacuum), ε_r (relative permittivity of the medium), Θ (quadrupole moment), r (distance between cation and adsorbed molecule), and z (valence of the cation).[228] It is noteworthy from the last three variables found in eqn (2.25) that cations with a smaller radius and a high valence may lead to stronger interactions. Indeed, Hasegawa and Matsumoto,[229] when studying CO_2 adsorption over low silica X zeolites verified that the interaction between Li^+ and CO_2 was the strongest and gave the highest differential heat of adsorption. By careful analysis of the cations found in this study (Li^+, Na^+, and Ca^{2+}), one may expect a similar interaction between Li^+ and Ca^{2+} due to their radius and valence, respectively. However, it appeared that Ca-LSX did not behave as a divalent Ca^{2+}, but rather as monovalent $Ca(OH)^-$ sites and, therefore, did not induce a strong electrostatic field.

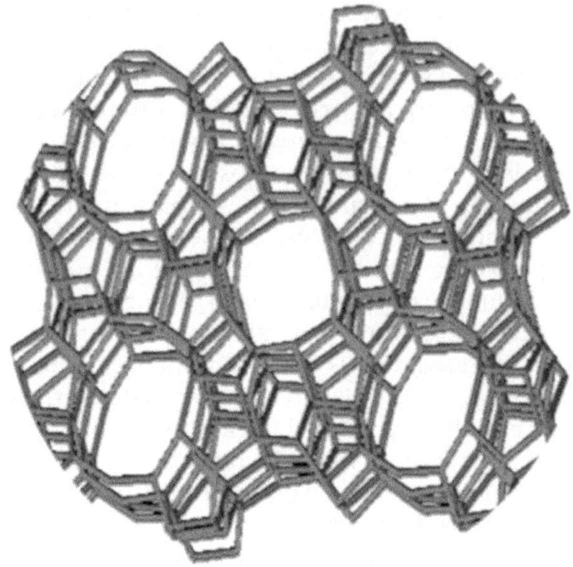

Figure 2.9 MFI zeolite framework with pore sizes ranging between 5.1 and 5.6 Å.

The polarizing power of the cations is inversely proportional to their ionic radius ($Li^+ > Na^+ > K^+ > Rb^+ > Cs^+$). Katoh *et al.*[230] observed that the amount of CO_2 strongly adsorbed in ZSM-5 zeolite (Figure 2.9) diminished while increasing the size of the cation. This is in agreement with the study of Yamazaki *et al.*[231] dealing with the energy of the interaction between CO_2 and the cationic sites of alkali metal-exchanged ZSM-5 zeolites.

Unfortunately, discrepancies can be found in the literature while attempting to establish the trend between the heat of adsorption of CO_2 and the zeolite structure. For instance, the capacity of adsorption at high coverage in CHA (Figure 2.10), followed the *a priori* expected trend: $Li^+ > Na^+ > K^+$[220,232,233] as well as $Li^+ > Na^+ > K^+ > Rb^+ > Cs^+$ in X zeolite.[219,234]

In contrast, by taking into account the interaction at low pressure in X and Y zeolites, a stronger interaction between CO_2 and Rb or Cs cations was found, becoming similar to Li. At low pressures, the interaction should be guided mainly by the basic character; so the lithium cation being less alkaline should exhibit a lower ion–quadrupole interaction due to its large charge density, comparable to the acid–base effects of CsX.[219] The same behaviour was observed for CHA by Zhang *et al.*, where Ba-CHA demonstrated a higher sorption capacity at lower pressures.[235]

In the case of BEA zeolite (Figure 2.11), the equilibrium adsorption capacity for CO_2 increased according to the following order: $Mg^{2+} < Cs^+ \sim Ca^{2+} < Ba^{2+} < Li^+ < Na^+ < K^+$, being (again) in agreement with the variation in the electrostatic field related to the ionic radius,[236] as already mentioned in former paragraphs.

An important observation was made in the latter study regarding CO_2–N_2 selectivities: at low adsorption capacities, this ratio was higher, whilst for

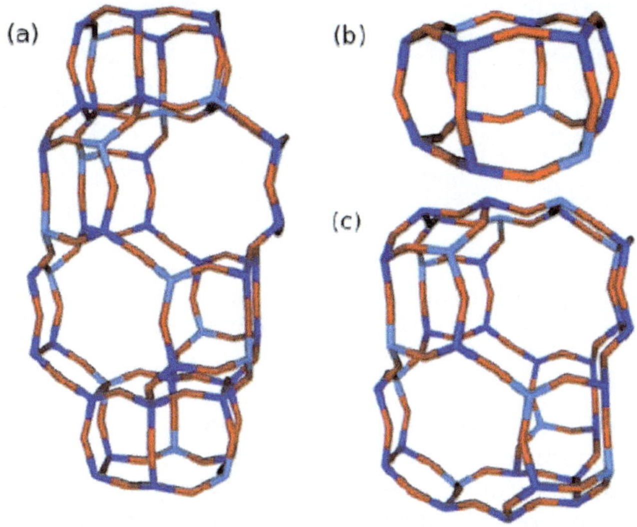

Figure 2.10 CHA zeolite framework with a pore size of 3.8 Å; (a–c) different channel systems.

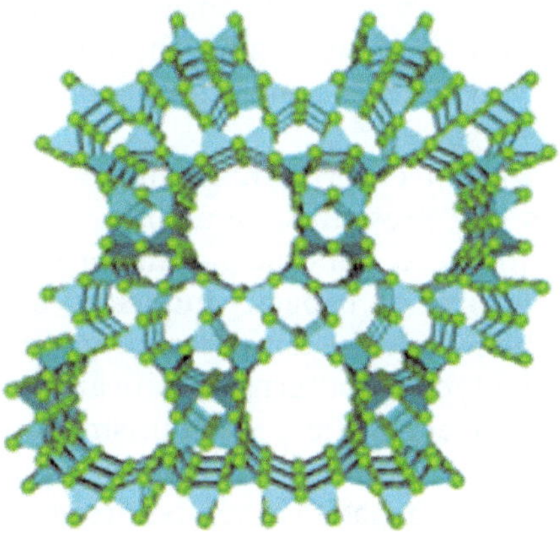

Figure 2.11 BEA zeolite framework with pore sizes of 5.6 Å and 6.6 × 7.7 Å.

K^+- and Na^+-exchanged zeolites, the adsorption capacity was higher but at the expense of CO_2–N_2 selectivity.[236] This may be explained by the higher interaction of the quadrupole moment of N_2 and CO_2 molecules with smaller cations. With an increase in the surface coverage, this trend was maintained but the values for the heat of adsorption diminished due to saturation of strong adsorption sites at first.

Regarding MCM-22 (MWW structure), shown in Figure 2.12 (left), changes were noticed in the sorption capacity order: K > Li > Cs > Na, which are even more pronounced in FER (Figure 2.12, right) with Na ~ K > Li > Cs.[237–239]

To summarize, the former part highlights that CO_2 adsorption on alkali metal cations in zeolites cannot always be explained by simple

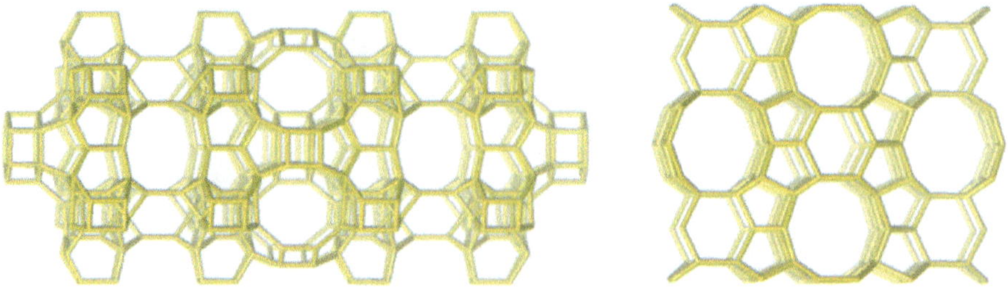

Figure 2.12 (Left) MWW framework with pore sizes of 4.0 × 5.5 Å and 4.1 × 5.1 Å (view normal to [001]); (right) FER framework with 4.2 × 5.4 Å and 3.5 × 4.8 Å (view along [001]).

electrostatic effects. Indeed, the coordination of an individual cation to zeolite oxygen atoms depends strongly on the size of the respective cation and on its location within the zeolite channels. In general, it was observed that 'alkaline' zeolites (thanks to an alkali cation) possess a strong electropositivity, thus leading to a higher adsorption capacity for molecules exhibiting an acidic character. Likewise, molecules with a permanent quadrupole moment also present a high affinity with those 'alkaline' surfaces. However, it is necessary to pay attention to the fact that a thermally stable chemisorption of CO_2 could hinder the regeneration of the adsorbent and significantly reduce the sorption capacity during successive adsorption–desorption cycles.[240]

By conducting adsorption experiments using a gravimetric method at different temperatures and pressures, Hauchhum concluded that the CO_2 adsorption isotherm followed general gas adsorption behavior, demonstrating that the CO_2 adsorption capacity increased while the pressure was increased and diminished while the temperature was increased. The adsorption isotherm followed a type I isotherm classification, according to IUPAC, representing a monolayer adsorption mechanism. Among the adsorbents tested, zeolite 13X led to the highest adsorption capacity, whilst activated charcoal provided the lowest capacity at temperatures ranging from 25 to 60 °C and pressures up to 1 bar.[241] The Langmuir adsorption model showed the best fit with the 13X and 4A zeolites, suggesting a monolayer adsorption of carbon dioxide molecules on surface active sites.[241] These results have been extended to other zeolites – EMC-1, SAPO-37 and USY – exhibiting all the FAU topology, by Pirngruber *et al.*[242]

The influence of the zeolite structure is also important with respect to its adsorption properties by the cations. As an example, one may cite the study from Pirngruber *et al.*,[243] which highlighted five adsorption sites in the FAU zeolite, as shown in Figure 2.13.

The preferred cationic site for adsorption, site II, is located in the supercage, just above the hexagonal window between the supercage and the sodalite cage. An occupation of this site minimizes the electrostatic repulsion between the cations. After a full occupation, the remaining cations are

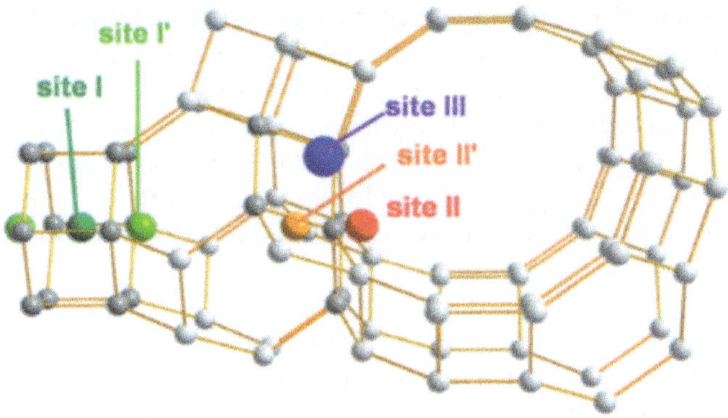

Figure 2.13 Adsorption sites on cation locations identified in the FAU zeolite. Adapted from ref. 243 with permission from the PCCP Owner Societies.

distributed between sites I and I′, located in the centre of the double 6-ring (D6R) and above the hexagonal window connecting the sodalite cage with the D6R, respectively. Although Li^+ and Na^+ can be distributed in sites I and I′, CO_2 cannot penetrate the sodalite cage, which does not have an impact on the adsorption properties. It was then concluded for LiY and NaY that the adsorption of CO_2 follows a linear trend. In CsY, the preferred adsorption results from the interaction with the cation and O-atoms of the 12-ring window of the FAU framework. For KY, two competitive and stable adsorption modes were found: a dual-cation site and one involving framework oxygens. A similar study performed by Thang *et al.*[244] led to different conclusions. While the former considered the order of the most stable CO_2 adsorption as Li-FAU > K-FAU > Na-FAU, the latter reported the following order: K-FAU > Na-FAU > Li-FAU. This discrepancy was attributed to different dispersion corrections used in DFT calculations.

Pringruber *et al.*[243] also demonstrated that these data cannot be extrapolated to X zeolite, since the cations are located in more exposed positions in the supercage, thus forming weaker bonds with the framework. Walton *et al.* came to the same conclusion with X and Y zeolites. Although these zeolites possess the same FAU topology, X zeolite contains more aluminium in the structure, which results in a higher amount of cations.[219] Wirawan *et al.* also verified that CO_2 could barely be adsorbed in silicalite-1 (MFI structure without Al, Figure 2.9), and slightly increased its adsorption in H-ZSM-5.[245] However, when the H^+ cation was exchanged by Na^+ and Ba^{2+}, the adsorption became higher, especially at a low CO_2 partial pressure. Further evidence was provided by Frantz *et al.*[246] and Chen *et al.*,[247] while investigating the effect of the Si–Al ratio in Na-ZSM-5 and 13X, respectively. They concluded that the CO_2 adsorption capacity increased with an increase in the aluminium content, *i.e.*, lower Si–Al ratio. However, it is noteworthy that the comparison between the zeolites was made at different specific surface area (SSA) values, which also have an impact on the adsorption capacity; the higher the SSA, the higher the sorption capacity. In that case, the charge differences become less significant on the adsorption than SSA values. Also, CO_2–CO_2 interactions become weaker than CO_2–zeolite interactions. Hence, CO_2 molecules tend to occupy the most energetically favourable position and the number of attractive sites is higher in smaller pores.[233] On the other hand, Ohlin *et al.*[248] verified that CO_2 had a higher affinity for site II in Na-ZSM-5, corresponding to adsorption on a few high-energy sites. For site I, a weaker adsorption was supposed, probably with the pore walls.

By combination of XRD and adsorption techniques, Pham *et al.*[233] identified two adsorption sites for CO_2 on CHA zeolites: (A) located in the 8 MR window and (B) located in the ellipsoid cages. Site A was identified as the main CO_2 adsorption site in pure-silica CHA due to the high number of close contacts with framework oxygen atoms, resulting in high van der Waals and quadrupolar interactions. The remaining free-space in Li-CHA and Na-CHA is larger, consequently, also the pore volume, than in K-CHA, which accounts for a larger adsorption capacity for the first two at a higher pressure.[235]

In cation-exchanged zeolites, the cation acts as a non-framework Lewis acid site and the framework oxygen atoms, bearing a partial negative charge, behave as basic sites, enhancing the electron density around the framework. The basic strength of these sites increases with the electropositivity of alkali metal cations, meaning $Li^+ < Na^+ < K^+ < Rb^+ < Cs^+$, based on local hard–soft acid base descriptors (HSAB).[249,250] Based on the Pearson theory, the interaction between the Lewis acid site (C-atom in CO_2) and the basic zeolite oxygen atoms remains the highest in zeolites exchanged by Cs^+ and Cu^{2+}. Extra-framework cations are considered as hard Lewis acid sites, following the order $Li^+ > Na^+ > K^+ > Cu^{2+}$, while O-atoms in CO_2 behave as hard basic sites. Taking into account that hard–hard and soft–soft interactions dominate the hard–soft interactions as demonstrated by the HSAB theory, the end-on CO_2 coordination with these cations ($M^{n+}\cdots O{=}C{=}O$) is more stable in Li^+ than in any other cation-exchanged zeolites. Such an ion–dipole interaction was further assessed by Jacobs and co-workers.[251]

Hasegawa *et al.* observed that CO_2 is more strongly adsorbed near the pore entrance in Li-LSX, thus hindering diffusion within inner cavities and resulting therefore in a smaller breakthrough time than Na-LSX.[229] However, there was a faster drop in the heat of adsorption with the CO_2 loading in Li-CHA than in Na-CHA, which was attributed to a higher dependence between the interaction energy on Li-exchanged sorbents.[233]

Another aspect pointed out by Katoh *et al.*[230] for Li-ZSM-5 zeolite was the higher amount of strongly adsorbed CO_2 while enhancing the pressure. This can be partially explained by the existence of two types of sinusoidal channels. The smaller ones (5.1 Å) may hinder the diffusion of CO_2 at low pressure. However, at a higher pressure, CO_2 may interact with Li^+ by entering deeply into the narrow channels. For larger cations, although the CO_2 molecules can also enter the pores deeply with an increase in pressure, it will be in a smaller quantity leading to a reduced amount of CO_2 adsorbed.

In several cationic zeolites, the chemical adsorption of CO_2 is accompanied by the formation of carbonates including very stable monodentate or unidentate carbonates and bidentate carbonates, at the surface, due to the interaction between CO_2 and the oxygen from the zeolite framework (Figure 2.14). The most common form is a highly asymmetric carbonate species observed initially by Bertsch and Habgood.[213]

Busca *et al.* have shown that the more strongly bound carbonates are bound through bi-coordination, as shown in Figure 2.15.[252]

The presence of these strongly adsorbed carbonates might decrease the accessibility of CO_2 to the major part of the zeolite surface sites, thereby contributing to the limiting of its adsorption. Bertsch *et al.* also reported that a small amount of water allowed a faster adsorption of CO_2 on zeolite Na-X at low relative partial pressures.[213] Due to that, Arletti *et al.*[240] related

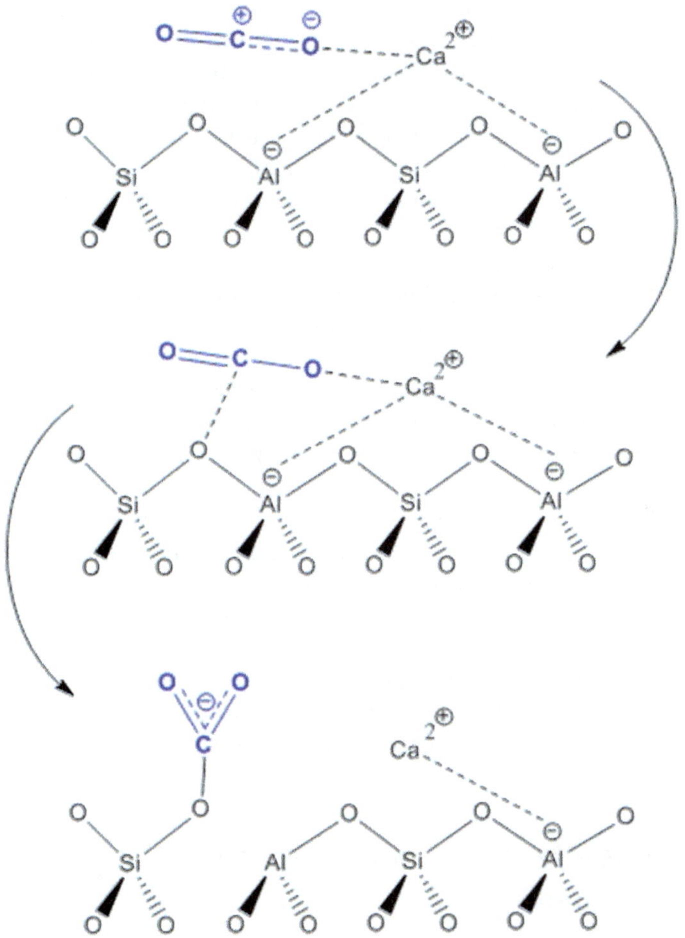

Figure 2.14 Carbonate species formed within the zeolite framework.

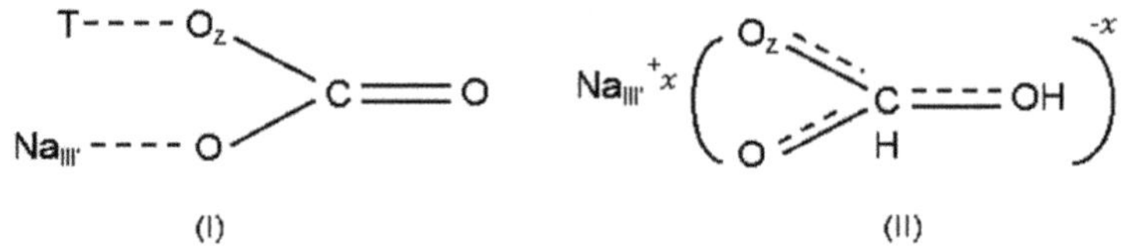

Figure 2.15 Two types of carbonates associated *via* bi-coordination.

the adsorption and desorption of CO_2 molecules in the porosity of Na–Y zeolite to the presence of water molecules. FAU is built by eight supercages, each of them can allocate eight CO_2 molecules, but only six were experimentally observed. They suggested that this difference could be caused by the presence of four H_2O molecules per supercage. Chen *et al.*[247] and Ohlin *et al.*[248] confirmed that the presence of water significantly diminished the CO_2 adsorption capacities in 13X and Na-ZSM-5 zeolites, respectively. Moreover, the presence of cations with a coordination sphere totally filled by water molecules usually induces a linear conformation of adsorbed CO_2. Through synchrotron studies, Arletti and co-workers found that only at a low water loading (two molecules H_2O per unit cell) does electron donation favor the formation of bent CO_2.[240] This implies that accessible dehydrated cations, direct CO_2–cation interactions, are required to allow the formation of bent CO_2 species.

By using an elegant DRIFT spectroscopy technique, Hunger *et al.* showed the absence of a linearly adsorbed carbon dioxide in an *end-on* configuration to a Cs^+ and Na^+ cation in X and Y zeolites. The carbonate structures associated to the bands obtained were adsorbed CO_2 coordinated to an extra-framework cation and a framework oxygen atom. Indeed, according to Pirngruber *et al.*,[243] some carbonate species were observed with a high intensity in CsY and KY, in which the interaction between the CO_2 molecule and the framework was discussed previously.

According to a theoretical study by Larin,[253] the formation of hydrogenocarbonate species (formed from CO_2 and H_2O) stabilized in all alkali-cation exchanged Y is possible, leading to a strong modification in the parameters of CO_2 adsorption in K- and Cs–Y zeolites. K^+ and Cs^+ cations would, in that case, occupy sites II, III and II′ in FAU supercages (Figure 2.13).

To get more clarity and quantitative values for CO_2 adsorption in ion-exchanged zeolites, Table 2.8 summarizes data found in the literature, in terms of heat of CO_2 adsorption and capacity of CO_2 adsorption (given in mmol CO_2 per gram of zeolite).

2.6.3.3 *Zeolite Acid–Base Properties: How Do They Impact the Sorption Features?*

As mentioned, the sorption properties have to be related to the higher affinity between *an acid* and *a base*. Zeolites exchanged with alkali or alkaline earth cations were shown to increase the basic character and therefore favored

Table 2.8 Heat of CO_2 adsorption and capacities of CO_2 sorption on different ion-exchanged zeolites.

Cation	Zeolite structure	Qst (kJ mol^{-1})	Capacity (mmol g^{-1})	Reference
Li^+	BEA	26.7–53		236
	Y	34.4	2.0	243,244
	X	49.9–50.0		229,234
	CHA	34.0–41.0		235
Na^+	BEA	29.1–56.5	1.1	236
	Y	28.0–36.0	1.5	225,233,243,244,317
	X	45.7–55.0		229,234
	ZSM-5	54.0	1.2–2.8	246,248
	CHA	37.0–43.0		233,235
	MCM-22	31.0		237
K^+	BEA	35.5–59.6	1.6	236
	Y	34.0	2.0	243,244
	X	42.2		234
	MCM-22	41.0		237
Cs^+	BEA	33.7–89.6	0.74	236
	Y	45.5		243
	X	40.6		234
	MCM-22	37.5		237

the interaction with the acidic carbon dioxide molecule. In the early 80 s, Scokart and Rouxhet showed that the adsorption of pyrrole on metal oxides and zeolites could be a suitable way to determine the basic strength.[254,255] The red shifting of the N–H stretching bond can therefore be used to probe the strength of the interaction between the N-atom and the cation. Later on, Barthomeuf extended this method and reported numerous contributions that shed light on the basic character of solids like zeolites.[256–258] She was able to demonstrate that the basicity could be ascribed to the net charge of the framework O-atom and that K-exchanged high Al-content FAU exhibited the highest basicity.[256] A basicity scale could therefore be developed while using several probe molecules over several cation-exchanged zeolites. In stark contrast, low Al-content MFI zeolites did not exhibit basic sites,[259] being in agreement with the low negative charge carried by O-atoms. Barthomeuf has set the relationship between the Sanderson electronegativity rule and the zeolite basicity.[256] Hence, the electrostatic environment of the MFI zeolite (ZSM-5) matrix, due to the presence of aluminium and counter cations, as well as the pore topology and sizes (both from the cation and the pores) allow us to guide the reactants toward the active sites by electrostatic interactions.[260–265]

In 1963, Bertsch and Habgood[213] had already studied the adsorption of weak Lewis acid CO_2 over alkali-exchanged FAU zeolites and confirmed the formation of carbonate-like species with the zeolite O-atom, thus acting as a weak Lewis base. According to the basicity scale (Figure 2.16), one may therefore *a priori* select K^+, Na^+ or Li^+ as a suitable alkali-cation to enhance the interaction with the CO_2 molecule in a high Al-content zeolite framework.

Figure 2.16 Alkali-exchanged FAU zeolites' basicity scale determined by pyrrole and pyridine adsorption measurements.

We have now seen that the cation's nature and size strongly influence the electric field present within the pores. Besides, the zeolite pore architecture and its available pore volume provide a convenient means for tuning the adsorption properties of these microporous materials.

2.6.3.4 CO_2 Sorption in Alkali-exchanged Zeolites: the 'Li' Paradox

The aim of this section is to tentatively rationalize the behavior of alkali-doped zeolites (described in former sections) and to show that lithium appears as a main player in carbon dioxide adsorption. Walton and co-workers[219] studied CO_2 adsorption over FAU zeolites, both Y and X, after exchange with Li, K, Na, Rb and Cs. As *a priori* expected, they found that the highest CO_2 sorption capacity was achieved over Li-promoted zeolites. The size of the small Li^+ cation may facilitate its strong interaction with the CO_2 molecule that possesses a quadrupole moment. In contrast, Díaz *et al.*[227] observed better performances over Na–Y with respect to Cs–Y zeolite at relatively high temperatures, suggesting that the sole basicity scale cannot explain the zeolite reactivity.

Regarding high-silica zeolites, Yamazaki *et al.*[231] and Wirawan *et al.*[245,266] investigated CO_2 adsorption over cation-exchanged ZSM-5 zeolite. Amongst Li, Na, K, Rb, Cs, H and Ba, it appeared that the latter led to the highest CO_2 adsorption capacity. In addition, adsorbed species also demonstrated a higher thermal stability. A study with small-pore CHA structures (Figure 2.10) showed that Na-CHA and Ca-CHA led to similar performances[235] at higher temperatures, whilst NaX exhibited a superior performance at lower temperatures.

Li-rich RHO zeolite (pore diameter 3.6 Å, Si–Al = 3.9) has been assessed as efficient for the separation of CO_2 from CH_4 in dry gas streams in terms of capacity, selectivity, and kinetics.[267] The authors' explanation was that enough empty windows were present to allow rapid motion of CO_2 molecules throughout the structure because Li^+ cations preferentially occupy 6R sites in LTA cages, away from the 8R windows. Besides the different alkali cations, Li^+ seems to play a major role in terms of 'affinity' with carbon dioxide. Though the sorption performances depend on numerous parameters (zeolite pore topology, alkali element loading and localization, operating conditions), this small and hard acid cation often led to a higher CO_2 sorption capacity, at least during the first cycle. The flipside might be that the affinity between the two protagonists is too strong, which may hinder an efficient cycling performance. A parallel can therefore be drawn between zeolites and silicates in terms of Li-doping.

Based on all the aforementioned considerations, it appears that lithium-exchanged zeolites are of prime importance for CO_2 sorption applications. In spite of a lack of generalization but rather a case-by-case study, some features of Li^+ can be drawn: a small radius, high polarizing power, hard Lewis acidity, the ability to locate in particular positions within small-pore zeolites. These peculiarities led this element to exhibit a strong affinity with molecules with a quadrupolar moment like CO_2. Hence, the presence of Li^+, which is mainly suitable for small-pore zeolites (8 MR), induces a strong electrostatic field, thus allowing faster gas diffusion than larger cations. Moreover, Li^+ is ready to exchange up to high loadings in high Al-content zeolites.

According to Barrer's findings,[268] further crystallographic studies have demonstrated that synthetic lithium silicate with small pore openings, for instance Li-ABW (3.2×4.1 Å), exhibited a tendency to 'breathe', inducing, to some extent, unit cell swelling, hence demonstrating high potential as an adsorbent.[269]

Theoretical calculations act as powerful tools to investigate adsorption enthalpies at different site locations depending on the zeolite structure.[270] Recently, Mothé-Esteves and co-workers reported an elegant DFT study using the finite cluster/supermolecule approach and shed light on CO_2 and N_2 adsorption in CHA zeolites.[271] Such a strategy may help to pave the way for designing new generations of adsorbents.

2.6.4 Miscellaneous Parameters

2.6.4.1 Impact of the Presence of Water

Though traces of water have been shown to enhance CO_2 uptake in zeolites,[213,227] a gas stream containing a high concentration of water, such as flue gas, leads to its selective adsorption within zeolite pores. Hence, the uptake of other gases is, in turn, usually hindered. A study by Gallei *et al.*[214] on carbon dioxide's physical adsorption on Ca–Y and Ni–Y illustrates CO_2/H_2O competitive adsorption with almost no CO_2 sorption. Similar findings were reported over X-zeolite.[272,273] Besides such competitive adsorption in hydrophilic zeolites, a dealumination phenomenon occurs, causing extensive zeolite surface modifications as well as degradation of zeolite-binder composite pellets.[274] The presence of these extra-framework aluminium species, or more exactly debris within the pores, seriously impacts gas transport through the zeolite bed.

As high Al-content zeolites (Figure 2.7) are hydrophilic, their efficient use in carbon dioxide capture should be done after upstream dehydration of flue gases. The other option may be to use low Al-content zeolites, ZSM-5 for instance, which behave as hydrophobic materials. Zeolites with Si–Al > 10 suffer limited damage caused by strong interactions with water. In contrast, the sorption capacity remains very low. Recently, numerous studies dealing with metal–organic framework and organic–inorganic hybrid materials have reported high selective CO_2 adsorptions, outperforming zeolites and activated carbons at high pressures.[23] However, critical issues, such as material stability in the presence of air or steam, are barely available in the literature.[210]

2.6.4.2 Impact of Sorbent Particle Size and Morphology

Regarding the effect of zeolite particle size, a higher sorption capacity may follow the faster diffusion observed in smaller particles, according to the Thiele–Weisz modulus.[275,276] The tricks to design an efficient zeolite catalyst can therefore be applied to develop an efficient sorbent: hierarchical porosity, presence of numerous surface sites, appropriate particle size and shape.[277]

In terms of the impact of crystal morphology, it is more complex, and such information can barely be found in the literature. However, one may expect zeolites to behave in a similar manner to silicate-based sorbents. Table 2.9 summarizes the sorption performances reached over Li_4SiO_4 materials depending on their morphologies. Li_4SiO_4, lithium silicate, has a theoretical maximal sorption capacity of 36.7 wt%. It is therefore better to compare sorbents with different morphologies according to their utilization ratio. Conventionally prepared Li_4SiO_4 exhibited 88% efficiency, whilst highly porous counterparts led to 93 and 99% utilization ratios for macroporous and mesoporous, respectively. In terms of stability, the higher cycling capacities were achieved over macroporous Li_4SiO_4 (6% loss after 10 cycles), platelet sample that did not deactivate during five cycles;[278] the best was the material prepared with SBA-15 mesoporous silica, which led to only a 1% loss after 20 cycles.[20] It is noteworthy that the particle morphology strongly influenced both the sorbent capacity and its stability; with these two features following the same trend. One may therefore argue that the higher the porosity of the siliceous precursor, the better the sorption behavior. If the same trend can be found for zeolite crystals, whose morphology can readily be modified at will,[279,280] it should be possible to further enhance the potentiality of

Table 2.9 Influence of Li_4SiO_4 morphology on high temperature CO_2 adsorption capacity and recyclability (550–650 °C).

Morphology	% CO_2	CO_2 sorption performance (wt%)	Utilization ratio (%)	Relative loss in capacity (%) (x cycles)	Reference
Conventional	100	32.3	88	12 (15)	318
Coral	15	32.4	88	66 (25)	319
Macroporous	100	34.2	93	6 (10)	15
Platelet	100	35.0	95	0 (5)	278
Mesoporous	100	36.3	99	1 (20)	20

zeolite-based low-temperature sorbents. Hence, a rational design strategy could help to bridge the gap between low- and high-temperature sorbents.

2.6.4.3 Selectivity

Besides a high capacity of sorption, adsorption-driven CO_2 separation from flue gas processes require a high selectivity with respect to N_2.[211] Zeolite ZK-4 (Si–Al ~ 1.3), possessing an LTA structure, demonstrated a high CO_2 capacity of 4.85 mmol g^{-1} (273 K, 1 bar) in its Na^+ form. The partial exchange of 26% of the cations by K^+ led to approximately a 10% loss in sorption capacity.[281] However, the nitrogen uptake became negligible (<0.03 mmol g^{-1}). This molecular sieving effect has been partly explained by computer modeling, suggesting hindered diffusion of nitrogen while increasing the K^+ content within the cages.

2.6.4.4 Possibility of Dual-site Adsorption?

It has often been seen, especially when dealing with small molecules, that individual extra-framework (exchangeable) cations behave as adsorption site for molecules coming from the gas phase to form the corresponding adsorption complex. However, Nachtigall *et al.*[282] have highlighted, by combining variable temperature infrared spectroscopy (VTIR) and periodic DFT calculations, that some types of adsorption sites can involve two or more cations. Adsorption complexes formed on these dual cationic adsorption sites differ in terms of structure and stability from an adsorption on a single cationic site. The capacity of cationic sites to attract, adsorb and activate CO_2 is strongly dependent on the cation size, its polarizing power, pore topology, and Si–Al ratio. Nonetheless, according to Nachtigall's studies, some subtle 'details' can have a strong bearing on both the structure of the adsorption complex and the corresponding carbon dioxide–zeolite surface interaction energy.[283–285]

VTIR spectra of CO_2 adsorbed on a K-FER zeolite (Si–Al = 28) presented two distinct IR absorption bands appearing at 2346 and 2355 cm^{-1} attributed to the asymmetric stretching modes of two different kinds of adsorbed CO_2 complexes.[284] By sophisticated DFT/CC calculations, taking into consideration *adsorbate–adsorbent* dispersion interactions, the importance of the latter dispersion forces was ascertained, thanks to the match with the zeolite pore diameter, accounting for half of the overall interaction energy between CO_2 and K-FER.

The adsorption of CO_2 in Li- and Na-FER was investigated by combining experimental measurements (IR spectroscopy and volumetric adsorption isotherms) with theoretical calculations.[283,285] For the CO_2/Na-FER system, different heat adsorptions were attributed to the existence of dual sites at higher Na^+ concentrations (low Si–Al ratio). In contrast, for the CO_2/Li-FER system, it was shown that smaller Li^+ cations are unlikely to be close enough (to each other) to generate dual cation sites.

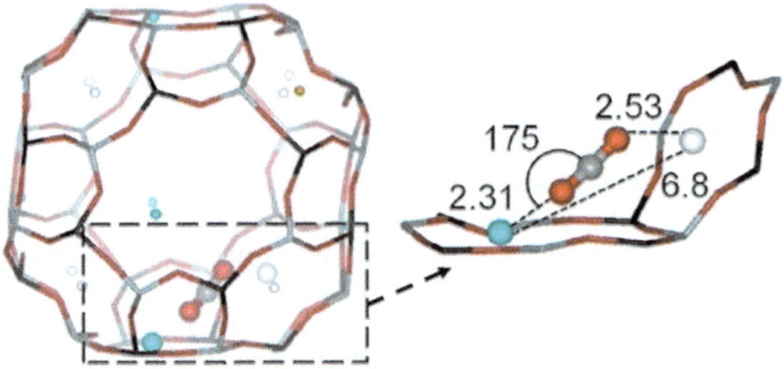

Figure 2.17 CO_2 adsorption complex formed on a dual cation site in Na-LTA.

CO_2 adsorption on alkali-metal exchanged ZSM-5 zeolites was investigated by a combination of volumetric adsorption isotherms and calorimetry.[286] The adsorption heats found for Li-, Na-, and K-ZSM-5 samples (Si–Al = 30) were 54, 49 and 43 kJ mol^{-1}, respectively. These values are in agreement with previous reports by Dunne *et al.*[287] and Bonelli *et al.*,[288] and they were assigned to the formation of adsorption complexes on single cationic sites.

The differences between CO_2 adsorption enthalpy when comparing high-silica FER and ZSM-5 zeolites could be explained by an enhanced interaction of the adsorbed molecule over M-ZSM-5 (where M stands for the alkali) when compared to M-FER. This effect can be related to the channel intersection sites in ZSM-5 where cations are less coordinatively saturated than in FER channel intersection sites.

To illustrate the possible adsorption of CO_2 molecules with a dual site, Figure 2.17 shows how a LTA zeolite adsorption site can be stabilized significantly by the presence of a secondary extra-framework sodium cation (adapted from Zukal).[289] It can be observed that the distances between the two O-atoms from carbon dioxide with the two Na^+ are different – 2.53 and 2.31 Å. The two Na^+ cations are located in neighboring cages and are separated by roughly 7 Å.

2.6.5 Zeolite-like Materials as Precursors to Design Performant Li-silicate Sorbents: How to Bridge the Gap Between High Affinity and High Recyclability?

Lithium silicates (or silicoaluminates), including Li_4SiO_4, $Li_{4+x}(Si_{1-x}Al_x)O_4$, Li_8SiO_6, $Li_6Si_2O_7$, $Li_2Si_2O_5$, and $Li_2Si_3O_7$, have been studied for CO_2 sorption.[78] Li_4SiO_4 and Li_8SiO_6 emerged as the most promising owing to their higher CO_2 sorption capacities and sorption/desorption kinetics. Li_4SiO_4 had already been reported to possess outstanding CO_2 sorption behavior 20 years ago,[290] thanks to its reversible reaction with CO_2, producing Li_2SiO_3

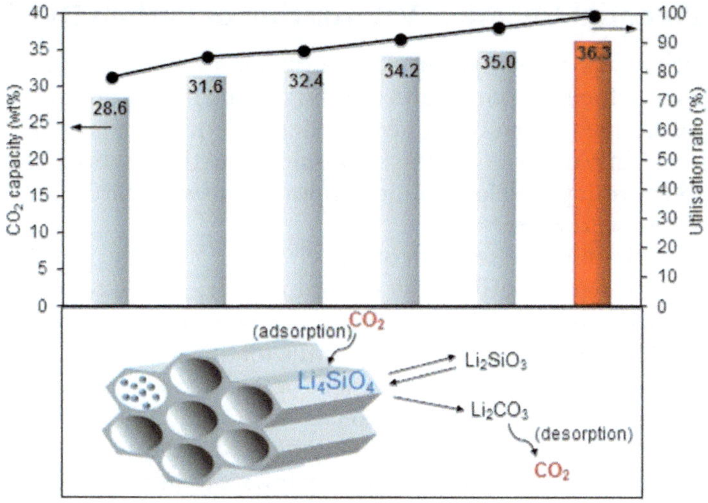

Figure 2.18 Impact of tailored mesoporosity of the silica precursor (SBA-15) on the CO_2 sorption capacity.

and Li_2CO_3. Several investigations devoted to the improvements of the CO_2 uptake capacity and cycling stability of Li_4SiO_4 relied on the use of alternative silicon sources loaded with lithium to design unique particle morphologies. Figure 2.18 highlights (in red) the highest capacity and efficiency in terms of utilization ratio achieved over SBA-15, a zeolite-related material.[20] Amazingly, the lithium silicate prepared with a regular array of mesopores led to 99% efficiency and did not lead to any deactivation after 20 cycles. The other materials described were prepared either with barely porous or microporous sources. Again, one may intend to draw a parallel between lithium silicates and zeolite catalysts. Indeed, the smart strategy of Wang's group to design powerful sorbents[20] confirms former results related to zeolite crystallization from mesoporous amorphous silica sources to design performant catalysts.[291,292]

2.7 Outline: Towards an Efficient Chemical Transformation of CO_2 into Fuels

Down-stream processes do not compete with CO_2 capture processes. In contrast, its chemical transformation has to be seen as a way to introduce added-value to the sequestration step. In addition, depending on the technology chosen, the adsorption–desorption equilibrium can be shifted and thus favor further CO_2 capture according to Le Châtelier's principle. However, the quantities we are talking about are huge; one may therefore convert carbon dioxide into molecules produced on a large scale as fuels; *methane* and *methanol* appearing as appropriate targets.[293,294]

2.7.1 Potential Chemical Valorization of Carbon Dioxide

The use of CO_2 as a raw carbon source opens up strategic perspectives to produce either fuels or important building blocks for industrial chemistry, with the subsequent need to reduce our dependence on fossil energies. The economic viability of these processes should be further ascertained by significant support from politicians while supporting new research: EU Emissions Trading Scheme, lower taxation,...

In spite of the known benefits of the chemical transformation of carbon dioxide, technological bottlenecks still inhibit large-scale feasibility. Firstly, the main block is the high stability of the carbon dioxide molecule. According to Gibbs' free energy, both *entropy* and *enthalpy* terms are unfavorable to C=O double bond breaking.[295] Somorjai and Dubois estimated that roughly 2% of CO_2 was able to dissociate into CO and O_2 at 2000 °C.[296] Such a thermodynamic limitation led to the development of catalysts able to overcome the low kinetics of CO_2 activation. These catalysts require a high resistance to deactivation by feed impurities, together with easy regeneration at moderate costs.

2.7.2 Synthesis of Energy Carriers

To allow the subsequent reduction of CO_2 emissions, its large-scale conversion into fuels should be warranted. Indeed, the world consumption of fuels is about two orders of magnitude superior to the one for chemical products.

During the last thirty years, numerous scientists have chosen a strategy to perform the hydrogenation of carbon dioxide into *methane* and *methanol*. Besides the catalytic conversion of CO_2 into hydrocarbons, a resurgence of interest in electrochemical, photochemical and biological routes has occurred, thanks to the constant need to reduce greenhouse gas emissions. While electrochemistry and photochemistry routes suffer from major drawbacks: (i) poor yield, (ii) absence of long chain hydrocarbons produced,[297] (iii) necessity to use a sacrificial electron donor, as shown by Nobel laureate Sauvage,[298] both biological and catalytic routes are still prominent. In this way, plants and micro-algae have been able to fix carbon dioxide by performing photosynthesis for millions of years.[299] The biomass can then be converted by either fermentation, digestion by an anaerobic process, or (thermo) chemical processes (gasification, pyrolysis, transesterification) into ethanol, methane, charcoal, or biodiesel. Researchers are constantly attempting to get inspiration from Nature.

Figure 2.19 presents the already established processes, at least at the pilot scale, of the thermochemical conversion of carbon dioxide into energetic carriers.[300] Interestingly, one may cite the dry reforming (DR) of methane, which allows its reaction with CO_2 to produce carbon monoxide and hydrogen in a 1 : 1 ratio (eqn (2.26)):

$$CO_2 + CH_4 \rightarrow 2\ CO + 2\ H_2\ \left(\Delta H_{r,25^\circ C} = +247\ \text{kJ mol}^{-1}\right) \quad (2.26)$$

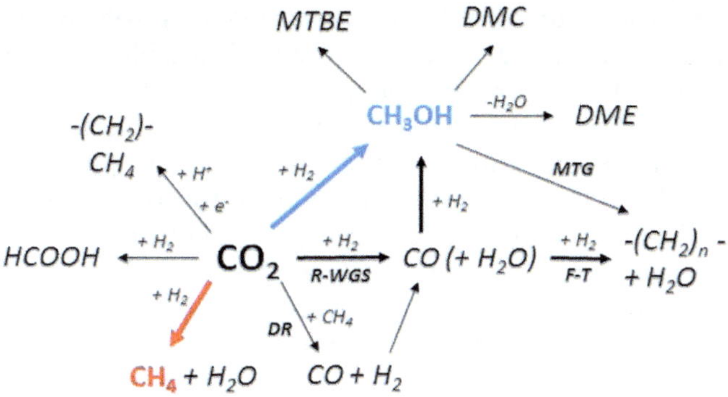

Figure 2.19 Conversion of carbon dioxide into energetic carriers.

The DR process' main advantage is to simultaneously consume two greenhouse gases. It is also a convenient way to convert the natural gas in syngas. Pilot plants based in Japan are already producing more than 500 barrels per day.[301] Unfortunately, this highly endothermic reaction occurs at temperatures above 700 °C, thus incurring significant energy costs.

In order to synthesize fuels, it is usually expected to produce a syngas with a $CO–H_2 < 1$ composition. The reverse water gas shift reaction (R-WGS), occurring simultaneously with the DR, diminishes this ratio below unity (eqn (2.27)):

$$CO_2 + H_2 \rightarrow CO + H_2O \left(\Delta H_{r,25°C} = +41 \text{ kJ mol}^{-1}\right) \tag{2.27}$$

It is then straightforward, while reaching $0.5 < CO–H_2 < 1$, to produce liquid fuels, either alkanes (and/or alkenes) according to Fischer–Tropsch synthesis (eqn (2.28)), or directly methanol (eqn (2.29)):

$$n\text{CO} + 2n\,\text{H}_2 \rightarrow (-\text{CH}_2)_n + n\,\text{H}_2\text{O} \left(\Delta H_{r,300°C} = -166 \text{ kJ mol}^{-1}\right) \tag{2.28}$$

$$CO + 2\,H_2 \rightarrow CH_3OH \left(\Delta H_{r,25°C} = -91 \text{ kJ mol}^{-1}\right) \tag{2.29}$$

Two remaining key options to directly convert CO_2 and H_2 into methane and methanol are marked in red and blue, respectively, in Figure 2.19. One might seriously question the viability of the expensive hydrogen source. This may seriously hinder the economic viability of the two processes. To render the hydrogenation of carbon dioxide energetically and economically attractive, the hydrogen has to originate from non-carbon sources such as wind, solar, hydroelectric or nuclear plants.[293] Electricity produced in a continuous (nuclear plants) or discontinuous (renewable energies) manner can permit the dissociation of water into dioxygen and dihydrogen *via* electrolysis. Since electricity can be barely stored (in capacitors),[302] one may transform this excess into chemical energy.

Methanol is a key industrial product, used as a solvent, fuel additive (MTBE) or a fuel by itself (fuel cells), or as a building block to produce high-added-value products (formaldehyde, acetic acid, methyl methacrylate, monomers). World demand increased to above 40 million tons in 2007.[293] Originated from the coal industry, the first methanol plant starting from syngas was built in 1923 by BASF. The conversion of carbon dioxide into methanol is limited by thermodynamics and high energy demands (eqn (2.30)). For instance, the maximum CO_2 conversion that may be achieved is nearly 30% at 300 °C and 50 bar pressure.[303] Another disadvantage relies on the selectivity, often below 50%, due to the presence of CO as the main by-product.[304,305]

$$CO_2 + 3\,H_2 \rightarrow CH_3OH + H_2O\left(\Delta H_{r,25^\circ C} = -49\ \text{kJ mol}^{-1}\right) \tag{2.30}$$

Methanol can be used as a convenient means of energy storage as well as an energy carrier. It can be dehydrated easily to dimethylether (DME), which has already replaced CFCs in aerosols, and is also a potential alternative to diesel fuels.[306]

Finally, carbon dioxide can easily be hydrogenated into methane, according to the Sabatier–Senderens reaction,[307] which was discovered in 1902 (eqn (2.31)):

$$CO_2 + 4H_2 \rightarrow CH_4 + 2\,H_2O\left(\Delta H_{r,25^\circ C} = -165\ \text{kJ mol}^{-1}\right) \tag{2.31}$$

This last option for reducing carbon dioxide and supplying methane presents the main advantage of a distribution network that already exists.

Although non-negligible progress has been made in carbon dioxide hydrogenation to various C_1 chemicals, it is still a great challenge to synthesize value-added products containing C_{5+} carbon atoms, such as gasoline. Recently, Gao and co-workers were able to achieve a 78% selectivity to C_{5+} hydrocarbons, using a bifunctional In_2O_3/zeolite catalyst at an appreciable CO_2 conversion (13%).[308]

2.7.3 Future Prospects

Solid sorbents are seen as good alternatives to aqueous amine scrubbing solutions for post-combustion CO_2 capture. The reversible adsorption of CO_2 at low temperatures can be successfully achieved on several materials. (Poly) amine-supported materials have attracted attention over the past decade due to the robustness of amine functionality and similarity to the alkanolamine scrubbing technology currently in use.

Several promoted silicate or alumino-silicate materials have been extensively considered in the literature, and are now close to a mature technology. The loss in selectivity in the presence of steam in the flow still needs to be fixed for alkali-doped zeolites. However, their industrial potential is superior to those of the MOF or ZIF families, which suffer from stability issues. Likewise, the carbonation kinetics is in favor of zeolite-type sorbents with respect to the oxides.

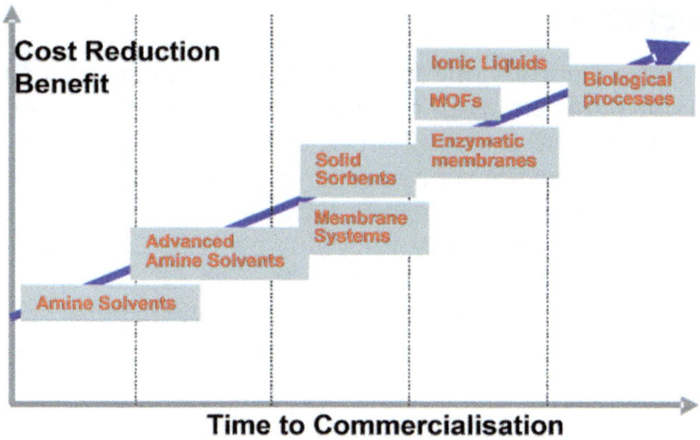

Figure 2.20 Cost reduction as a function of the time to put on the market prediction for carbon dioxide capture technology.

R&D efforts are necessary to understand the effect of steam on CO_2 adsorption and particularly the role of the interaction with the surface, since this could significantly affect the energy requirement for regenerative operations. Furthermore, it is obvious that R&D should be more closely focused on the design of the system and reactor than it has been. Sorbent design should be based indeed on system and reactor requisites rather than finding the highest possible capacity or lower adsorption energy, as may often be seen in the literature.

Figure 2.20 gives a qualitative idea of the benefits of sorbent technologies as a function of time to commercialization. Solid sorbents, mainly alkali-based silicates (and aluminosilicates), that are close to being brought to market, represent a good compromise, leading to greater benefits than advanced amine scrubbing technology. In contrast, the practical applicability of MOFs, and enzymatic and biological processes may take much longer as several issues must be overcome.

To date, one can confidently stress that there is no universal sorbent material; its choice relies on the operating conditions: CO_2 concentration, flue gas nature, temperature, pressure...

Rather, a proper design is obtained following case-by-case studies. One may, however, claim that clays and zeolites that did not appear to be promising at first have since been found to be potentially useful.

The last section of the chapter covered several ways to convert CO_2, appearing as an ubiquitous carbon source, into fuels. Besides the interest in carbon dioxide conversion reactions, the latter also acts as a driving force for continuously re-capturing CO_2.[309] The challenge of its sequestration, from industrial sources (flue gas from power plants) or even from ambient air, and utilization as a reagent to produce fuels (CCU) relies on a strategy to mitigate anthropogenic CO_2 emissions in the atmosphere. In this context, the combination of capture and utilization induces a lower energy demand, while avoiding intermediate desorption and compression steps for captured CO_2.[310]

2.8 Conclusion

Whilst hundreds of porous materials have been evaluated for CO_2 capture, either in pre- or post-combustion applications, this research field remains challenging. Indeed, the design of an efficient adsorbent for the direct capture of CO_2 from air has not yet been found.[311] However, we have attempted throughout this chapter to put the necessity for optimizing existing sorbents in terms of CO_2 uptake/selectivity, re-cycling operations, resistance to water, temperature and pressure at the forefront. Besides this, the main families of silicate-based sorbents were discussed (clays, alkali-silicates, zeolites), compared with amine-grafted mesoporous silica, with the aim of investigating the impact of the electronic structure, polarity, pore size/geometry, acidity, and cation nature/loading/localization. It is important to mention that the sorbent nanopores play a crucial role since they need to be designed to allow neither too strong an adsorption (high energy consumption for regeneration), nor too weak a bond (loss in selectivity). Regarding post-combustion technology, the main issue relies on the separation of CO_2 from N_2 in power plant flue-gas purification. Flue gases usually contain 3–15% CO_2, 5–7% H_2O and more than 70% N_2.[311]

Nature is solely able to carry out atmospheric CO_2 capture *via* photosynthesis. For instance, ribulose-1,5-bisphosphate carboxylase/oxygenase (Rubisco), the most abundant protein on Earth (approximately 10 kg per human being),[312] demonstrates a CO_2–O_2 selectivity of roughly 1500 with three molecules of CO_2 bound per molecule of O_2. Such a value remains a challenge for any synthetic material and this spectacular selectivity is due to synergetic effects between the Mg^{2+} metal site and the amino group of lysine. This may inspire researchers in their quest to develop new families of sorbents. Hence, the final target for selective CO_2 capture would be to design a CO_2–N_2 molecular sieve with a small area accessible to nitrogen adsorption, but a high CO_2 polarization power provided by enzyme-like 'pockets' with multiple binding contributions to keep the adsorption strong but reversible, along with a high stability in the presence of water.

A second parallel can be made with Nature and the presence of well-defined active sites in the pockets from enzymes: when pores in zeolites or related materials are close to the kinetic diameter of molecules, like CO_2, a perfect match can guarantee a high selectivity. Nevertheless, when free space in the pores disappears, the diffusion may become extremely hindered in microporous solids.[313] The introduction of additional transport (meso)pores favors the diffusion of gases in the zeolites but at the expense of selectivity. Once well-chosen, the zeolite sorbents usually exhibit high CO_2 uptakes at low pressure and moderate temperatures thanks to their basicity, the electrostatic field in the channels and cavities, and the higher number of cations able to promote the adsorption by charge–quadrupole interactions.[246] The main drawback is the higher affinity to water adsorption; the term *zeolite* means 'boiling stone' in Greek, thus inducing a drastic drop in their CO_2 sequestration potential or requiring a water purification step prior to the capture. This issue, once solved, may render zeolites the materials of choice for post-combustion CO_2 capture technology.

Finally, the technological realization of selective CO_2 capture from relevant sources (flue gas or air) will start after mastering the chemistry and physics of solid sorbent materials, but will then continue with engineering considerations: heat management, cost, safety issues, process operating conditions, reactor design...

Acknowledgements

The authors are grateful to Dr Catherine Hulot-Laugel for offering part of her precious time to correct this chapter. RB thanks St Gobain-Norpro and Region Grand Est for her PhD grant. BL is indebted to the Cercle Gutenberg for being awarded from the Guy Ourisson Prize.

References

1. United Nations Climate Change, Environment Program, COP21, Paris, 2015, http://unfccc.int/meetings/paris_nov_2015/session/9057.php, (accessed Jan 2018).
2. F. Princiotta, *Global Climate Change - The Technology Challenge*, Springer, Netherlands, 2011.
3. A. Álvarez, A. Bansode, A. Urakawa, A. V. Bavykina, T. A. Wezendonk, M. Makkee, J. Gascon and F. Kapteijn, *Chem. Rev.*, 2017, **117**, 9804–9838.
4. M. Aresta, *Carbon Dioxide Recovery and Utilization*, Springer, Netherlands, 1st edn, 2003.
5. X. Xiaoding and J. A. Moulijn, *Energy Fuels*, 1996, **10**, 305–325.
6. M. A. Wójtowicz, J. R. Pels and J. A. Moulijn, *Fuel Process. Technol.*, 1993, **34**, 1–71.
7. Recent Monthly Average Mauna Loa CO_2, National Oceanic and Atmospheric Administration - Earth System Research Laboratory - Global Monitoring Division, https://www.esrl.noaa.gov/gmd/ccgg/trends/, (accessed Jan 2018).
8. M. Oppenheimer, *Nature*, 1998, **393**, 325.
9. The Economics of Climate Change - A Primer, Morgan Stanley Research Europe, October 3, 2007, Economics, http://greensantaclarita.com/files/2012/10/2.-The-Economics-of-Climate-Change-A-Primer.pdf, (accessed Jan 2018).
10. R. C. Willson, *Science*, 1997, **277**, 1963–1965.
11. O. Wild, *The Straight Facts on Forests, Carbon, and Global Warming*, Oregon Wild, 2007.
12. Y. Cao, Y. Zhao, F. Song and Q. Zhong, *J. Energy Chem.*, 2014, **23**, 468–474.
13. Q. Wang, H. H. Tay, Z. Zhong, J. Luo and A. Borgna, *Energy Environ. Sci.*, 2012, **5**, 7526–7530.

14. J. Wang, X. Mei, L. Huang, Q. Zheng, Y. Qiao, K. Zang, S. Mao, R. Yang, Z. Zhang, Y. Gao, Z. Guo, Z. Huang and Q. Wang, *J. Energy Chem.*, 2015, **24**, 127–137.
15. H. Kim, H. D. Jang and M. Choi, *Chem. Eng. J.*, 2015, **280**, 132–137.
16. Q. Zheng, L. Huang, Y. Zhang, J. Wang, C.-Z. Zhao, Q. Zhang, W. Zheng, D. Cao, D. O'Hare and Q. Wang, *J. Mater. Chem. A*, 2016, **4**, 12889–12896.
17. F. N. Ridha, V. Manovic, Y. Wu, A. Macchi and E. J. Anthony, *Int. J. Greenhouse Gas Control*, 2013, **16**, 21–28.
18. M. Y. Veliz-Enriquez, G. Gonzalez and H. Pfeiffer, *J. Solid State Chem.*, 2007, **180**, 2485–2492.
19. E. Ochoa-Fernández, M. Rønning, T. Grande and D. Chen, *Chem. Mater.*, 2006, **18**, 1383–1385.
20. Y. Pan, Y. Zhang, T. Zhou, B. Louis, D. O'Hare and Q. Wang, *Inorg. Chem.*, 2017, **56**, 7821–7834.
21. P. Luis, *Desalination*, 2016, **380**, 93–99.
22. R. A. Khatri, S. S. C. Chuang, Y. Soong and M. Gray, *Energy Fuels*, 2006, **20**, 1514–1520.
23. Q. Wang, J. Luo, Z. Zhong and A. Borgna, *Energy Environ. Sci.*, 2011, **4**, 42–55.
24. X. Ren, H. Li, J. Chen, L. Wei, A. Modak, H. Yang and Q. Yang, *Carbon*, 2017, **114**, 473–481.
25. Z. Chen, S. Deng, H. Wei, B. Wang, J. Huang and G. Yu, *Front. Environ. Sci. Eng.*, 2013, **7**, 326–340.
26. X. Wang, H. Li and X.-J. Hou, *J. Phys. Chem. C*, 2012, **116**, 19814–19821.
27. Y. Lin, C. Kong and L. Chen, *RSC Adv.*, 2016, **6**, 32598–32614.
28. D. P. Bezerra, F. W. M. d. Silva, P. A. S. d. Moura, A. G. S. Sousa, R. S. Vieira, E. Rodriguez-Castellon and D. C. S. Azevedo, *Appl. Surf. Sci.*, 2014, **314**, 314–321.
29. N. Hashimah Alias, K. Kamarudin, N. Ghazali, T. A. Tengku Mohd, A. Sauki and M. Redwan Jaafar, *Carbon Dioxide Separation Using Amine Modified Zeolite in Pressure Swing Adsorption System*, 2014.
30. C. Chen, J. Kim and W.-S. Ahn, *Korean J. Chem. Eng.*, 2014, **31**, 1919–1934.
31. C. Chen, S. Zhang, K. H. Row and W.-S. Ahn, *J. Energy Chem.*, 2017, **26**, 868–880.
32. E. E. Ünveren, B. Ö. Monkul, Ş. Sarıoğlan, N. Karademir and E. Alper, *Petroleum*, 2017, **3**, 37–50.
33. C. T. Kresge, M. E. Leonowicz, W. J. Roth, J. C. Vartuli and J. S. Beck, *Nature*, 1992, **359**, 710.
34. J. S. Beck, J. C. Vartuli, W. J. Roth, M. E. Leonowicz, C. T. Kresge, K. D. Schmitt, C. T. W. Chu, D. H. Olson, E. W. Sheppard, S. B. McCullen, J. B. Higgins and J. L. Schlenker, *J. Am. Chem. Soc.*, 1992, **114**, 10834–10843.
35. Q. Huo, D. I. Margolese and G. D. Stucky, *Chem. Mater.*, 1996, **8**, 1147–1160.
36. D. Zhao, Q. Huo, J. Feng, B. F. Chmelka and G. D. Stucky, *J. Am. Chem. Soc.*, 1998, **120**, 6024–6036.

37. S. Che, A. E. Garcia-Bennett, T. Yokoi, K. Sakamoto, H. Kunieda, O. Terasaki and T. Tatsumi, *Nat. Mater.*, 2003, **2**, 801.
38. P. T. Tanev and T. J. Pinnavaia, *Science*, 1995, **267**, 865.
39. S. A. Bagshaw, E. Prouzet and T. J. Pinnavaia, *Science*, 1995, **269**, 1242.
40. S. Shen, A. E. Garcia-Bennett, Z. Liu, Q. Lu, Y. Shi Yan, C. Yu, W. Liu, Y. Cai, O. Terasaki and D. Zhao, *J. Am. Chem. Soc.*, 2005, **127**, 6780–6787.
41. F. Kleitz, S. Hei Choi and R. Ryoo, *Chem. Commun.*, 2003, 2136–2137.
42. F. Kleitz, D. Liu, G. M. Anilkumar, I.-S. Park, L. A. Solovyov, A. N. Shmakov and R. Ryoo, *J. Phys. Chem. B*, 2003, **107**, 14296–14300.
43. A. P. Wight and M. E. Davis, *Chem. Rev.*, 2002, **102**, 3589–3614.
44. G. Kickelbick, *Angew. Chem., Int. Ed.*, 2004, **43**, 3102–3104.
45. M. F. David, E. S. Eric and F. S. Daniel, *Nanotechnology*, 2005, **16**, S458.
46. H. Yoshitake, *New J. Chem.*, 2005, **29**, 1107–1117.
47. F. Hoffmann, M. Cornelius, J. Morell and M. Froba, *Angew. Chem., Int. Ed.*, 2006, **45**, 3216–3251.
48. X. Xu, C. Song, J. M. Andresen, B. G. Miller and A. W. Scaroni, *Energy Fuels*, 2002, **16**, 1463–1469.
49. X. Xu, C. Song, J. M. Andresen, B. G. Miller and A. W. Scaroni, *Microporous Mesoporous Mater.*, 2003, **62**, 29–45.
50. H. Y. Huang, R. T. Yang, D. Chinn and C. L. Munson, *Ind. Eng. Chem. Res.*, 2003, **42**, 2427–2433.
51. J. C. Hicks and C. W. Jones, *Langmuir*, 2006, **22**, 2676–2681.
52. J. C. Hicks, R. Dabestani, A. C. Buchanan and C. W. Jones, *Chem. Mater.*, 2006, **18**, 5022–5032.
53. C.-H. Yu, C.-H. Huang and C.-S. Tan, *Aerosol Air Qual. Res.*, 2012, **12**, 745–769.
54. X. Xu, C. Song, B. G. Miller and A. W. Scaroni, *Ind. Eng. Chem. Res.*, 2005, **44**, 8113–8119.
55. A. Sayari, Y. Belmabkhout and R. Serna-Guerrero, *Chem. Eng. J.*, 2011, **171**, 760–774.
56. A. Heydari-Gorji and A. Sayari, *Chem. Eng. J.*, 2011, **173**, 72–79.
57. G. Qi, Y. Wang, L. Estevez, X. Duan, N. Anako, A.-H. A. Park, W. Li, C. W. Jones and E. P. Giannelis, *Energy Environ. Sci.*, 2011, **4**, 444–452.
58. A. Sayari and Y. Belmabkhout, *J. Am. Chem. Soc.*, 2010, **132**, 6312–6314.
59. W.-J. Son, J.-S. Choi and W.-S. Ahn, *Microporous Mesoporous Mater.*, 2008, **113**, 31–40.
60. C. Chen, W.-J. Son, K.-S. You, J.-W. Ahn and W.-S. Ahn, *Chem. Eng. J.*, 2010, **161**, 46–52.
61. X. Ma, X. Wang and C. Song, *J. Am. Chem. Soc.*, 2009, **131**, 5777–5783.
62. M. B. Yue, Y. Chun, Y. Cao, X. Dong and J. H. Zhu, *Adv. Funct. Mater.*, 2006, **16**, 1717–1722.
63. M. B. Yue, L. B. Sun, Y. Cao, Y. Wang, Z. J. Wang and J. H. Zhu, *Chem. Eur. J.*, 2008, **14**, 3442–3451.
64. M. B. Yue, L. B. Sun, Y. Cao, Z. J. Wang, Y. Wang, Q. Yu and J. H. Zhu, *Microporous Mesoporous Mater.*, 2008, **114**, 74–81.

65. D. M. D'Alessandro, B. Smit and J. R. Long, *Angew. Chem., Int. Ed.*, 2010, **49**, 6058–6082.
66. C.-H. Huang, K.-P. Chang, C.-T. Yu, P.-C. Chiang and C.-F. Wang, *Chem. Eng. J.*, 2010, **161**, 129–135.
67. O. Leal, C. Bolívar, C. Ovalles, J. J. García and Y. Espidel, *Inorg. Chim. Acta*, 1995, **240**, 183–189.
68. A. C. C. Chang, S. S. C. Chuang, M. Gray and Y. Soong, *Energy Fuels*, 2003, **17**, 468–473.
69. F.-Y. Chang, K.-J. Chao, H.-H. Cheng and C.-S. Tan, *Sep. Purif. Technol.*, 2009, **70**, 87–95.
70. R. Serna-Guerrero, Y. Belmabkhout and A. Sayari, *Adsorption*, 2010, **16**, 567–575.
71. G. P. Knowles, J. V. Graham, S. W. Delaney and A. L. Chaffee, *Fuel Process. Technol.*, 2005, **86**, 1435–1448.
72. J. Wei, J. Shi, H. Pan, W. Zhao, Q. Ye and Y. Shi, *Microporous Mesoporous Mater.*, 2008, **116**, 394–399.
73. G. P. Knowles, S. W. Delaney and A. L. Chaffee, *Ind. Eng. Chem. Res.*, 2006, **45**, 2626–2633.
74. N. Hiyoshi, K. Yogo and T. Yashima, *Microporous Mesoporous Mater.*, 2005, **84**, 357–365.
75. Y. Duan and D. C. Sorescu, *J. Chem. Phys.*, 2010, **133**, 074508.
76. B. Zhang, Y. Duan and K. Johnson, *J. Chem. Phys.*, 2012, **136**, 064516.
77. Y. Duan, D. Luebke and H. Henry Pennline, *Int. J. Clean Coal Energy*, 2012, **01**, 1–11.
78. J. Wang, L. Huang, R. Yang, Z. Zhang, J. Wu, Y. Gao, Q. Wang, D. O'Hare and Z. Zhong, *Energy Environ. Sci.*, 2014, **7**, 3478–3518.
79. M. Wang and C.-G. Lee, *Energy Convers. Manage.*, 2009, **50**, 636–638.
80. H. Gupta and L.-S. Fan, *Ind. Eng. Chem. Res.*, 2002, **41**, 4035–4042.
81. G. Tilekar, K. Shinde, K. Kale, R. Raskar and A. Gaikwad, *Front. Chem. Sci. Eng.*, 2011, **5**, 477–491.
82. M. T. Rodríguez and H. Pfeiffer, *Thermochim. Acta*, 2008, **473**, 92–95.
83. R. Rodríguez-Mosqueda and H. Pfeiffer, *J. Phys. Chem. C*, 2013, **117**, 13452–13461.
84. S. Kumar and S. K. Saxena, *Mater. Renew. Sustain. Energy*, 2014, **3**, 30.
85. M. Kato, K. Nakagawa, K. Essaki, Y. Maezawa, S. Takeda, R. Kogo and Y. Hagiwara, *Int. J. Appl. Ceram. Technol.*, 2005, **2**, 467–475.
86. R. Xiong, J. Ida and Y. S. Lin, *Chem. Eng. Sci.*, 2003, **58**, 4377–4385.
87. M. Kato, S. Yoshikawa and K. Nakagawa, *J. Mater. Sci. Lett.*, 2002, **21**, 485–487.
88. Y. Duan, H. Pfeiffer, B. Li, I. C. Romero-Ibarra, D. C. Sorescu, D. R. Luebke and J. W. Halley, *Phys. Chem. Chem. Phys.*, 2013, **15**, 13538–13558.
89. K. Essaki, M. Kato and H. Uemoto, *J. Mater. Sci.*, 2005, **40**, 5017–5019.
90. E. Ochoa-Fernández, T. Zhao, M. Rønning and D. Chen, *J. Environ. Eng.*, 2009, **135**, 397–403.
91. R. Quinn, R. J. Kitzhoffer, J. R. Hufton and T. C. Golden, *Ind. Eng. Chem. Res.*, 2012, **51**, 9320–9327.

92. R. Pacciani, J. Torres, P. Solsona, C. Coe, R. Quinn, J. Hufton, T. Golden and L. F. Vega, *Environ. Sci. Technol.*, 2011, **45**, 7083–7088.
93. M. Seggiani, M. Puccini and S. Vitolo, *Int. J. Greenhouse Gas Control*, 2011, **5**, 741–748.
94. T. Tang, Z. Zhang, J.-B. Meng and D.-L. Luo, *Fusion Eng. Des.*, 2009, **84**, 2124–2130.
95. S. Shan, S. Li, Q. Jia, L. Jiang, Y. Wang and J. Peng, *Ind. Eng. Chem. Res.*, 2013, **52**, 6941–6945.
96. D. Cruz and S. Bulbulian, *J. Am. Ceram. Soc.*, 2005, **88**, 1720–1724.
97. V. L. Mejía-Trejo, E. Fregoso-Israel and H. Pfeiffer, *Chem. Mater.*, 2008, **20**, 7171–7176.
98. K. Wang, X. Wang, P. Zhao and X. Guo, *Chem. Eng. Technol.*, 2014, **37**, 1552–1558.
99. X. Wu, Z. Wen, X. Xu, X. Wang and J. Lin, *J. Nucl. Mater.*, 2009, **392**, 471–475.
100. I. C. Romero-Ibarra, J. Ortiz-Landeros and H. Pfeiffer, *Thermochim. Acta*, 2013, **567**, 118–124.
101. H. Xu, W. Cheng, X. Jin, G. Wang, H. Lu, H. Wang, D. Chen, B. Fan, T. Hou and R. Zhang, *Ind. Eng. Chem. Res.*, 2013, **52**, 1886–1891.
102. H. Pfeiffer, E. Lima and P. Bosch, *Chem. Mater.*, 2006, **18**, 2642–2647.
103. H. Pfeiffer, C. Vázquez, V. H. Lara and P. Bosch, *Chem. Mater.*, 2007, **19**, 922–926.
104. M. Seggiani, M. Puccini and S. Vitolo, *Int. J. Greenhouse Gas Control*, 2013, **17**, 25–31.
105. J. F. Stebbins, Z. Xu and D. Vollath, *Solid State Ionics*, 1995, **78**, L1–L8.
106. K. Jackowska and A. R. West, *J. Mater. Sci.*, 1983, **18**, 2380–2384.
107. C. Gauer and W. Heschel, *J. Mater. Sci.*, 2006, **41**, 2405–2409.
108. J. Ortiz-Landeros, C. Gómez-Yáñez, L. M. Palacios-Romero, E. Lima and H. Pfeiffer, *J. Phys. Chem. A*, 2012, **116**, 3163–3171.
109. F. Duran-Munoz, I. C. Romero-Ibarra and H. Pfeiffer, *J. Mater. Chem. A*, 2013, **1**, 3919–3925.
110. S. Mukherjee, *The Science of Clays - Applications in Industry, Engineering, and Environment*, Springer, Netherlands, 2013.
111. F. Bergaya, B. K. G. Theng and G. Lagaly, *Developments in Clay Science*, Elsevier, Oxford, 2006.
112. J. A. Delgado, M. A. Uguina, J. L. Sotelo, B. Ruíz and M. Rosário, *J. Nat. Gas Chem.*, 2007, **16**, 235–243.
113. J. Pires, M. Bestilleiro, M. Pinto and A. Gil, *Sep. Purif. Technol.*, 2008, **61**, 161–167.
114. A. Azzouz, N. Platon, S. Nousir, K. Ghomari, D. Nistor, T. C. Shiao and R. Roy, *Sep. Purif. Technol.*, 2013, **108**, 181–188.
115. C. Chen, D.-W. Park and W.-S. Ahn, *Appl. Surf. Sci.*, 2013, **283**, 699–704.
116. A. Azzouz, S. Nousir, N. Platon, K. Ghomari, T. C. Shiao, G. Hersant, J.-Y. Bergeron and R. Roy, *Int. J. Greenhouse Gas Control*, 2013, **17**, 140–147.
117. W. Wang, J. Xiao, X. Wei, J. Ding, X. Wang and C. Song, *Appl. Energy*, 2014, **113**, 334–341.

118. W. H. Chan, M. N. Mazlee, Z. A. Ahmad, M. A. M. Ishak and J. B. Shamsul, *J. Mater. Cycles Waste Manage.*, 2017, **19**, 1–14.
119. A. Gil, R. Trujillano, M. A. Vicente and S. A. Korili, in *Studies in Surface Science and Catalysis*, ed. P. L. Llewellyn, F. Rodriquez-Reinoso, J. Rouqerol and N. Seaton, Elsevier, 2007, vol. 160, pp. 327–334.
120. R. R. Pawar, H. A. Patel, G. Sethia and H. C. Bajaj, *Appl. Clay Sci.*, 2009, **46**, 109–113.
121. K. J. Shah, T. Imae, M. Ujihara, S.-J. Huang, P.-H. Wu and S.-B. Liu, *Chem. Eng. J.*, 2017, **312**, 118–125.
122. K. J. Shah, T. Imae and A. Shukla, *RSC Adv.*, 2015, **5**, 35985–35992.
123. M. Irani, M. Fan, H. Ismail, A. Tuwati, B. Dutcher and A. G. Russell, *Nano Energy*, 2015, **11**, 235–246.
124. E. Vilarrasa-García, J. A. Cecilia, M. Bastos-Neto, C. L. Cavalcante, D. C. S. Azevedo and E. Rodríguez-Castellón, *Appl. Surf. Sci.*, 2017, **410**, 315–325.
125. M. Jaroniec, K. P. Gadkaree and J. Choma, *Colloids Surf., A*, 1996, **118**, 203–210.
126. S. Sircar, *Ind. Eng. Chem. Res.*, 2006, **45**, 5435–5448.
127. A. Azzouz, D. Messad, D. Nistor, C. Catrinescu, A. Zvolinschi and S. Asaftei, *Appl. Catal., A*, 2003, **241**, 1–13.
128. L. Stevens, K. Williams, W. Y. Han, T. Drage, C. Snape, J. Wood and J. Wang, *Chem. Eng. J.*, 2013, **215–216**, 699–708.
129. S. Nousir, N. Platon, K. Ghomari, A. S. Sergentu, T. C. Shiao, G. Hersant, J. Y. Bergeron, R. Roy and A. Azzouz, *J. Colloid Interface Sci.*, 2013, **402**, 215–222.
130. A. Azzouz, E. Assaad, A.-V. Ursu, T. Sajin, D. Nistor and R. Roy, *Appl. Clay Sci.*, 2010, **48**, 133–137.
131. A. Azzouz, A.-V. Ursu, D. Nistor, T. Sajin, E. Assaad and R. Roy, *Thermochim. Acta*, 2009, **496**, 45–49.
132. E. A. Roth, S. Agarwal and R. K. Gupta, *Energy Fuels*, 2013, **27**, 4129–4136.
133. J. C. Hicks, J. H. Drese, D. J. Fauth, M. L. Gray, G. Qi and C. W. Jones, *J. Am. Chem. Soc.*, 2008, **130**, 2902–2903.
134. R. G. Leliveld, T. G. Ros, A. J. van Dillen, J. W. Geus and D. C. Koningsberger, *J. Catal.*, 1999, **185**, 513–523.
135. A. Gil, M. A. Vicente and S. A. Korili, *J. Catal.*, 2005, **229**, 119–126.
136. W. Wang, X. Wang, C. Song, X. Wei, J. Ding and J. Xiao, *Energy Fuels*, 2013, **27**, 1538–1546.
137. E. Vilarrasa-García, J. A. Cecilia, D. C. S. Azevedo, C. L. Cavalcante and E. Rodríguez-Castellón, *Microporous Mesoporous Mater.*, 2017, **249**, 25–33.
138. A. Gil, L. M. Gandia and M. A. Vicente, *Catal. Rev.: Sci. Eng.*, 2000, **42**, 145–212.
139. A. Gil and L. M. Gandía, *Chem. Eng. Sci.*, 2003, **58**, 3059–3075.
140. W. J. J. Huijgen and R. N. J. Comans, *Carbon Dioxide Sequestration by Mineral Carbonation: Literature Review*, Energy Research Centre of the Netherlands, 2003.

141. W. J. J. Huijgen and R. N. J. Comans, *Carbon Dioxide Sequestration by Mineral Carbonation, Literature Review Update 2003–2004*, Energy Research Centre of the Netherlands, 2005.
142. J. Sipilä, S. Teir and R. Zevenhoven, *Carbon Dioxide Sequestration by Mineral Carbonation – Literature Review Update 2005–2007.*, Heat Engineering Laboratory, Faculty of Technology, Åbo Akademi University, 2008.
143. A. A. Olajire, *J. Pet. Sci. Eng.*, 2013, **109**, 364–392.
144. F. Goff and K. S. Lackner, *Environ. Geosci.*, 1998, **5**, 89–101.
145. B. Metz, O. Davidson, H. C. de Coninck, M. Loos and L. A. Meyer, *IPCC Special Report on Carbon Dioxide Capture and Storage*, Cambridge University Press, Cambridge, United Kingdom and New York, NY, USA 2005.
146. K. S. Lackner, C. H. Wendt, D. P. Butt, E. L. Joyce and D. H. Sharp, *Energy*, 1995, **20**, 1153–1170.
147. H. Bearat, M. J. McKelvy, A. V. G. Chizmeshya, R. Nunez and R. W. Carpenter, *Investigations of the Mechanisms that Govern Carbon Dioxide Sequestration via Aqueous Olivine Mineral Carbonation*, 2003.
148. A. Foucault and J.-F. Raoult, *Dictionnaire de géologie*, Paris, Milan, 2005.
149. F. Wypych, W. H. Schreiner and E. Richard, *J. Colloid Interface Sci.*, 2004, **276**, 167–173.
150. M. Marcello, *Am. Mineral.*, 1982, **67**, 587–598.
151. J. C. Rucklidge and J. Zussman, *Acta Crystallogr.*, 1965, **19**, 381–389.
152. N. Koukouzas, V. Gemeni and H. J. Ziock, *Int. J. Miner. Process.*, 2009, **93**, 179–186.
153. M. M. Maroto-Valer, D. J. Fauth, M. E. Kuchta, Y. Zhang and J. M. Andrésen, *Fuel Process. Technol.*, 2005, **86**, 1627–1645.
154. S. J. Gerdemann, D. C. Dahlin and W. K. O'Connor, *Carbon Dioxide Sequestration by Aqueous Mineral Carbonation of Magnesium Silicate Mineral*, Elsevier Science, Amsterdam, 2003.
155. W. K. O'Connor, D. C. Dahlin, D. N. Nilsen, R. P. Walters and P. C. Turner, *Carbon Dioxide Sequestration by Direct Aqueous Mineral Carbonation*, Clearwater, Florida, USA, 2006.
156. P. Baláž, E. Turianicová, M. Fabián, R. A. Kleiv, J. Briančin and A. Obut, *Int. J. Miner. Process.*, 2008, **88**, 1–6.
157. M. Fabian, M. Shopska, D. Paneva, G. Kadinov, N. Kostova, E. Turianicová, J. Briančin, I. Mitov, R. A. Kleiv and P. Baláž, *Miner. Eng.*, 2010, **23**, 616–620.
158. T. A. Haug, R. A. Kleiv and I. A. Munz, *Appl. Geochem.*, 2010, **25**, 1547–1563.
159. D. Kelland, *IEEE Trans. Magn.*, 1973, **9**, 307–310.
160. D. J. Fauth, P. M. Goldberg, J. P. Knoer, Y. Soong, W. K. O'Connor, D. C. Dahlin, D. N. Nilsen, R. P. Walters, K. S. Lackner, H. J. Ziock, M. J. McKelvy and Z. Y. Chen, *Div. Fuel Chem. Am. Chem. Soc.*, 2000, **45**, 708–712.
161. W. K. O'Connor, D. C. Dahlin, D. N. Nilsen, S. J. Gerdemann, G. E. Rush, L. R. Penner, R. P. Walters and P. C. Turner, *Research Status on the Sequestration of Carbon Dioxide by Direct Aqueous Mineral Carbonation,* Newcastle Australia, 2001.

162. S. Eloneva, S. Teir, H. Revitzer, J. Salminen, A. Said, C.-J. Fogelholm and R. Zevenhoven, *Steel Res. Int.*, 2009, **80**, 415–421.
163. R. D. Balucan, E. M. Kennedy, J. F. Mackie and B. Z. Dlugogorski, *Greenhouse Gases: Sci. Technol.*, 2011, **1**, 294–304.
164. G. W. Brindley and J. Zussman, *Am. Mineral.*, 1957, **42**, 461–474.
165. A. Cattaneo, A. F. Gualtieri and G. Artioli, *Phys. Chem. Miner.*, 2003, **30**, 177–183.
166. C. Jolicoeur and D. Duchesne, *Can. J. Chem.*, 1981, **59**, 1521–1526.
167. C. Schmidt and B. Romanosky, *Program Overviews: CO_2 Mineral Sequestration*, NETL, USA, 2011.
168. L. Marini, *Geological Sequestration of Carbon Dioxide: Thermodynamics, Kinetics, and Reaction Path Modeling* - 1st edn, Elsevier, Amsterdam, The Netherlands, 2006.
169. T. Kojima, A. Nagamine, N. Ueno and S. Uemiya, *Energy Convers. Manage.*, 1997, **38**, S461–S466.
170. S. J. Gerdemann, W. K. O'Connor, D. C. Dahlin, L. R. Penner and H. Rush, *Environ. Sci. Technol.*, 2007, **41**, 2587–2593.
171. M. Hänchen, V. Prigiobbe, G. Storti, T. M. Seward and M. Mazzotti, *Geochim. Cosmochim. Acta*, 2006, **70**, 4403–4416.
172. I. A. Munz, J. Kihle, Ø. Brandvoll, I. Machenbach, J. W. Carey, T. A. Haug, H. Johansen and N. Eldrup, *Energy Procedia*, 2009, **1**, 4891–4898.
173. K. S. Lackner, D. P. Butt and C. H. Wendt, *Energy Convers. Manage.*, 1997, **38**, S259–S264.
174. M. Kakizawa, A. Yamasaki and Y. Yanagisawa, *Energy*, 2001, **26**, 341–354.
175. J. H. Kwak, J. Z. Hu, D. W. Hoyt, J. A. Sears, C. Wang, K. M. Rosso and A. R. Felmy, *J. Phys. Chem. C*, 2010, **114**, 4126–4134.
176. R. Zevenhoven, S. Teir and S. Eloneva, *Energy*, 2008, **33**, 362–370.
177. J. Fagerlund, J. Highfield and R. Zevenhoven, *RSC Adv.*, 2012, **2**, 10380–10393.
178. R. Zevenhoven and J. Kohlmann, *CO_2 Sequestration by Magnesium Silicate Mineral Carbonation,* Genève, Switzerland, 2002.
179. J. Fagerlund, E. Nduagu, I. Romão and R. Zevenhoven, *Energy*, 2012, **41**, 184–191.
180. J. Fagerlund, E. Nduagu and R. Zevenhoven, *Energy Procedia*, 2011, **4**, 4993–5000.
181. J. Fagerlund, E. Nduagu, I. Romão and R. Zevenhoven, *Front. Chem. Eng. China*, 2010, **4**, 133–141.
182. E. Nduagu, J. Bergerson and R. Zevenhoven, *Energy Convers. Manage.*, 2012, **55**, 116–126.
183. E. Nduagu, T. Björklöf, J. Fagerlund, E. Mäkilä, J. Salonen, H. Geerlings and R. Zevenhoven, *Miner. Eng.*, 2012, **30**, 87–94.
184. E. Nduagu, T. Björklöf, J. Fagerlund, J. Wärnå, H. Geerlings and R. Zevenhoven, *Miner. Eng.*, 2012, **30**, 75–86.
185. G. P. Assima, F. Larachi, G. Beaudoin and J. Molson, *Ind. Eng. Chem. Res.*, 2012, **51**, 8726–8734.
186. D. T. Beruto and R. Botter, *J. Eur. Ceram. Soc.*, 2000, **20**, 497–503.
187. K. J. Fricker and A.-H. A. Park, *Chem. Eng. Sci.*, 2013, **100**, 332–341.

188. S. Kwon, M. Fan, H. F. M. DaCosta and A. G. Russell, *J. Environ. Sci.*, 2011, **23**, 1233–1239.
189. F. Larachi, I. Daldoul and G. Beaudoin, *Geochim. Cosmochim. Acta*, 2010, **74**, 3051–3075.
190. F. Larachi, J. P. Gravel, B. P. A. Grandjean and G. Beaudoin, *Int. J. Greenhouse Gas Control*, 2012, **6**, 69–76.
191. V. Manovic and E. J. Anthony, *Ind. Eng. Chem. Res.*, 2010, **49**, 9105–9110.
192. S. Kwon, M. Fan, H. F. M. Dacosta, A. G. Russell and C. Tsouris, *J. Phys. Chem. A*, 2011, **115**, 7638–7644.
193. V. Nikulshina, M. E. Gálvez and A. Steinfeld, *Chem. Eng. J.*, 2007, **129**, 75–83.
194. J. Pronost, G. Beaudoin, J. Tremblay, F. Larachi, J. Duchesne, R. Hébert and M. Constantin, *Environ. Sci. Technol.*, 2011, **45**, 9413–9420.
195. P. V. Brady and S. R. Gíslason, *Geochim. Cosmochim. Acta*, 1997, **61**, 965–973.
196. S. R. Gislason, E. H. Oelkers and Á. Snorrason, *Geology*, 2006, **34**, 49–52.
197. D. S. Goldberg, D. V. Kent and P. E. Olsen, *Proc. Natl. Acad. Sci. U. S. A.*, 2010, **107**, 1327–1332.
198. T. K. Flaathen and S. R. Gislason, *Geochim. Cosmochim. Acta*, 2007, **71**, A283.
199. B. P. McGrail, H. T. Schaef, A. M. Ho, Y.-J. Chien, J. J. Dooley and C. L. Davidson, *J. Geophys. Res.: Solid Earth*, 2006, **111**, 1–13.
200. J. M. Matter, W. S. Broecker, M. Stute, S. R. Gislason, E. H. Oelkers, A. Stefánsson, D. Wolff-Boenisch, E. Gunnlaugsson, G. Axelsson and G. Björnsson, *Energy Procedia*, 2009, **1**, 3641–3646.
201. S. R. Gislason, D. Wolff-Boenisch, A. Stefansson, E. H. Oelkers, E. Gunnlaugsson, H. Sigurdardottir, B. Sigfusson, W. S. Broecker, J. M. Matter, M. Stute, G. Axelsson and T. Fridriksson, *Int. J. Greenhouse Gas Control*, 2010, **4**, 537–545.
202. P. B. Kelemen and J. Matter, *Proc. Natl. Acad. Sci. U. S. A.*, 2008, **105**, 17295–17300.
203. E. H. Oelkers, S. R. Gislason and J. Matter, *Elements*, 2008, **4**, 331–335.
204. T. K. Mlambo, F. J. Doucet, E. M. Van der Merwe and W. Altermann, *The Application of Mineral Carbonation Engineering Principles to Geological CO_2 Sequestration: a Conceptual Approach to Improved Reservoir Integrity*, Turku, Finland, 2010.
205. A. Ridgwell and R. E. Zeebe, *Earth Planet. Sci. Lett.*, 2005, **234**, 299–315.
206. S.-I. Yang, D.-Y. Choi, S.-C. Jang, S.-H. Kim and D.-K. Choi, *Adsorption*, 2008, **14**, 583–590.
207. Y. Wang, Y. Zhou, C. Liu and L. Zhou, *Colloids Surf., A*, 2008, **322**, 14–18.
208. R. V. Siriwardane, M.-S. Shen, E. P. Fisher and J. A. Poston, *Energy Fuels*, 2001, **15**, 279–284.
209. C. Lu, H. Bai, B. Wu, F. Su and J. F. Hwang, *Energy Fuels*, 2008, **22**, 3050–3056.
210. S. Keskin, T. M. van Heest and D. S. Sholl, *ChemSusChem*, 2010, **3**, 879–891.
211. O. Cheung and N. Hedin, *RSC Adv.*, 2014, **4**, 14480–14494.

212. D. Bonenfant, M. Kharoune, P. Niquette, M. Mimeault and R. Hausler, *Sci. Technol. Adv. Mater.*, 2008, **9**, 013007.
213. L. Bertsch and H. W. Habgood, *J. Phys. Chem.*, 1963, **67**, 1621–1628.
214. E. Gallei and G. Stumpf, *J. Colloid Interface Sci.*, 1976, **55**, 415–420.
215. D. Ko, R. Siriwardane and L. T. Biegler, *Ind. Eng. Chem. Res.*, 2003, **42**, 339–348.
216. J. A. C. Silva, K. Schumann and A. E. Rodrigues, *Microporous Mesoporous Mater.*, 2012, **158**, 219–228.
217. F. Akhtar and L. Bergström, *J. Am. Ceram. Soc.*, 2011, **94**, 92–98.
218. F. Akhtar, Q. Liu, N. Hedin and L. Bergstrom, *Energy Environ. Sci.*, 2012, **5**, 7664–7673.
219. K. S. Walton, M. B. Abney and M. Douglas LeVan, *Microporous Mesoporous Mater.*, 2006, **91**, 78–84.
220. F. N. Ridha, Y. Yang and P. A. Webley, *Microporous Mesoporous Mater.*, 2009, **117**, 497–507.
221. M. M. Lozinska, E. Mangano, J. P. S. Mowat, A. M. Shepherd, R. F. Howe, S. P. Thompson, J. E. Parker, S. Brandani and P. A. Wright, *J. Am. Chem. Soc.*, 2012, **134**, 17628–17642.
222. R. M. Barrer and R. M. Gibbons, *Trans. Faraday Soc.*, 1965, **61**, 948–961.
223. Z. Zhao, X. Cui, J. Ma and R. Li, *Int. J. Greenhouse Gas Control*, 2007, **1**, 355–359.
224. C. Graham, D. A. Imrie and R. E. Raab, *Mol. Phys.*, 1998, **93**, 49–56.
225. P. J. E. Harlick and F. H. Tezel, *Microporous Mesoporous Mater.*, 2004, **76**, 71–79.
226. G. Calleja, J. Pau and J. A. Calles, *J. Chem. Eng. Data*, 1998, **43**, 994–1003.
227. E. Díaz, E. Muñoz, A. Vega and S. Ordóñez, *Ind. Eng. Chem. Res.*, 2008, **47**, 412–418.
228. J. Israelachvili, *Intermolecular and Surface Forces* - 3rd edn, Amsterdam, 2011.
229. K. Hasegawa and A. Matsumoto, *AIP Conf. Proc.*, 2017, **1865**, 020002.
230. M. Katoh, T. Yoshikawa, T. Tomonari, K. Katayama and T. Tomida, *J. Colloid Interface Sci.*, 2000, **226**, 145–150.
231. T. Yamazaki, M. Katoh, S. Ozawa and Y. Ogino, *Mol. Phys.*, 1993, **80**, 313–324.
232. F. N. Ridha and P. A. Webley, *Sep. Purif. Technol.*, 2009, **67**, 336–343.
233. T. D. Pham, M. R. Hudson, C. M. Brown and R. F. Lobo, *ChemSusChem*, 2014, 7, 3031–3038.
234. R. S. Pillai, S. A. Peter and R. V. Jasra, *Microporous Mesoporous Mater.*, 2012, **162**, 143–151.
235. J. Zhang, R. Singh and P. A. Webley, *Microporous Mesoporous Mater.*, 2008, **111**, 478–487.
236. S.-T. Yang, J. Kim and W.-S. Ahn, *Microporous Mesoporous Mater.*, 2010, **135**, 90–94.
237. A. Zukal, J. Pawlesa and J. Čejka, *Adsorption*, 2009, **15**, 264–270.
238. J. Pawlesa, A. Zukal and J. Čejka, *Adsorption*, 2007, **13**, 257–265.

239. I. Domínguez, J. Pawlesa, A. Zukal and J. Čejka, in *Studies in Surface Science and Catalysis*, ed. A. Gédéon, P. Massiani and F. Babonneau, Elsevier, 2008, vol. 174, pp. 603–606.
240. R. Arletti, L. Gigli, F. di Renzo and S. Quartieri, *Microporous Mesoporous Mater.*, 2016, **228**, 248–255.
241. L. Hauchhum and P. Mahanta, *Int. J. Energy Environ. Eng.*, 2014, **5**, 349–356.
242. E. J. García, J. Pérez-Pellitero, G. D. Pirngruber, C. Jallut, M. Palomino, F. Rey and S. Valencia, *Ind. Eng. Chem. Res.*, 2014, **53**, 9860–9874.
243. G. D. Pirngruber, P. Raybaud, Y. Belmabkhout, J. Cejka and A. Zukal, *Phys. Chem. Chem. Phys.*, 2010, **12**, 13534–13546.
244. H. V. Thang, L. Grajciar, P. Nachtigall, O. Bludský, C. O. Areán, E. Frýdová and R. Bulánek, *Catal. Today*, 2014, **227**, 50–56.
245. S. K. Wirawan and D. Creaser, *Microporous Mesoporous Mater.*, 2006, **91**, 196–205.
246. T. S. Frantz, W. A. Ruiz, C. A. da Rosa and V. B. Mortola, *Microporous Mesoporous Mater.*, 2016, **222**, 209–217.
247. H. Chen, W. Wang, J. Ding, X. Wei and J. Lu, *Energy Procedia*, 2017, **105**, 4370–4376.
248. L. Ohlin, P. Bazin, F. Thibault-Starzyk, J. Hedlund and M. Grahn, *J. Phys. Chem. C*, 2013, **117**, 16972–16982.
249. R. C. Deka, R. Kinkar Roy and K. Hirao, *Chem. Phys. Lett.*, 2000, **332**, 576–582.
250. T.-L. Ho, *Chem. Rev.*, 1975, **75**, 1–20.
251. P. A. Jacobs, F. H. van Cauwelaert, E. F. Vansant and J. B. Uytterhoeven, *J. Chem. Soc., Faraday Trans. 1*, 1973, **69**, 1056–1068.
252. T. Montanari and G. Busca, *Vib. Spectrosc.*, 2008, **46**, 45–51.
253. A. V. Larin, *Microporous Mesoporous Mater.*, 2016, **228**, 182–195.
254. P. O. Scokart and P. G. Rouxhet, *J. Chem. Soc., Faraday Trans. 1*, 1980, **76**, 1476–1489.
255. P. O. Scokart and P. G. Rouxhet, *Bull. Soc. Chim. Belg.*, 1981, **90**, 983–984.
256. D. Barthomeuf, *J. Phys. Chem.*, 1984, **88**, 42–45.
257. D. Barthomeuf, in *Studies in Surface Science and Catalysis*, ed. G. Öhlmann, H. Pfeifer and R. Fricke, Elsevier, 1991, vol. 65, pp. 157–169.
258. D. Barthomeuf, in *Zeolite Microporous Solids: Synthesis, Structure, and Reactivity*, ed. E. G. Derouane, F. Lemos, C. Naccache and F. R. Ribeiro, Springer, Amsterdam, 1992, vol. 352, pp. 193–223.
259. M. Derewinski, J. Haber, J. Ptaszynski, J. A. Lercher and G. Rumplmayr, in *Studies in Surface Science and Catalysis*, ed. Y. Murakami, A. Iijima and J. W. Ward, Elsevier, 1986, vol. 28, pp. 957–964.
260. C. M. Zicovich-Wilson, A. Corma and P. Viruela, *J. Phys. Chem.*, 1994, **98**, 10863–10870.
261. J. Fraissard, in *Studies in Surface Science and Catalysis*, ed. B. Imelik, C. Naccache, Y. B. Taarit, J. C. Vedrine, G. Coudurier and H. Praliaud, Elsevier, 1980, vol. 5, pp. 343–350.

262. E. G. Derouane and C. D. Chang, *Microporous Mesoporous Mater.*, 2000, **35–36**, 425–433.
263. A. Corma, in *Studies in Surface Science and Catalysis*, ed. H. K. Beyer, H. G. Karge, I. Kiricsi and J. B. Nagy, Elsevier, 1995, vol. 94, pp. 736–747.
264. D. Barthomeuf, in *Studies in Surface Science and Catalysis*, ed. H. Chon, S.-K. Ihm and Y. S. Uh, Elsevier, 1997, vol. 105, pp. 1677–1706.
265. M. Wallau and U. Schuchardt, *J. Braz. Chem. Soc.*, 1995, **6**, 393–403.
266. S. K. Wirawan and D. Creaser, *Sep. Purif. Technol.*, 2006, **52**, 224–231.
267. M. M. Lozinska, E. Mangano, A. G. Greenaway, R. Fletcher, S. P. Thompson, C. A. Murray, S. Brandani and P. A. Wright, *J. Phys. Chem. C*, 2016, **120**, 19652–19662.
268. R. M. Barrer and E. A. D. White, *J. Chem. Soc.*, 1951, 1267–1278.
269. I. S. Kerr, *Z. Kristallogr.*, 1974, **139**, 186–195.
270. E. Jaramillo and M. Chandross, *J. Phys. Chem. B*, 2004, **108**, 20155–20159.
271. A. C. P. Barbosa, P. M. Esteves and M. A. C. Nascimento, *J. Phys. Chem. C*, 2017, **121**, 19314–19320.
272. S. U. Rege and R. T. Yang, *Chem. Eng. Sci.*, 2001, **56**, 3781–3796.
273. F. Brandani and D. M. Ruthven, *Ind. Eng. Chem. Res.*, 2004, **43**, 8339–8344.
274. M. Suckow, W. Lutz, J. Kornatowski, M. Rozwadowski and M. Wark, *Gas Sep. Purif.*, 1992, **6**, 101–108.
275. M. Baerns, H. Hofmann and A. Renken, *Chemische Reaktionstechnik* - 3rd edn, Georg Thieme Verlag, Stuttgart/New York, 1999.
276. J. Kärger and D. M. Ruthven., *Diffusion in Zeolites and Other Microporous Solids*, John Wiley & Sons, Inc., New York, 1992.
277. V. Blay, B. Louis, R. Miravalles, T. Yokoi, K. A. Peccatiello, M. Clough and B. Yilmaz, *ACS Catal.*, 2017, **7**, 6542–6566.
278. P. V. Subha, B. N. Nair, P. Hareesh, A. P. Mohamed, T. Yamaguchi, K. G. K. Warrier and U. S. Hareesh, *J. Mater. Chem. A*, 2014, **2**, 12792–12798.
279. B. Louis, E. S. Gomes, P. Losch, G. Lutzweiler, T. Coelho, A. Faro, J. F. Pinto, C. S. Cardoso, A. V. Silva and M. M. Pereira, *ChemCatChem*, 2017, **9**, 2065–2079.
280. B. Louis, G. Laugel, P. Pale and M. M. Pereira, *ChemCatChem*, 2011, **3**, 1263–1272.
281. O. Cheung, Z. Bacsik, P. Krokidas, A. Mace, A. Laaksonen and N. Hedin, *Langmuir*, 2014, **30**, 9682–9690.
282. P. Nachtigall, M. R. Delgado, D. Nachtigallova and C. O. Arean, *Phys. Chem. Chem. Phys.*, 2012, **14**, 1552–1569.
283. A. Zukal, A. Pulido, B. Gil, P. Nachtigall, O. Bludsky, M. Rubes and J. Cejka, *Phys. Chem. Chem. Phys.*, 2010, **12**, 6413–6422.
284. C. Otero Areán, M. Rodríguez Delgado, G. Fiol Bibiloni, O. Bludský and P. Nachtigall, *ChemPhysChem*, 2011, **12**, 1435–1443.
285. A. Pulido, P. Nachtigall, A. Zukal, I. Domínguez and J. Čejka, *J. Phys. Chem. C*, 2009, **113**, 2928–2935.
286. R. Bulánek, K. Frolich, E. Frýdová and P. Čičmanec, *J. Therm. Anal. Calorim.*, 2011, **105**, 443–449.

287. J. A. Dunne, M. Rao, S. Sircar, R. J. Gorte and A. L. Myers, *Langmuir*, 1996, **12**, 5896–5904.
288. B. Bonelli, B. Civalleri, B. Fubini, P. Ugliengo, C. O. Areán and E. Garrone, *J. Phys. Chem. B*, 2000, **104**, 10978–10988.
289. A. Zukal, C. O. Arean, M. R. Delgado, P. Nachtigall, A. Pulido, J. Mayerová and J. Čejka, *Microporous Mesoporous Mater.*, 2011, **146**, 97–105.
290. K. Nakagawa and T. Ohashi, *J. Electrochem. Soc.*, 1998, **145**, 1344–1346.
291. F. Ocampo, H. S. Yun, M. M. Pereira, J. P. Tessonnier and B. Louis, *Cryst. Growth Des.*, 2009, **9**, 3721–3729.
292. B. Louis, F. Ocampo, H. S. Yun, J. P. Tessonnier and M. M. Pereira, *Chem. Eng. J.*, 2010, **161**, 397–402.
293. G. A. Olah, A. Goeppert and G. K. S. Prakash, in *Beyond Oil and Gas: The Methanol Economy*, Wiley-VCH Verlag GmbH & Co. KGaA, Weinheim, Germany, 2009, DOI: 10.1002/9783527627806.index, pp. 327–334.
294. M. M. Pereira and B. Louis, in *New and Future Developments in Catalysis. Catalysis for Remediation and Environmental Concerns*, ed. S. L. Suib, Elsevier, Amsterdam, The Netherlands, 2013, pp. 535–562.
295. M. Jacquemin, A. Beuls and P. Ruiz, *Catal. Today*, 2010, **157**, 462–466.
296. L. H. Dubois and G. A. Somorjai, *Surf. Sci.*, 1983, **128**, L231–L235.
297. H. Noda, S. Ikeda, Y. Oda, K. Imai, M. Maeda and K. Ito, *Bull. Chem. Soc. Jpn.*, 1990, **63**, 2459–2462.
298. J. P. Collin and J. P. Sauvage, *Coord. Chem. Rev.*, 1989, **93**, 245–268.
299. B. Wang, Y. Li, N. Wu and C. Q. Lan, *Appl. Microbiol. Biotechnol.*, 2008, **79**, 707–718.
300. L. K. Rihko-Struckmann, A. Peschel, R. Hanke-Rauschenbach and K. Sundmacher, *Ind. Eng. Chem. Res.*, 2010, **49**, 11073–11078.
301. F. Yagi, S. Hodoshima, S. Wakamatsu, R. Kanai, K. Kawazuishi, Y. Suehiro and M. Shimura, in *Studies in Surface Science and Catalysis*, ed. F. Bellot Noronha, M. Schmal and E. Falabella Sousa-Aguiar, Elsevier, 2007, vol. 167, pp. 385–390.
302. M. Specht, F. Baumgart, B. Feigl, V. Frick, B. Stürmer and U. Zuberbühler, *FVEE AEE Topics*, 2009, 69–78.
303. N. Tsubaki, M. Ito and K. Fujimoto, *J. Catal.*, 2001, **197**, 224–227.
304. M. Kilo, J. Weigel, A. Wokaun, R. A. Koeppel, A. Stoeckli and A. Baiker, *J. Mol. Catal. A: Chem.*, 1997, **126**, 169–184.
305. F. Arena, K. Barbera, G. Italiano, G. Bonura, L. Spadaro and F. Frusteri, *J. Catal.*, 2007, **249**, 185–194.
306. S. Ivanova, E. Vanhaecke, B. Louis, S. Libs, M.-J. Ledoux, S. Rigolet, C. Marichal, C. Pham, F. Luck and C. Pham-Huu, *ChemSusChem*, 2008, **1**, 851–857.
307. P. Sabatier and J. B. Senderens, *C. R. Acad. Sci. Paris*, 1902, **134**, 514.
308. P. Gao, S. Li, X. Bu, S. Dang, Z. Liu, H. Wang, L. Zhong, M. Qiu, C. Yang, J. Cai, W. Wei and Y. Sun, *Nat. Chem.*, 2017, **9**, 1019–1024.
309. G. A. Olah, A. Goeppert and G. K. S. Prakash, *J. Org. Chem.*, 2009, **74**, 487–498.

310. J. Kothandaraman, A. Goeppert, M. Czaun, G. A. Olah and G. K. Surya Prakash, *Green Chem.*, 2016, **18**, 5831–5838.
311. M. Oschatz and M. Antonietti, *Energy Environ. Sci.*, 2018, **11**, 57–70.
312. G. Zhu, R. G. Jensen, R. B. Hallick and G. F. Wildner, *Plant Physiol.*, 1992, **98**, 764.
313. J. Perez-Ramirez, C. H. Christensen, K. Egeblad, C. H. Christensen and J. C. Groen, *Chem. Soc. Rev.*, 2008, **37**, 2530–2542.
314. R. Zevenhoven, J. Fagerlund, E. Nduagu, I. Romão, B. Jie and J. Highfield, *Energy Procedia*, 2013, **37**, 5945–5954.
315. F. Dufaud, I. Martinez and S. Shilobreeva, *Chem. Geol.*, 2009, **265**, 79–87.
316. W. J. J. Huijgen, G.-J. Witkamp and R. N. J. Comans, *Chem. Eng. Sci.*, 2006, **61**, 4242–4251.
317. G. Maurin, P. L. Llewellyn and R. G. Bell, *J. Phys. Chem. B*, 2005, **109**, 16084–16091.
318. K. Wang, P. Zhao, X. Guo, Y. Li, D. Han and Y. Chao, *Energy Convers. Manage.*, 2014, **81**, 447–454.
319. J. H. Lee, B. Moon, T. K. Kim, S. Jeoung and H. R. Moon, *Dalton Trans.*, 2015, **44**, 15130–15134.
320. J. Breu, W. Seidl and A. Stoll, *Z. Anorg. Allg. Chem.*, 2003, **629**, 503–515.
321. J. E. Post, D. L. Bish and P. J. Heaney, *Am. Mineral.*, 2007, **92**, 91–97.
322. G. Falini, E. Foresti, M. Gazzano, A. F. Gualtieri, M. Leoni, I. G. Lesci and N. Roveri, *Chem. – Eur. J.*, 2004, **10**, 3043–3049.

CHAPTER 3

Metal–Organic Framework (MOF)-based CO_2 Adsorbents

PILLAIYAR PUTHIARAJ, YU-RI LEE, SEENU RAVI, SIQIAN ZHANG AND WHA-SEUNG AHN*

Department of Chemistry and Chemical Engineering, Inha University, Incheon, 402-751, South Korea
*E-mail: whasahn@inha.ac.kr

3.1 Introduction

CO_2 capture from post-combustion flue gases has become an urgent research area because of global warming caused by the presence of greenhouse gases in the atmosphere.[1,2] The contribution of CO_2 to this environmental problem is about 60%, and CO_2 emissions continue to increase because of the combustion of fossil fuels. The International Energy Agency (IEA) recommends that CO_2 emissions into the atmosphere be cut to half the 2007 level by 2050; this can be achieved by implementing green CCS technologies. The IEA expects that about 100 CCS projects will be executed by 2020 and over 3000 by 2050 to limit climate destabilization.[3] CCS technology encompasses three different types of CO_2 capture: post-combustion

Inorganic Materials Series No. 2
Post-combustion Carbon Dioxide Capture Materials
Edited by Qiang Wang

Published by the Royal Society of Chemistry, www.rsc.org

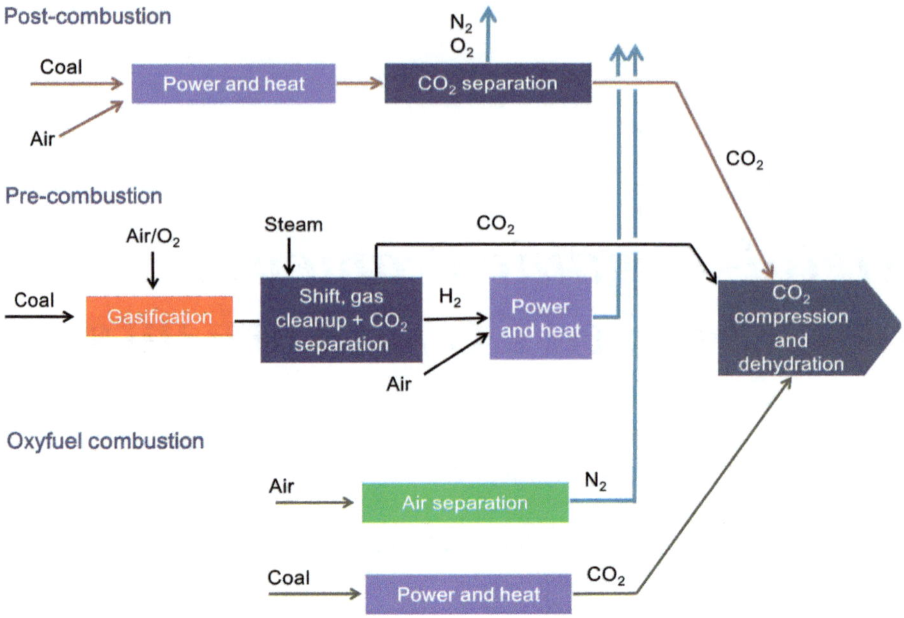

Figure 3.1 Technical options for CO_2 capture from coal power plants.[4]

capture, pre-combustion capture, and oxy-fuel combustion (see Figure 3.1).[4,5] Post-combustion CO_2 capture involves CO_2 removal from the flue gases generated by burning fossil fuels or other carbonaceous materials such as coal and biomass. For conventional post-combustion CO_2 capture, alkanolamines in aqueous solutions such as ethanolamine, diethanolamine, and methyldiethanolamine have been used, and these can capture up to 95% of the CO_2 in the flue gas. However, these absorption methods face several problems such as the high energy requirements and cost (1° > 2° > 3° amines) of amine regeneration, as well as the corrosion of pipelines.[6,7] Thus, many scientists involved in CCS have focused on research to develop effective alternative adsorbents that operate either *via* chemisorption or physisorption and have outstanding CO_2 capture performances, easier reversibility, and lower energy requirements. For this purpose, various porous materials such as zeolites, silica, carbons, metal–organic frameworks (MOFs), and porous covalent organic polymers have been extensively investigated as adsorbents for post-combustion CO_2 capture.[8–12] Of these materials, the lower surface areas and destructive effect of water on zeolites and silica limit their use for CO_2 capture; in contrast, carbon materials and covalent organic polymers have high surface areas but suffer from relatively low adsorption selectivities for CO_2.[9,12,13]

MOFs (also known as porous coordination polymers) have emerged as a new class of porous materials and were first developed near the end of 1990.[14] MOFs have regular pore sizes ranging from micro to mesopores and

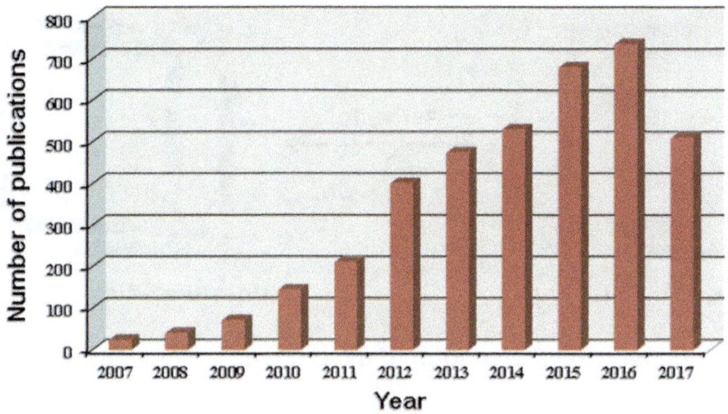

Figure 3.2 The number of publications between 2007 and 2017 concerning the use of MOFs for CO_2 adsorption (from Web of Science up to August 2017).

high surface areas (up to 8000 $m^2 g^{-1}$), large pore volumes, and significant structural flexibility. Importantly, MOFs can be prepared by rational design. To date, more than 20 000 MOF structures have been reported. MOFs are self-assembled from metal ions or metal clusters (also known as secondary building units or SBUs) with bi-, tri-, or multi-dentate organic linkers *via* coordination bonding and other weak interactions, such as van der Waals forces. In addition to their stable geometries and diverse structures attained *via* coordination bonding between the metal ions and organic linkers,[15,16] their pores can be easily tuned by pre- or post-synthetic modification for potential applications in CO_2 capture.[17] Recently, increasing efforts have been made to construct novel MOFs with enhanced CO_2 capture capacities (see Figure 3.2).[2,7] The insertion of CO_2 binding sites with high affinity into MOFs is critical for their application to CO_2 capture, and high affinity binding sites can be added by various methods, such as (i) the use of metal ions with coordination unsaturated open metal sites, (ii) the attachment of functional groups, (iii) the preparation of mixed-ligand MOFs, and (iv) the use of flexible MOFs, (v) interpenetrated MOFs, and (vi) MOF composites.[18] In this chapter, CO_2 adsorption studies carried out up to August 2017 using MOFs with different active sites are summarized.

3.2 CO_2 Adsorption by MOFs With Open Metal Sites

Open metal sites (OMSs, also referred to as coordinatively unsaturated metal sites) in MOFs play a vital role in the specific recognition of gas molecules by their Lewis acidic character and can be generated by removing metal-bound solvent or water molecules (see Figure 3.3). OMSs in MOFs were first characterized by Chen *et al.*[19] in MOF-11. In this MOF, 1,3,5,7-adamantanetetracarboxylates are connected to the dicopper paddle wheel SBU, which is partially coordinated with water molecules. The rigid MOF-11 structure was subjected to thermal treatment to complete the removal

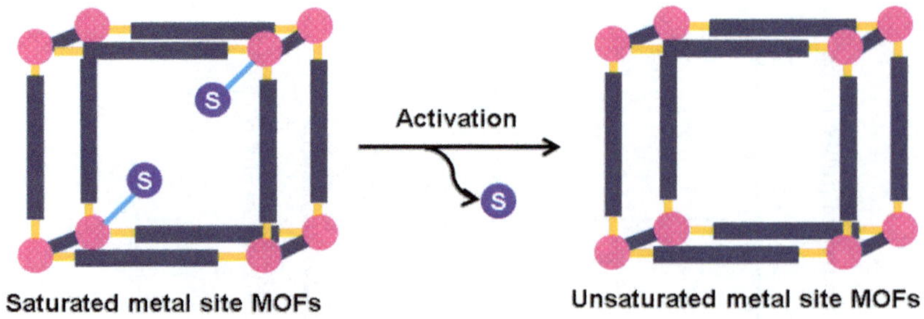

Figure 3.3 Schematic representation of the preparation of MOFs with OMSs.

of water molecules from the copper sites to yield a stable 3D framework with coordinatively unsaturated copper sites; the absence of water molecules was confirmed by single crystal X-ray analysis. Soon after, many different MOF structures with OMSs such as [Cu(H_2salphdc)]·H_2O ((H_4salphdc) = *N*,*N*′-phenylenebis(salicylideneimine)dicarboxylic acid),[20] Cu-BTC (also referred to as HKUST-1, where BTC is benzene-1,3,5-tricarboxylic acid),[21] MIL-100(Cr, Fe),[22,23] MIL-101(Cr, Fe),[24,25] UMCM-150,[26] MOF-505,[27] and UiO-66[28] were synthesized.

Yaghi *et al.* prepared $Zn_4O(CO_2)_6$ units joined with either one or two kinds of organic linkers, such as 1,4-benzenedicaboxylic acid (H_2BDC), hydropyrenedicarboxylic acid (H_2HPDC), cyclobutylbenzenedicarboxylic acid ($H_2C_2H_4BDC$), 4,4′,4″-benzene-1,3,5-triyl-tribenzoic acid (H_3BTB), 4,4′,4″-(benzene-1,3,5-triyl-tris(ethyne-2,1-diyl))tribenzoic acid (H_3BTE), and 4,4′,4″-(benzene-1,3,5-triyl-tris(benzene-4,1-diyl))tribenzoic acid (H_3BBC) to obtain six different MOFs: IRMOF-1, IRMOF-11, IRMOF-6, MOF-177, MOF-180, and MOF-200, respectively. Among them, MOF-200 has a high Brunauer–Emmett–Teller (BET) surface area and pore volume of 4530 m^2 g^{-1} and 3.59 cm^3 g^{-1}, respectively, and a high CO_2 uptake of up to 2400 mg g^{-1} at 50 bar and 298 K.[29,30]

One of the most widely studied MOFs with OMSs is Cu-BTC (also referred to as HKUST-1), which is synthesized from a mixture of $Cu(NO_3)_2.H_2O$ and BTC in H_2O–EtOH (v/v, 50:50) at 180 °C for 12 h. Cu-BTC is composed of dimeric cupric tetracarboxylate units with a short Cu–Cu internuclear separation of 2.628(2) Å, and the 12 carboxylate oxygens from the two BTC ligands bind to four coordination sites for each of the three Cu^{2+} ions to give a pseudo-octahedral complex with an axial water ligand opposite the Cu–Cu moiety. Cu-BTC with open copper sites can be generated by the dehydration of Cu-BTC at 100 °C, which is accompanied by a color change from turquoise to dark blue.[31] The dehydrated Cu-BTC has a BET surface area of 1781 m^2 g^{-1} and a CO_2 uptake of 471 mg g^{-1} at 35 bar and 298 K.[30] Nachtigall *et al.* investigated the interactions of CO_2 with Cu-BTC in both the low- and high-coverage regions by experiments at 1 bar and modeling by density functional theory (DFT). They showed that only one CO_2 molecule interacts with each open Cu^{2+} site at low coverage, whereas CO_2 molecules are adsorbed at the open Cu^{2+} sites, as well as positioned in the center of small and large cages, at high coverages; these results are in good agreement with the DFT theoretical calculations.[32]

Ferey and co-workers reported MIL-53, −100, and −101(M) with BTC and terephthalic acid organic linkers (M = Cr,[23] Fe,[22,33] and Al[34]). Among these, MIL-100(Cr) and MIL-101(Cr) are highly stable and active for CO_2 adsorption. These materials are made of trimers of Cr octahedral units with terminal (H_2O, F) ligands and rigid organic carboxylates. The terminal water molecules in MIL-100 and MIL-101 can be removed by thermal treatment under vacuum at 423 K, which generates two coordinatively unsaturated open Cr^{3+} sites. The high heats of adsorption of CO_2 at zero coverage (45 and 63 kJ mol^{-1} for MIL-101 and MIL-100, respectively) are attributed to the coordination of CO_2 molecules onto the Lewis acid Cr^{3+} sites.[35] He *et al.* examined the influence of flue gas contaminants (H_2O, NO, and SO_2) on CO_2 adsorption by MIL-101(Cr), as well as the adsorbent regeneration conditions. The trace flue gas contaminants were found to have a minimal impact on the CO_2 adsorption capacity by MIL-101(Cr), and complete regeneration was observed at 328 K within 10 min by temperature-swing adsorption with N_2-stripping at 0.2 bar. The adsorption–desorption cycles showed that the CO_2 adsorption capacity of MIL-101(Cr) was maintained after five cycles under a gas flow of 10 vol% CO_2, 100 ppm SO_2, and 100 ppm NO in 10% relative humidity (RH).[36]

Many studies have also been conducted on the CO_2 adsorption of the MOF-74 series. MOF-74(Zn) (also referred to as CPO-27) was first reported by Yaghi and co-workers in 2005.[37] In this MOF, the tetraanionic 2,5-dihydroxy-1,4-benzenedicarboxylic acid (dobdc) ligand is coordinated to Zn^{2+} ions. It was found that the as-synthesized hexagonal channels of the MOF were occupied with solvent molecules (dimethylformamide (DMF) and water) and that coordinately unsaturated open Zn sites (square pyramidal geometry) were generated by the thermal evacuation of the solvent molecules. These sites are responsible for binding various guest species. MOF synthesis has been expanded to prepare MOFs containing Co, Ni, Mg, Mn, Fe, Cd, and Cu species.[38–46] Matzger *et al.* synthesized the MOF-74 series (Zn, Co, Ni, and Mg) for CO_2 adsorption.[42] Among them, MOF-74(Mg) has a high BET surface area of 1495 m^2 g^{-1} and CO_2 uptakes of 236 and 352 mg g^{-1} at 0.1 and 1 bar, respectively. Blom *et al.* reported CO_2 storage and separation by MOF-74(Mg and Ni), in which the maximum CO_2 adsorption of 509 mg g^{-1} (21.7 bar) and 629 mg g^{-1} (32.6 bar) at 298 K were observed for MOF-74(Ni) and MOF-74(Mg) (corresponding to 1.92/1.90 molecules CO_2 per Ni/Mg atoms and 34.5/34.0 molecules CO_2 per unit cell), respectively. This result proved that the total CO_2 mass uptake is higher for Mg than Ni, but the actual amount of CO_2 adsorbed by each species is roughly equal for both materials.[47] Recently, computational simulations have suggested that the binding affinity of CO_2 for MOF-74 can be tuned by metal substitution (Mg, Ca, Sc, Ti, V, Cr, Mn, Fe, Co, Ni, Cu, and Zn), in which MOF-74(Ti) and MOF-74(V) have higher CO_2 binding affinities than the others.[48,49] The effects of long-term storage and simulated flue gas conditions on the CO_2 adsorption capacities of MOF-74(Mg and Ni) were also studied. The long storage stability (one year) experiments showed that MOF-74(Mg) is less stable than MOF-74(Ni) because of the stronger oxidizing capability of the Mg ions, which reduces the number of OMSs in the framework. The flue gas (containing a trace amount of SO_2 and NO) only slightly affected the CO_2 adsorption of MOF-74(Ni) at room temperature (RT), which indicates that the acidic gas does not react extensively with the unsaturated Ni sites in MOF-74(Ni) at RT.[50] Yaghi and co-workers studied the separation of CO_2 over CH_4 by MOF-74(Mg) using breakthrough experiments with an 80% CH_4–20% CO_2 mixture at RT and atmospheric pressure, which showed that the complete separation of CO_2 and the dynamic capacity (89 mg g^{-1}) was slightly higher than that of NaX zeolite (85 mg g^{-1}). In addition, the material could be regenerated under mild conditions.[51] A microporous MOF based on Zn and 2,4-di(3′,5′-dicarboxylphenyl)benzoic acid was also developed for CO_2 adsorption and separation, and this MOF was found to have a CO_2 uptake of 168 mg g^{-1} with a selectivity of 95.2 at 273 K and 1 bar.[52] The Hupp group synthesized the carborane-based MOF, $[Zn_3(OH)(p\text{-}CDC)_{2.5}(DEF)_4]_n$ (*p*-CDC = 1,12-dihydroxycarbonyl-*closo*-dodecaborane-1,12-dicarboxylic acid; DEF = *N*,*N*′-diethylformamide), which was activated at 300 °C to create open Zn sites in the material, which showed enhanced CO_2 separation from a CO_2–CH_4 mixture compared to the MOF without OMSs, demonstrating that OMSs in MOFs can support the separation of (quadru)polar (CO_2)–nonpolar (CH_4) pairs.[53,54] Mirkin *et al.* also suggested that OMSs improve the CO_2–CH_4 selectivity in carborane-based Co(II)-MOFs by comparing the samples with and without OMSs. Snurr and co-workers used both experiments and simulations to screen a diverse set of MOFs for low-pressure CO_2 capture.[55,56] Representative MOFs with OMSs considered for CO_2 adsorption are listed in Table 3.1.

3.3 CO_2 Adsorption by Amine-functionalized MOFs

Lewis basic sites, such as -NH_2, alkylamines, and arylamines, have high CO_2 adsorption affinities and are recognized to be effective for the selective capture of CO_2 accompanied by a high enthalpy of CO_2 adsorption. Generally, amine-functionalized MOFs (see Figure 3.4) are prepared by following three methods: (i) *in situ* synthesis (amine-functionalized ligands are used for the synthesis), (ii) post-synthesis amine functionalization of MOFs *via* covalent or coordination bonding, and (iii) physical incorporation of amine species into unmodified MOFs. Among these methods, *in situ* synthesis is most widely used because of the easy synthesis protocols and wide commercial availability of the organic linkers. Examples of the CO_2 uptake values of amine-functionalized MOFs are summarized in Table 3.2.

Table 3.1 CO_2 uptakes of selected MOFs with OMSs.

Samples[a]	BET area (m^2 g^{-1})	Temp. (K)	Pressure (bar)	CO_2 capture (mg g^{-1})	$-Q_{st}$ (kJ mol^{-1})	Ref.
CuBTC	1400	293	1	198	25.5	57
CuBTC	—	298	1	184	—	58
CuBTC	—	295	1	183	—	56
CuBTC	1781	198	1	152	—	30
	1781	298	35	471	—	30
CuBTC	1482	295	1	150	—	59
CuBTC	—	313	1	62	—	60
CuBTC	—	298	0.8	106	—	61
	921	298	20	418	—	61
CuBTC	857	295	1	62	—	59
Mg-HKUST-1	—	298	5	836	—	62
Co-HKUST-1	—	298	5	520	—	62
Zn-HKUST-1	—	298	5	484	—	62
Ni-HKUST-1	—	298	5	492	—	62
UMCM-150	—	298	1	102	—	56
UMCM-150	—	298	1	108	—	56
Cu-BTTri	1770	298	1	143	19	63
UIO-66	1105	298	2	119	26	64
MIL-101(Cr)	2674	319	1	42	24	65
MIL-101(Cr)	—	304	5	1760	45	35
	—	304	5	792	63	35
MOF-74(Zn)	—	296	0.15	76	—	49
MOF-74(Zn)	816	296	1	244	—	42
	816	298	35	457	—	30
MOF-74(Mg)	1495	296	0.1	191	47	42
MOF-74(Mg)	1800	303	0.15	206	—	56
	—	313	0.15	189	—	56
	—	323	0.15	167	—	56
	—	333	0.15	145	—	56
MOF-74(Mg)	1495	296	1	352	47	42
MOF-74(Mg)	1174	298	1	275	—	66
MOF-74(Mg)	1542	298	32.6	629	—	47
MOF-74(Ni)	936	298	0.15	169	—	56
	936	298	1	239	—	56
MOF-74(Ni)	1070	296	0.1	104	41	42
MOF-74(Ni)	639	298	1	227	—	61
MOF-74(Ni)	1083	303	1	226	—	67
MOF-74(Ni)	1218	298	21.7	509	—	47
MOF-74(Co)	957	298	0.15	142	—	56
	957	298	1	249	—	56
MOF-74(Co)	1080	296	1	234	37	42
NOTT-140	2620	293	1	117	—	68
Zn-BTC	408	295	1	76	—	69
NJU-Bai12	3058	298	20	873	—	70
MOF-2	345	298	35	140	—	30
MOF-177	4508	298	35	1474	—	30
MOF-200	4530	298	50	2400	—	29
$[H_2N(Me)_2]_2$ $[Zn_4(L)_2(H_2O)_{1.5}]\cdot$ $5DMF\cdot H_2O$ (1DMF)	544	195	1	314	—	52

[a]HKUST = Hong Kong University of Science and Technology; UMCM = University of Michigan Crystalline Material; H_3BTTri = 1,3,5-tris(1*H*-1,2,3-triazol-5-yl)benzene; UIO = University of Oslo; MIL = matériauxs de l'Institut Lavoisier; NOTT = Nottingham University; NJU-Bai = Nanjing University Bai group; L1 = 2,4-di(3′,5′-dicarboxylphenyl)benzoic acid.

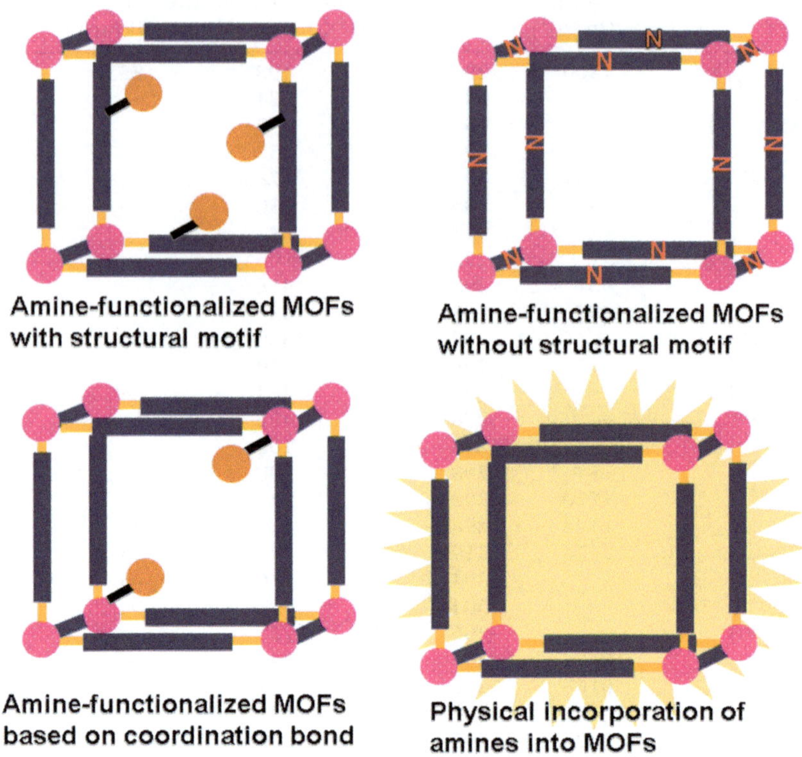

Figure 3.4 Schematic representations of amine-functionalized MOFs.

3.3.1 *In situ* Synthesized Amine-functionalized MOFs

3.3.1.1 Amine-functionalized MOFs With a Structural Motif

Mostly, hydrothermal and solvothermal synthesis schemes are used for the *in situ* preparation of amine-functionalized MOFs.[71] However, some amine functional groups in the ligand may also coordinate easily with metal ions, producing unwanted solid materials. Representative MOFs with *in situ* amine-functionalization are shown in Figure 3.5.

Yaghi *et al.* reported a robust 3D amine-functionalized MOF, MOF-46, which was constructed from 2-amino-1,4-benzenedicarboxylic acid (ABDC) and Zn^{2+} ions.[72] The same group also reported IRMOF-3, which is identical to the MOF-5 structure but has high stability and porosity (BET surface area of 2160 $m^2\ g^{-1}$). IRMOF-3 is prepared from a mixture of ABDC and $Zn(NO_3)_2.4H_2O$ in diethylformamide (DEF) under solvothermal conditions.[31] IRMOF-3 has been widely studied for catalysis,[73–75] but rarely for gas adsorption.[31,76] The CO_2 adsorption capacity of IRMOF-3 is 823 mg g^{-1} at 298 K and 35 bar.[31] Walton *et al.* compared IRMOF-3 and MOF-5 (without amine functionalization), for CO_2 adsorption both under experimental and molecular simulation conditions, and the results showed that IRMOF-3 has a higher

Table 3.2 Amine-functionalized MOFs for CO_2 adsorption.

Amine-functionalized MOFs[a]	BET area ($m^2\ g^{-1}$)	CO_2 (bar) Pressure 0.1	1.0	Temp. (°C)	$-Q_{st}$ (kJ mol^{-1})	Ref.
CAU-1	1268	—	316	273	48	83
NH_2-MIL-101(Cr)	1675	—	140	273	52	77
NH_2-MIL-53(Al)	—	—	48	298	56	124
NH_2-MIL-53(Al)	—	—	132	298	50	125
NH_2-MIL-53(Al)	937	—	80	298	60	80
USO-1-Al-A	960	—	120	298	50	125
USO-2-Ni-A	1530	—	140	298	—	125
Cd(NH_2-bdc)(bphz)$_{0.5}$	298	—	190	195	38	126
NH_2-UIO-66	930–1040	185	—	273	26.9	90
NH_2-MIL-125	1203	—	132	273	—	85
IRMOF-74- III-CH_2NH_2	2310	—	140	298	—	127
IRMOF-74- III-CH_2NHBoc	2170	—	93	298	—	127
IRMOF-74- III-CH_2NHMe	2250	—	120	298	—	127
IRMOF-74-III-CH_2NHMeBoc	2220	—	81	298	—	127
IRMOF-74- III-NH_2	2720	—	135	298	—	127
Cu-(NH_2BTC)	1834	—	231	298	—	91
ZJNU-54	2134	—	336	273	25	92
ZJNU-41	2530	—	188	298	24	93
ZJNU-43	2243	—	200	296	—	94
ZJNU-44	2314	—	222	296	—	94
ZJNU-45	2232	—	207	296	—	94
CPF-13	642	—	228	273	28.2	128
MAF-66	1014	—	275	273	26	97
Zn(Atz)$_2$	782	—	191 (1.2 bar)	273	40.8	129
Mn(pmdc)	410	—	73 (0.9 bar)	293	—	130
Pd(2-pymo)$_2$	600	—	73 (0.86 bar)	293	—	131
Cu(2-pymo)$_2$	350	—	38 (0.89 bar)	298	—	131
Al(OH)(bpdc)	2160	—	62	298	—	132
BioMOF-11	1040	—	264	273	45	95
FMOF-2	378	—	45	298	—	133
NTU-105	3543	67		273	—	134
SNU-50	2300	29.9		298	25.8	135
IFMC-1a	780	30		298	—	136
en-CuBTTri		16 (0.06 bar)		298	90	63
mmen-CuBTTri	870	104	184	298	96	110
en-Mg-MOF-74	469	66 (0.0004 bar)		298	—	111
mmen-Mg-MOF-74	70	137	169	298	75	114
dmen-Mg-MOF-74	675	165	—	298	—	112
mmen-Mg_2(dondc)	99	181	—	298	—	113
$Cu_3(btc)_2$(3-picolylamine)$_2$	—	—	140	273	28.2	137
DETA-MIL-101(Cr)	1560	—	79	298	6942.5	108
TEPA-MIL-101(Cr)	1553	—	124	273	42.5	107
PEI-MIL-101–50	1802	104	—	298	—	138

(*continued*)

Table 3.2 (*continued*)

Amine-functionalized MOFs[a]	BET area ($m^2 g^{-1}$)	CO_2 (bar) Pressure 0.1	1.0	Temp. (°C)	$-Q_{st}$ (kJ mol^{-1})	Ref.
PEI-MIL-101–75	1112	159	—	298	—	138
PEI-MIL-101–100	608	176	—	298	—	138
PEI-MIL-101–125	182	157	—	298	—	138

[a]CAU = Christian-Albrechts-University; bphz = 1,2-bis(4-pyridyl-methylene)hydrazine; ZJNU = Zhejiang Normal University; CPF = crystalline porous framework; Atz = 3-amino-1,2,4- triazole; pmdc = pyrimidine-4,6-dicarboxylic acid; pymo = 2-pyrimidinolate; bpdc = 4,4′-biphenyldicarboxylate; SNU = Seoul National University; IFMC - Institute of Functional Material Chemistry

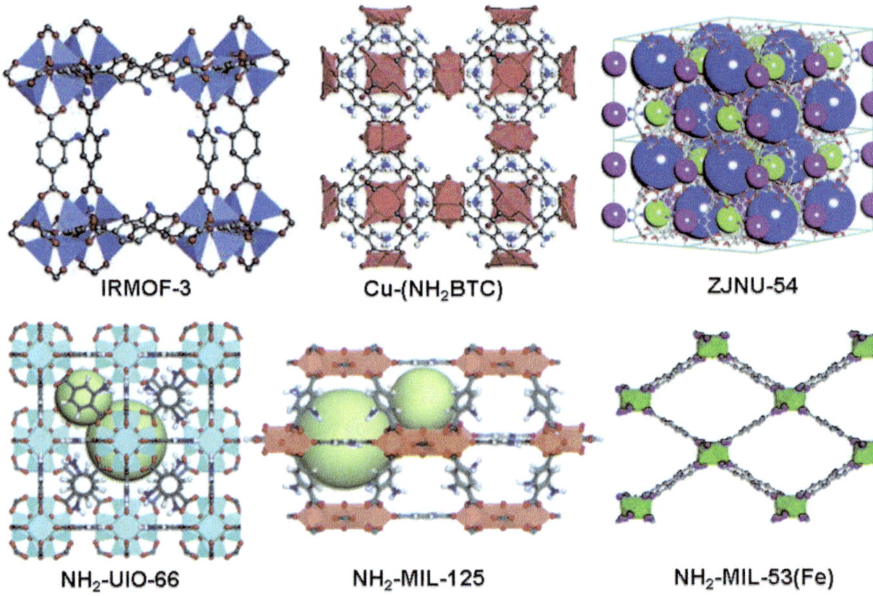

Figure 3.5 Structures of some *in situ* amine-functionalized MOFs.

CO_2 adsorption capacity than MOF-5 at low pressure because of the amine functionality of the framework.[76]

Subsequently, researchers expanded the investigation of amine-functionalized MOFs with different metal ions and amine-functionalized linkers. Lin *et al.* synthesized amine-functionalized MIL-101(Cr) using ABDC, which has a BET surface area of 1675 $m^2 g^{-1}$ and a small particle size of around 50 nm. NH_2-MIL-101(Cr) has a higher CO_2 adsorption capacity at low pressure than MIL-101(Cr), whereas the CO_2 adsorption capacity is inferior at high pressure (up to 660 mg g^{-1} at 298 K and 25 bar) because of the decreased surface area caused by the bulkier ligand.[77] Jiang *et al.* reported a multi-scale modeling study of X-MIL-101 functionalized with a series of functional groups (X = –NH_2, –Cl, –CH_3, –CN, and –NO_2) for CO_2 capture and the separation of CO_2

from N_2. The CO_2 uptake increased in the order of MIL-101 < CN-MIL-101 < NO_2-MIL-101 < Cl-MIL-101 < CH_3-MIL-101 < NH_2-MIL-101 in the low-pressure region, suggesting that the –NH_2 groups interact more strongly with the CO_2 molecules. However, in the high-pressure region, MIL-101 functionalized with smaller groups exhibited higher CO_2 uptakes (CN-MIL-101 > CH_3-MIL-101 > NH_2-MIL-101 > Cl-MIL-101 > NO_2-MIL-101). The CO_2–N_2 selectivity and breakthrough time also increased in the order of MIL-101 < NO_2-MIL-101 < CN-MIL-101 < Cl-MIL-101 < CH_3-MIL-101 < NH_2-MIL-101, and the longest breakthrough time by NH_2-MIL-101 was almost twice that of MIL-101, indicating the effect of amine-functionalization in MIL-101 for the improved separation of CO_2 from N_2.[78]

Kapteijn *et al.* synthesized NH_2-MIL-53(Al) *via* the reaction between ABDC and $Al(NO_3)_3.9H_2O$ in DMF under solvothermal conditions, which exhibited higher CO_2 adsorption and more selective separation from CH_4 than MIL-53(Al) owing to the presence of amine groups in the framework, which enhanced the affinity for CO_2.[79] Kim *et al.* also prepared NH_2-MIL-53(Al) *via* a solvothermal route. The synthesized MOF, NH_2-MIL-53(Al), achieved a high CO_2 adsorption capacity (82 and 302 mg g^{-1} at 0.1 and 3.0 MPa, respectively), and a high selectivity against N_2 (>30 at 2.0 MPa) was shown at 298 K. The heat of adsorption for CO_2 near zero coverage by NH_2-MIL-53(Al) was *ca.* 60 kJ mol^{-1} compared to *ca.* 50 kJ mol^{-1} by MIL-53(Al). Six repeated pressure swing adsorption–desorption cycles between 0.5 and 2.0 MPa were carried out at 298 and 323 K and no deterioration in the CO_2 adsorption capacity was observed; furthermore, reversible adsorbent regeneration was maintained for a total of 240 min.[80] The mechanism of the high CO_2 adsorption efficiency of NH_2-MIL-53(Al) was also examined by a combination of experimental and theoretical studies. Spectroscopic and DFT investigations revealed that the amine moieties play an indirect role in CO_2 adsorption, and CO_2 is not adsorbed by the formation of a direct chemical bond.[81] Hadjiivanov *et al.* studied the adsorption of CO_2 on MIL-53(Al) and NH_2-MIL-53(Al) by Fourier transform infrared (FTIR) spectroscopy at different temperatures and pressures.[82]

When the aluminium source of $Al(NO_3)_3$ was replaced by $AlCl_3·6H_2O$, NH_2-MIL-101(Al) was obtained instead, which has a BET surface area of 1268 m^2 g^{-1}, a CO_2 uptake of 317 mg g^{-1} at 1 bar, and a high CO_2 selectivity to CO_2 over N_2 (101 : 1) at 273 K.[83] Kapteijn *et al.* also reported the CO_2 separation from a CH_4 mixture using NH_2-MIL-101(Al), which exhibited higher CO_2–CH_4 selectivity than MIL-101(Cr), although the CO_2 adsorption capacity was lower.[84] Blom *et al.* reported amine-functionalized MOFs based on Al, In, and Ni ions, and reported CO_2 adsorption capacities of up to 140 mg g^{-1} at 298 K and 1 bar. Another amine-functionalized MOF in the MIL series was NH_2-MIL-125, which has a quasi-cubic tetragonal structure and was solvothermally synthesized from titanium-oxo-hydroxo clusters and ABDC.[85] The NH_2-MIL-125 was prepared under microwave heating conditions and applied for CO_2 adsorption. The CO_2 adsorption by NH_2-MIL-125 exhibited a significantly higher capacity than MIL-125 over the entire temperature range of 273–303

K, and the CO_2 breakthrough over NH_2-MIL-125 occurred later than that of MIL-125.[86]

NH_2-UiO-66 (UiO = University of Oslo), which has a BET surface area of 1112 m^2 g^{-1}, was prepared from the structural motif of UiO-66[28] from the treatment of $ZrCl_4$ and ABDC under solvothermal conditions in DMF.[87] A series of functionalized X-UiO-66 (X = –Br, $-NH_2$, $-NO_2$, $-(CF_3)_2$, $-(OH)_2$, $-SO_3H$, $-(CH_3)_2$, and –COOH) with a wide range of surface polarities was synthesized, and the effects of the functionalization on the CO_2 adsorption and CO_2–CH_4 separation performance were studied. The CO_2 adsorption capacity of NH_2-UiO-66 was *ca.* 350 mg g^{-1} (303 K and 30 bar), and the CO_2 adsorption enthalpies increased in the order of $-(CH_3)_2 > -SO_3H > -COOH > -NH_2 > -OH > -NO_2 > -Br > -CF_3$, indicating that the enthalpy increases as the functional group becomes bulkier and the pore volume decreases.[88,89] Recently, Su *et al.* synthesized a metal–organic gel based on NH_2-UiO-66 nanoparticles for CO_2 adsorption and conversion. The CO_2 uptake capacity of the dried NH_2-UiO-66 gel nanoparticles (185 mg g^{-1} at 273 K, 0.1 MPa) is higher than that of crystalline NH_2-UiO-66, which may be due to the larger number of defects (the open Zr metal sites), high surface area, and amine functionality.[90] Froba *et al.* synthesized Cu-(NH_2BTC) (NH_2BTC = 2-amino-1,3,5-benzenetricarboxylic acid), which has a BET surface area of 1834 m^2 g^{-1}, and reported a CO_2 adsorption at 298 K under both low- and high-pressure conditions. The CO_2 adsorption at 45 bar was slightly lower, but the CO_2 adsorption was slightly improved at 1 bar compared to that of Cu-BTC.[91] He *et al.* prepared the aminopyrimidine-functionalized cage-based MOF, ZJNU-54, *via* the solvothermal reaction of 5,5′-(pyrimidin-5-amine-4,6-diyl)di(isophthalic acid) and $CuCl_2·2H_2O$ in a DMF–water–methanol mixture and used it for the selective adsorption of CO_2 over CH_4. The estimated dynamic CO_2 uptake breakthrough run from an equimolar CO_2–CH_4 gas mixture was found to be 96.3 mg g^{-1}.[92] In addition, the activated ZJNU-54 showed a CO_2 uptake of 336 and 232 mg g^{-1} at 1 bar at 278 and 295 K, respectively, which are higher than those of the copper diisophthalate frameworks owing to the presence of open Cu sites and uncoordinated nitrogen atoms on the pore surfaces.[93,94]

3.3.1.2 Amine-functionalized MOFs Without a Structural Motif

Some of the amine-functionalized MOFs have been synthesized without structural motifs such as in Bio-MOF-11, ZTF-1 (ZTF = zeolitic tetrazolate framework), and MAF-66 (MAF = metal azolate framework). The biocompatible Bio-MOF-11 was constructed from $Co(OAc)_2.4H_2O$ and adenine in DMF at 130 °C for 24 h. The biomolecular adenine ligand has rigid and multiple metal binding sites. The structure of Bio-MOF-11 is composed of cobalt-adeninate-acetate paddle-wheel clusters, in which two Co^{2+} are connected by two adeninates and two acetates. Bio-MOF-11 has a BET surface area of 1040 m^2 g^{-1}, CO_2 adsorption capacities of 264 and 180 mg g^{-1} (1 bar), and CO_2 over N_2 selectivities of 81 : 1 and 75 : 1 at 273 and 298 K, respectively. The high adsorption values are due to the presence of large numbers of Lewis basic amino and pyrimidine groups on the pore surfaces of the frameworks. The bio-compatibility and free amines of the adeninates in the framework support the importance of bio-MOFs in the amine-functionalized MOFs family.[95] Banerjee *et al.* synthesized an amine-functionalized ZTF-1 using $Zn(NO_3)_2.6H_2O$ and 5-aminotetrazole, which exhibited *ca.* 246 mg g^{-1} CO_2 uptake at 273 K and 1 bar.[96] MAF-66 is synthesized from 3-amino-1,2,4-triazole and Zn nodes and has a BET surface area of 1014 m^2 g^{-1}. MAF-66 shows high CO_2 uptakes of 276 and 194 mg g^{-1} at 273 and 298 K, respectively, and 1 bar and a high Henry's law-based CO_2–N_2 selectivity of 403 and 225 at 273 and 298 K, respectively, because of the high density of uncoordinated N atoms on the framework.[97] UTSA-49 (UTSA = University of Texas at San Antonio) was prepared using $Zn(NO_3)_2{\cdot}6H_2O$ and 5-methyl-1*H*-tetrazole *via* solvothermal synthesis at 90 °C for 24 h. UTSA-49 shows a slightly lower CO_2 uptake (213 mg g^{-1} at 273 K) than ZTF-1 and MAF-66 because of the narrower pore size and lower pore volume compared with ZTF-1 and MAF-66.[98] The amine-functionalized $[Zn_2(CN_5H_2)_3(H_2O)_3]{\cdot}6H_2O$ MOF was synthesized from $Zn(NO_3)_2.6H_2O$ and 5-amino-1*H*-tetrazole ligand in DMF by Chen *et al.* In this MOF, each organic ligand is connected to two zinc nodes in the framework, producing 1D square channels, and each channel layer is connected *via* a hydrogen bond. Free amine is located around the framework cavities. This framework exhibited 105 mg g^{-1} CO_2 uptake at 273 K and 1 bar.[99] NTU-111, 112, and 113 (NTU = Nanyang Technological University) MOFs were constructed from "click"-extended tetracarboxylates and $Cu(NO_3)_2$. The structural analyses revealed that all three MOFs have both open Cu sites and accessible nitrogen-rich triazole units. NTU-113 exhibits a higher CO_2 uptake compared to NTU-111 and NTU-112 because NTU-113 contains more nitrogen atoms and the two triazole units are located close together in the framework. The breakthrough experiment for the separation of CO_2 from N_2 showed a breakthrough time of 13.8 units for N_2 and 181.2 units for CO_2, and the large difference in time demonstrated the efficiency of NTU-113 for post-combustion CO_2 capture.[100] He *et al.* developed a series of isostructural copper-based ZJNU-43, ZJNU-44, and ZJNU-45 using 5,5′-(quinoline-5,8-diyl)-diisophthalate, 5,5′-(isoquinoline-5,8-diyl)-diisophthalate, and 5,5′-(quinoxaline-5,8-diyl)-diisophthalate, respectively.[94] The as-synthesized MOFs were activated at 373 °C, and the activated ZJNU-43a, ZJNU-44a, and ZJNU-45a MOFs have CO_2 uptakes of 199, 224, and 207 mg g^{-1}, respectively, at 298 K and 1 bar. However, of the three MOFs, only ZJNU-44 has easily accessible nitrogen sites.

3.3.2 Post-synthesis Amine-functionalized MOFs

A number of amine-functionalized MOFs have also been prepared by the post-synthetic modification (PSM) method. Two different approaches have been used to incorporate amine moieties into the MOFs. In the first scenario, the organic linker of the MOF is modified with amine functional groups *via* covalent bonding in the framework. In the second scenario, amines or alkyl amines are incorporated into the framework by binding to the unsaturated metal sites *via* coordination bonding.

3.3.2.1 *Post-synthesis Functionalization of MOFs* via *Covalent Bonding*

The covalent PSM of MIL-101(Cr) was reported by Stock *et al.* The aromatic ring of the BDC ligand was nitrated using nitric acid followed by hydrogenation using $SnCl_2$ in ethanol to obtain NH_2-MIL-101(Cr) (see Figure 3.6).[101] NH_2-MIL-101(Cr) shows a higher CO_2 uptake than unsubstituted MIL-101(Cr), whereas the NO_2-MIL-101(Cr) shows a lower CO_2 sorption capacity than MIL-101(Cr), indicating that the enhancement of CO_2 uptake is due to the presence of amine groups in the pore framework.[102] NH_2-MIL-53(Al) was modified post-synthetically using diphosgene or thiophosgene to produce the isocyanate or isothiocyanate MOFs (NCO-MIL-53(Al) or NCS-MIL-53(Al)). Finally, these cyanate MOFs were reacted with amine and alcohols to produce (thio)carbamate- or (thio)urea-bearing MOFs. The resulting amide MOFs show high selectivity for CO_2 over N_2 but slightly lower CO_2 uptakes than NH_2-MIL-53(Al) because of the decrease in the surface area caused by the larger substituents in the pore structure.[103] The breathing properties of NH_2-DMOF-1 post-synthetically modified with anhydrides were studied. For this study, NH_2-DMOF-1 was constructed from Zn(II), ABDC, and 1,4-diazabicyclo[2.2.2]octane (DABCO), followed by treatment with linear alkyl anhydrides ($O[CO(CH_2)_nCH_3]_2$, $n = 0–5$) to obtain AM1-, AM2-, AM3-, AM4-, AM5-, and AM6-DMOF-1. The breathing behavior of the modified and unmodified NH_2-DMOF-1 was studied by N_2, Ar, and CO_2 adsorption. The unmodified NH_2-DMOF-1 and long alkyl chain in NH_2-DMOF-1 (AM4-, AM5-, and AM6-DMOF-1) did not show breathing properties during gas adsorption. However, for the short-alkyl-chain-modified AM1-DMOF-1, non-ideal type-I adsorption behavior with a small unambiguous hysteresis was detected, and

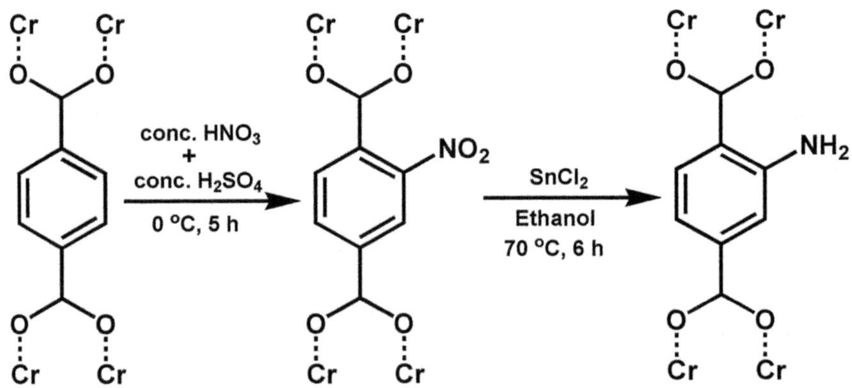

Figure 3.6 Synthesis of NH_2-MIL-101(Cr) from MIL-101(Cr) *via* covalent PSM.

AM2- and AM3-DMOF-1 showed breathing behavior with all the gases. The observed difference in adsorption was explained by structural transitions (different pore sizes reflect the different phases) in the framework; AM1- and AM2-DMOF-1 are in the narrow pore form, AM3-DMOF-1 is in the intermediate narrow and large pore form, and NH_2-DMOF-1, AM4-, AM5-, and AM6-DMOF-1 are in the large pore form.[104] Cao *et al.* studied the effects of amino and carboxyl groups in the MOFs on CO_2 capture. NH_2-UMCM-1 was synthesized using $Zn(NO_3)_2 \cdot 6H_2O$, ABDC, and H_3BTB and then carboxyl groups were introduced in NH_2-UMCM-1 by post-synthetic modification with maleic anhydride (NH_2-MA-UMCM-1). NH_2-UMCM-1 exhibited higher CO_2 uptakes in the high-pressure region (869 mg g^{-1} at 298 K and 18 bar) compared to UMCM-1 and NH_2-MA-UMCM-1. However, NH_2-MA-UMCM-1 showed higher CO_2 uptakes in the low-pressure region because of the decreased pore size compared to NH_2-UMCM-1.[105]

3.3.2.2 *Post-synthetic Functionalization of MOFs* via *Coordination Bonding*

The pore sizes in MOFs play an important role in the post-synthetic amine-functionalization *via* coordination bonding. This method of PSM can tune the porosity of the MOFs and incorporate amine molecules into the MOFs at the same time for CO_2 adsorption. Here, the MOFs were activated first to remove the solvent molecules for the creation of uncoordinated metal sites followed by functionalization with amine molecules. Férey and co-workers selected MIL-101(Cr) for the PSM of amines because of the open Cr sites, high surface area, and mesoporosity. In their study, MIL-101(Cr) was activated at 150 °C under vacuum for 12 h to remove the coordinated solvent molecules, and then ethylenediamine (EN) was grafted to produce the EN-grafted MIL-101(Cr) (see Figure 3.7). The formation of this MOF was confirmed by powder X-ray diffraction (XRD) and FT-IR spectroscopy. The results showed that one end of EN was anchored to an open Cr site of MIL-101(Cr) and the other end is free.[106] Subsequently, Li *et al.* grafted tetraethylenepentamine

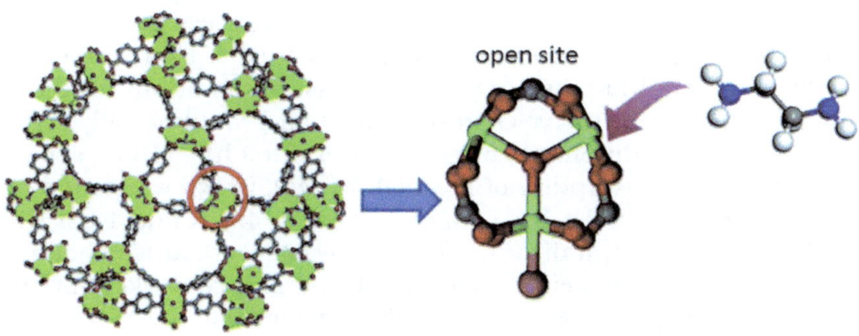

Figure 3.7 Schematic representation of ethylenediamine-grafted MIL-101(Cr).

(TEPA) to the coordinatively unsaturated Cr(III) sites of MIL-101 and applied it to the selective adsorption of CO_2 over CO. The resulting TEPA-MIL-101(Cr) showed a slightly higher CO_2 adsorption than the parent MIL-101(Cr), but the selectivity for CO_2 over CO increased dramatically from 1.77 to 70.2 at 298 K and 0.4 bar because the amine groups in TEPA capture CO_2 *via* Lewis acid–base interactions.[107] Kim *et al.* prepared diethylenetriamine (DETA)-grafted MIL-101(Cr) for CO_2 adsorption, which showed a higher adsorption under low-pressure conditions (<0.15 bar), indicating the strong binding affinity of CO_2 for amines.[108] The amine-functionalized MIL-100(Cr) was synthesized by grafting EN and *N,N'*-dimethylethylenediamine (MMEN), respectively. The resulting amine-grafted MIL-100(Cr) has a reduced BET surface area and pore volume compared with the bare MIL-100(Cr). The CO_2 uptakes of MMEN-MIL-100(Cr) and EN-MIL-100(Cr) at 308 K and 1 bar were 75 and 105 mg g^{-1}, respectively, 6% and 50% increases relative to that of MIL-100(Cr) (70 mg g^{-1}).[109] Similarly, EN was used for the PSM of tetrazolate-bridged CuBTTri, where EN was attached to the open Cu sites. After EN had been attached, the color changed from red to blue, indicating the successful grafting of EN. The EN-CuBTTri exhibited CO_2 uptakes of 55 mg g^{-1} at 298 K and 1 bar with a very high heat of adsorption ($-Q_{st}$) of 90 kJ mol^{-1}, and the desorption isotherm showed a slight hysteresis, indicating that the EN in CuBTTri has a strong affinity for CO_2.[63] Subsequently, MMEN was also immobilized to open the Cu site of CuBTTri using the same procedure. The gravimetric CO_2 uptake of MMEN-CuBTTri was 184 mg g^{-1} at 298 K and 1 bar, which is higher than those of CuBTTri and EN-CuBTTri. In addition, MMEN-CuBTTri exhibited higher CO_2 selectivity over N_2 (327) at 298 K in a mixture of 0.15 bar CO_2–0.75 bar N_2. The high adsorption and selectivity to CO_2 of MMEN-CuBTTri were ascribed to the large number of amines interacting with the CO_2 molecules because at least twice the number of amine moieties was present in MMEN-CuBTTri compared to EN-Cu-MOF.[110]

MOF-74(Mg) has a large number of open Mg sites in the framework. Mg-MOF-74 was desolvated by heating at 250 °C under vacuum for 5 h. Subsequently, EN was incorporated in the open Mg sites of MOF-74(Mg), which resulted in a reduction in the surface area from 1094 to 469 m^2 g^{-1} after functionalization. Although the CO_2 adsorption capacities of MOF-74(Mg) decreased (from 59.4 to 46.6 mg g^{-1}) by almost 20% after four cycles, EN-MOF-74(Mg) showed excellent stability (66.4, 66.0, 67.8, and 68.2 mg g^{-1} at 298 K under ultra-dilute CO_2 partial pressure conditions for the first, second, third, and fourth cycles, respectively) and could be fully regenerated under mild conditions (Ar purge at 110 °C for 3 h).[111] Hong *et al.* also measured the CO_2 adsorption of EN-MOF-74(Mg), which was found to be 201 mg g^{-1} at 298 K and 1 bar and 132 mg g^{-1} at 423 K and 1 bar with a $-Q_{st}$ of 49–51 kJ mol^{-1}, indicating that CO_2 was adsorbed *via* chemisorption, forming carbamic acid.[111] Recently, Hong and co-workers prepared DMEN-MOF-74(Mg) (DMEN = *N,N*-dimethylethylenediamine)[112] and homodiamine-grafted Mg_2(dondc) (dondc = 1,5-dioxido-2,6-naphthalenedicarboxylic acid)[113] frameworks for CO_2 adsorption. DMEN-MOF-74(Mg) took up CO_2 at temperatures and pressures below 348 K and 1 bar, similar to the adsorption properties of EN-MOF-74(Mg) and MMEN-MOF-74(Mg), both of which show no CO_2 adsorption above 393 K. This result indicates that there are weaker interactions between CO_2 and the amine groups in DMEN-MOF-74(Mg) than between CO_2 and the amines in EN-MOF-74(Mg) and MMEN-MOF-74(Mg).[112] Similarly, MMEN-Mg_2(dobpdc) (dobpdc = 4,4′-dioxido-3,3′-biphenyldicarboxylate) was prepared, in which the fully activated Mg_2(dobpdc) (activated at 420 °C for 65 min in vacuum conditions) was suspended in a hexane solution of MMEN to obtain MMEN-MOF-74(Mg). MMEN-Mg_2(dobpdc) shows a CO_2 uptake 15 times higher than that of Mg_2(dobpdc) at 0.39 mbar and 298 K.[114] Subsequently, various MMEN-functionalized M_2(dobpdc) (M = Mg, Mn, Fe, Co, Ni, and Zn) MOFs were prepared and applied for CO_2 adsorption. In the MMEN-M_2(dobpdc) series, one amine group from MMEN is strongly bound to the uncoordinated metal site. Powder XRD patterns of MMEN-Mn(dobpdc) were recorded at 100 K before and after adsorption of 5 mbar of CO_2, and the CO_2 adsorbed MMEN-Mn(dobpdc) showed large changes in the relative intensities of the diffraction peaks. To find the mechanism of CO_2 binding on MMEN-M_2(dobpdc), CO_2 adsorption isotherms for each material were measured at 298, 313, 323, and 348 K and 1 bar. All of the materials, except MMEN-Ni_2(dobpdc), showed normal Langmuir-type adsorption behavior and exhibited sharp isothermal steps shifted to higher pressures with increasing temperature. Carbamate was formed, and one oxygen atom in the carbamate had a close interaction with the N atom of a neighboring MMEN, resulting in a structure of chains of ammonium carbamate. The CO_2 adsorption isotherm at 298 K indicates a cooperative CO_2 adsorption mechanism. The CO_2 adsorption step position varied in the order Mg < Mn < Fe < Zn < Co at the measured temperatures, which matches the published series of MOF stabilities. The lack of a step in the adsorption isotherm for MMEN-Ni_2(dobpdc) even at high pressure was ascribed to the high stability of MMEN-Ni_2(dobpdc), which prevents carbamate insertion taking place under the conditions measured.[115] On the other hand, the CO_2 adsorption isotherms of Mg_2(dondc), EN-Mg_2(dondc), MMEN-Mg_2(dondc), and ppz-Mg_2(dondc) (ppz = piperazine) exhibited adsorption capacities of 282, 115, 181, and 138 mg g^{-1}, respectively, at 298 K and 1 bar, and all the amine-functionalized Mg_2(dondc) showed decreased adsorption capacities because of the reduced pore spaces in the framework.[113] Kortright *et al.* investigated the adsorption mechanism of CO_2 for MMEN-MOF-74(Mg) using *in situ* X-ray absorption spectroscopy and DFT calculations.[116] Chen *et al.* prepared the N_2H_4-grafted MOF-74(Mg) and compared its CO_2 adsorption at low-pressure conditions with that of MOF-74(Mg). The CO_2 uptake of N_2H_4-MOF-74(Mg) was significantly higher (201 mg g^{-1}) than that of MOF-74(Mg) (3.96 mg g^{-1}) at 0.4 mbar and 298 K. In addition, temperature swing adsorption for N_2H_4-MOF-74(Mg) revealed a working capacity of 154 mg g^{-1} between the adsorption in a 15 : 85 CO_2–N_2 (v/v) mixture at 313 K for 48 min and the desorption in a pure CO_2 flow at 413 K (optimized) for 54 min.[117] Recently, Hatton *et al.* reported the post-synthetic functionalization of MOF-74(Mg) with TEPA, and the resulting TEPA-MOF-74(Mg) showed a higher CO_2 adsorption (269 mg g^{-1}) than the MOF-74(Mg) at 298 K and 1 bar (234 mg g^{-1}).[118] The sterically-hindered secondary amine ppz was incorporated into the open Ni^{2+} sites of MOF-74(Ni), which exhibited a lower CO_2 uptake at 298 K and 1 bar but a higher CO_2 selectivity over N_2 at the pressure region of 0.1–0.15 bar compared to non-grafted MOF-74(Ni).[119] Wang *et al.* studied the effect of NH_4OH on the structure and CO_2 adsorption of nano-sized UiO-66. At 273 K, the CO_2 adsorption uptake of NH_4-UiO-66 at 0.2 bar was lower than that of bare UiO-66, but at 0.987 bar, NH_4-UiO-66 exhibited a slightly higher CO_2 uptake (379 mg g^{-1}) compared to UiO-66 (356 mg g^{-1}).[120]

3.3.3 Physical Incorporation of Amines into Unmodified MOFs

This is one of the most convenient methods to prepare amine-incorporated MOFs, because of the facile methodology and simple control over the amount of amine introduced. Polyamines are usually used for the incorporation. However, the polyamine chain length should be short because volatility and deep incorporation into the framework are desirable to generate small MOF pore sizes. In addition, the pore sizes of the MOFs should be sufficiently large and MOF crystals should be small for fast and facile polyamine incorporation. Chen and co-workers synthesized polyethylenimine (PEI)-impregnated MIL-101(Cr) adsorbents (see Figure 3.8). In the synthesis, the activated MIL-101(Cr) was added slowly to a methanol solution of PEI (different loadings) with stirring. The resulting gel was dried at room temperature overnight and the sample was treated at 110 °C under vacuum conditions for 12 h before CO_2 adsorption. After PEI impregnation, CO_2 uptake was enhanced dramatically at low pressures (185 mg g^{-1} at 298 K and 150 mg g^{-1} at 323 °C with 100% PEI-loading), and an ultrahigh selectivity for CO_2 over N_2 (770 at 298 K and 1200 at 323 °C) was achieved in a flue gas mixture with 0.15 bar CO_2 and 0.75 bar N_2.[121] In addition, dual amine-functionalized MIL-101(Cr)

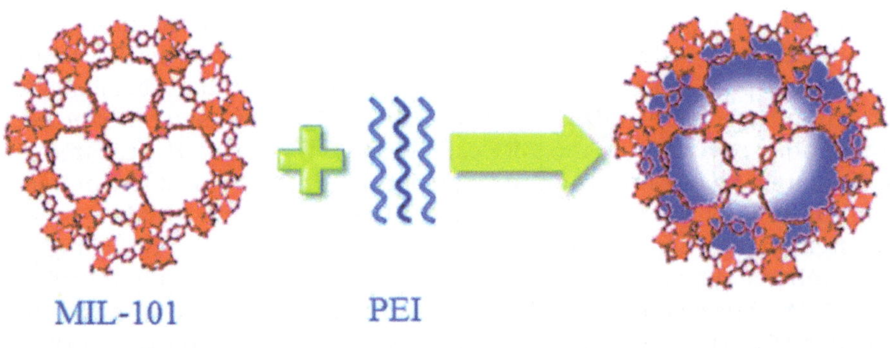

Figure 3.8 Schematic representation of the preparation of PEI-grafted MIL-101(Cr).

was prepared by the incorporation of different amounts of short length linear PEI (molecular weight of 300 g mol^{-1}) into the framework of the *in situ* amine-functionalized MIL-101 (NH_2-MIL-101(Cr)) for the adsorption of CO_2 and CH_4. The PEI-decorated NH_2-MIL-101(Cr) exhibited an outstanding CO_2 adsorption capacity at low pressures and a high CO_2 selectivity over CH_4.[122] A MIL-101(Cr) series incorporating PEIs with different average molecular weights (ranging from 300 to 10 000 g mol^{-1}) were prepared to identify the influence of PEI on the adsorption of CO_2. The adsorption results showed that the low-molecular-weight PEI/MIL-101(Cr) showed a better performance than others, probably because low-molecular-weight, short PEI chains are beneficial for their diffusion into MOFs.[123] Although the regeneration cost of the PEI/MOF adsorbent could be lower than that of a homogeneous amine solution, it is still expected to be high compared with that of *in situ* amine-functionalized MOFs.

The incorporation of amine groups into MOFs by physical impregnation and post-synthetic methods has some disadvantages such as the blocking of the OMSs, reduction in the surface area, blocking of pore windows, and reduction of the pore volume of the original MOFs, which often result in a decrease in the CO_2 adsorption capacity. In this respect, the *in situ* synthesis of amine-functionalized MOFs with OMSs using amine-containing ligands seems to be more effective for achieving CO_2 adsorption in the low-pressure region and high CO_2 selectivity over N_2 and CH_4 compared to other MOF modifications.

3.4 CO_2 Adsorption by Mixed-ligand-based MOFs

Mixed-ligand MOFs can be prepared using multiple ligands with different functionalities, sizes, shapes, and coordination features, and this can introduce new functionality and improve stability and textural properties. The different functionality created in the pore, adjustable pore size, and variable surface environment, coupled with the gate effect of a stable pillar ligand, can lead to synergistic CO_2 capture properties. Mixed-ligand-based MOFs for CCS applications can be classified into two groups based on the structural features of the mixed linkers (see Figure 3.9).

3.4.1 Pillared-layer Mixed-ligand MOFs (PL-MOFs)

Bharadwaj *et al.* synthesized a stable 3D MOF, {[Zn_3(2′,5′-dimethyl-*p*-terphenyl-4,4″-dicarboxylate)$_3$(DPB)$_{1:5}$]6DMF·H_2O}$_n$, pillared by 1,4-dipyridylbenzene (DPB) under solvothermal conditions. The ligand, which contains methyl groups, increases the hydrophobicity of the cavity of the MOF structure. The solvent molecules inside the cavity were removed at 120 °C, and the activated MOF was applied for CO_2, CH_4, H_2, and N_2 adsorption. The PL-MOF showed significant CO_2 adsorption (116 mg g^{-1}) at 273 K and 1 bar and moderate CO_2 selectivity over N_2 and CH_4 with an $-Q_{st}$ of around 21 kJ

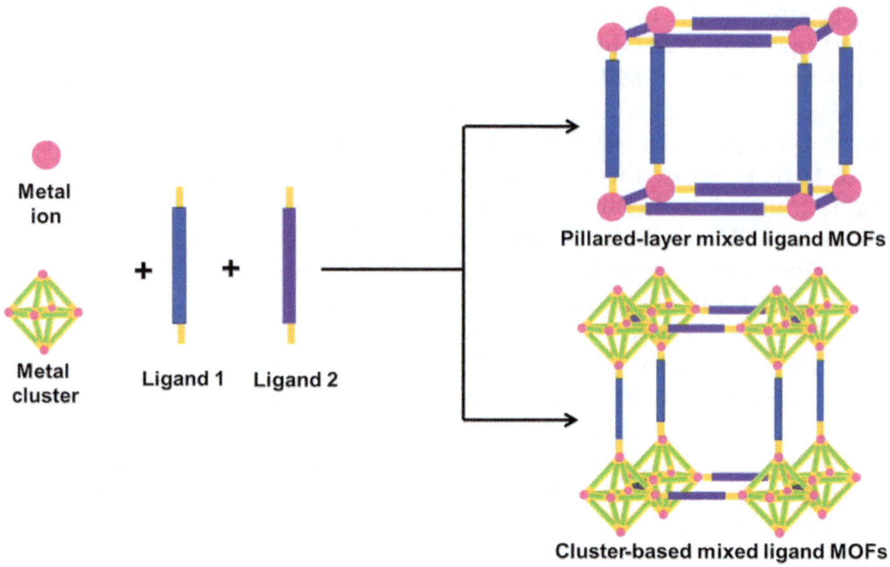

Figure 3.9 Types of mixed-ligand MOFs: (i) pillared-layered mixed-ligand MOFs and (ii) cluster-based mixed-ligand MOFs.

mol^{-1}, indicating a strong interaction of the CO_2 molecules with the framework walls.[139] Maldonado *et al.* synthesized two types of Cu-based pillared MOFs – Cu_2(pyrazine-2,3-dicarboxylate)$_2$-4,4′-bipyridine (CPL-2) and Cu_2(pyrazine-2,3-dicarboxylate)$_2$-1,2-di(4-pyridyl)ethylene (CPL-5). The CO_2 adsorption of CPL-2 at 298 K and 1 bar was slightly higher (72 mg g^{-1}) than that of CPL-5 (48 mg g^{-1}) because of the different structural properties of the pillared linkers. However, when the pressure was increased to 9 bar, CPL-5 adsorbed more CO_2 (*ca.* 224 mg g^{-1}) than CPL-2 (*ca.* 190 mg g^{-1}) because of the larger pore volume and flexible nature of the longer pillared linker, which enabled gallery expansion.[140] Farha *et al.* prepared two pillared MOFs with controllable pore size and functionality, TMU-34 and TMU-34(-2H), using (3,6-di(pyridin-4-yl)-1,4-dihydro-1,2,4,5-tetrazine) and (3,6-di(pyridin-4-yl)-1,2,4,5-tetrazine), respectively, for CO_2 adsorption and separation. TMU-34(-2H) showed higher CO_2 adsorption (*ca.* 83 mg g^{-1}) and $-Q_{st}$ (30 kJ mol^{-1}) than TMU-34 (CO_2 adsorption of *ca.* 54 mg g^{-1} and $-Q_{st}$ = 23 kJ mol^{-1}). In addition, the CO_2–N_2 selectivity, which was estimated by the ideal adsorption solution theory (IAST), of TMU-34(-2H) was higher (29.5) at low pressures compared to that of TMU-34 (15.5), increasing to 33.1 at high pressure.[141] The anionic sulfonate group-functionalized surface of [$Zn_3(OH)_3$(2-stp)(bpy)$_{1.5}$(H_2O)], which comprises 2-sulfonylterephthalate (2-stp) and 4,4′-bipyridyl (bpy) linkers, has also been reported. The CO_2 and CH_3OH sorption confirmed that the surface with the polarized sulfonyl group made the strong accommodation of gases *via* acid–base interactions possible.[142] Similarly, [Zn_2(sdb)$_2$(bpy)] (sdb = 4,4′-sulfonyldibenzoate) with low flexibility has been reported for gas adsorption.[143] The adsorption caused a structural change when gases were adsorbed at 195 and 298 K. The MOF showed a higher uptake of CO_2 (54 mg g^{-1}) compared to ethylene and ethane but no methane uptake at 195 K and 1 bar. A similar gas sorption pattern ($CO_2 > C_2H_4 > C_2H_6 \gg CH_4$) was also obtained at 298 K and 9 bar. These results were ascribed to the dual responsive gate opening by the guest gas molecules caused by the pillar motion. The flexible PL-MOF [Zn(ndc)(bpy)] (ndc = 2,7-naphthalenedicarboxylate) exhibited selective CO_2 adsorption properties. The framework adsorbed CO_2 (156 mg g^{-1}) at 195 K but did not adsorb N_2 and O_2 gases at 77 K and 1 bar. Similarly, the framework adsorbed negligible amounts of O_2 and N_2 but adsorbed CO_2 in the mixed component adsorption of CO_2, O_2, and N_2 with a ratio of 1 : 1 : 1 and at 10 bar, suggesting that the framework has high CO_2–N_2 and CO_2–O_2 selectivities with good recyclability.[144] Bu *et al.* prepared two zinc-containing PL-MOFs based on 1,1′-biphenyl-2,2′,6,6′-tetracarboxylic acid (H_4bpta) and two different bipyridine pillar ligands, {[Zn_4(bpta)2(4-pna-)$_2$(H_2O)$_2$]·4DMF·3H_2O}$_n$ and {[Zn_2(bpta)(bpy-ea)-(H_2O)]·2DMF·H_2O}*n* (where 4-pna = *N*-(4-pyridyl)-isonicotinamide and bpy-ea = 1,2-bis(4-pyridyl)ethane), and investigated their CO_2 adsorption properties.[145] The CO_2 adsorption capacities of these two PL-MOFs were 118 and 40 mg g^{-1} at 273 K and 1 bar for the complex of 4-pna and bpy-ea-based PL-MOFs, respectively. The IAST selectivity of CO_2 over CH_4 for the 4-pna-based PL-MOF decreased from 23 to 15 in the range of 0–1 bar; however, these values are still higher than those of Cu-BTC[146] and NJU-Bai3[147] under similar conditions. On the other hand, the CO_2–CH_4 selectivity of the bpy-ea-based PL-MOF increased from 13 to 19 owing to the flexible nature of the bpy-ea ligand in the PL-MOF. Later, two water-tolerant PL-MOFs, $M_2(BDC)_2$(DABCO) (M = Zn and Ni) were synthesized for CO_2 adsorption.[148] CO_2 adsorption capacities of 602 and 550 mg g^{-1} were measured for $Zn_2(BDC)_2$(DABCO) and $Ni_2(BDC)_2$(DABCO), respectively, at 298 K and 15 bar, and both PL MOFs show higher CO_2–N_2 selectivities than CO_2–CH_4 selectivities, which decreased gradually as the gas pressure increased. In addition, the MOF structures were highly stable even after O_2 and 30% relative humidity (RH) water vapor sorption at 298 K, but collapsed after 60% RH water vapor sorption at 298 K.

3.4.2 Cluster-based Mixed-ligand MOFs

The structures of metal clusters and two different organic linkers have attracted interest recently for the storage and release of small gas molecules because of their intriguing structures and different functionalities. Bordiga *et al.* prepared a series of mixed-ligand UiO-66 (different ABDC–BDC molar ratios were used: 0, 0.25, 0.50, 0.75, and 1.00) *via* low- (100 °C) and high-temperature (220 °C) synthesis for CO_2 adsorption to determine the role of the amine groups in the MOF materials.[149] Figure 3.10 shows that almost the same amount of CO_2 adsorption occurs at 1 bar for the MOFs prepared by both methods using 50% NH_2 content. In addition, a stronger affinity of the CO_2 molecule for amines was exhibited at greater

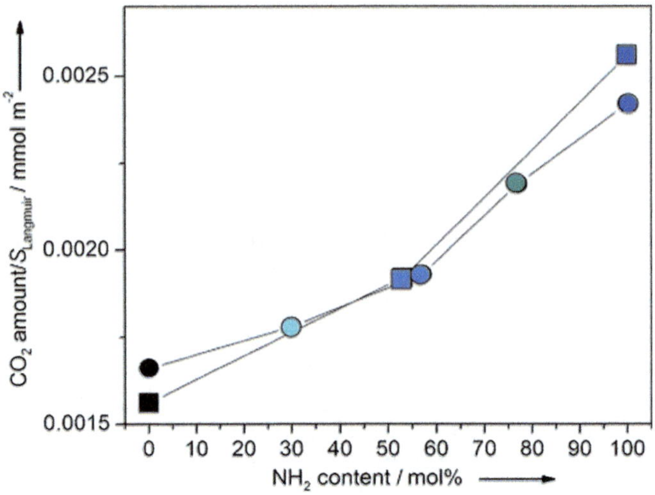

Figure 3.10 CO_2 capacities at 1 bar *vs.* NH_2 content. Circles: low-temperature mixed-ligand MOFs; squares: high-temperature mixed-ligand MOFs. Reproduced from ref. 149 with permission from John Wiley and Sons, © 2014 WILEY-VCH Verlag GmbH & Co. KGaA, Weinheim.

than 50% amine content in both cases, owing to the increased probability of the NH_2 sites being positioned next to each other at higher ABDC loadings. Wang *et al.* also prepared the mixed-ligand UiO-66-NO_2-N (N = $-NH_2$, $-(OH)_2$, $-(COOH)_2$) for enhanced CO_2 adsorption and selectivity for N_2. The methanol or chloroform activated UiO-66-NO_2-NH_2 showed much higher CO_2 adsorption compared to other samples, and the capacities were around 170 and 145 mg g^{-1} at 273 K and 0.76 bar for chloroform-activated and methanol-activated samples, respectively. In addition, a higher CO_2–N_2 selectivity (65) was obtained for the chloroform-activated UiO-66-NO_2-NH_2 compared to the other samples.[150]

3.5 CO_2 Adsorption by Flexible Ligand-based MOFs (FL-MOFs)

The unique properties of the more recent MOF materials stem from their flexible nature and may provide unique opportunities for their application in the separation of adsorbate mixtures.[151,152] It is, however, challenging to synthesize FL-MOFs because the flexible ligands with different configurations, which are a consequence of single bond rotation, can lead to different MOF structures. In addition, most of the FL-MOFs are fragile and lose their porosity after the evaporation of solvent molecules inside because of the ligand flexibility. As a consequence, the construction of stable FL-MOFs is more highly dependent on the synthetic conditions such as the temperature, pH, pressure, and time.[153] Although flexible chiral organic linkers, such as amino acids, peptides, and their derivatives, can be used for the construction of FL-MOFs with chiral centers, applicable for enantioselective separation and asymmetric catalysis, FL-MOFs may also offer "breathing" capability and the adaptive recognition of guest molecules.[154] For gas adsorption and separation, researchers have mostly focused on increasing the free pore volume, maintaining porosity after removal of solvent, and increasing the gas binding affinity of FL-MOFs. Representative CO_2 adsorption results for FL-MOFs are summarized in Table 3.3.

3.5.1 Increasing the Free Pore Volume in FL-MOFs

Both experimental and theoretical studies have indicated that a high surface area and large pore volume in MOFs are favorable for CO_2 adsorption.[155] The most common strategy to increase the pore volume is the use of long organic ligands in MOF synthesis, which results in FL-MOFs. For example, a series of non-interpenetrated mesoporous isoreticular FL-MOFs were prepared from 4,4′-(2,2-bis((4-carboxy-2-methoxyphenoxy)methyl)propane-1,3-diyl)bis(oxy)bis(3-methoxybenzoic acid)/3,3′-(4,4′-(2,2-bis((4-(2-carboxyvinyl)-2-methoxyphenoxy)methyl)propane-1,3-diyl)bis(oxy)bis-(3-methoxy-4,1-phenylene))-diacrylic acid/6,6′-(2,2-bis((6-carboxynaphthalen-2-yloxy)-methyl)propane-1,3-diyl)bis(oxy)di-2-naphthoic acid and $Zn(NO_3)_2{\cdot}6H_2O$. The resulting FL-MOFs possess quadrangular openings and two types of micro- and mesopore cages, the sizes of which increase gradually with the extension of the ligand in the framework. Each microporous cage is surrounded by eight mesoporous cages, resulting in very high porosities of 75.7%, 83.7%, and 84.6% in these three FL-MOFs. Very low N_2 uptakes were obtained at 77 K for these FL-MOFs because of the distorted framework after the removal of the internal solvent, whereas high CO_2 uptakes of up to 166 mg g^{-1} were obtained at 298 K and 1 bar with high CO_2 selectivity over N_2.[156] In addition, Zaworotko *et al.* used BTC to build $[Zn_3(BTC)_2\{Cr_3O(isonic)6(H_2O)_2(OH)\}]{\cdot}x$DMF (isonic = isonicotinic acid), which has a BET surface area of 1671 $m^2\ g^{-1}$ and a free pore volume fraction of more than 80% (calculated using the PLATON program). The resulting MOF exhibited a CO_2 uptake of 211 mg g^{-1} at 273 K and 1 bar and $-Q_{st}$ of 26 kJ mol^{-1} for CO_2.[157] Demiral and co-workers constructed a series of Zn-based FL-MOFs using 3,3′,5,5′-azobenzenetetracarboxylate and 1,2-, 1,3-, and 1,4-bis(imidazol-1-ylmethyl)benzenes. These MOFs are almost nonporous but show low CO_2 and N_2 adsorption because of the flexible nature of the MOF after activation.[158] Agarwal *et al.* synthesized flexible $\{[Ni_3(TBIB)_2(BTC)_2(H_2O)_6].5C_2H_5OH.9H_2O\}_n$ (TBIB = 1,3,5-tri(1*H*-benzo[*d*]imidazol-1-yl)benzene) for gas adsorption. There was no N_2 uptake at 77 K and a CO_2 uptake of 106 mg g^{-1} was observed at 273 K and 1 bar.[159]

Table 3.3 Flexible MOFs for CO_2 adsorption.

FL-MOF[a]	BET area ($m^2\ g^{-1}$)	Temp. (K)	Pressure (bar)	CO_2 uptakes ($mg\ g^{-1}$)	$-Q_{st}$ (kJ mol^{-1})	Ref.
$Zn_4O(L2)_{1.5}$	75.7	273/298	1	68/58	—	156
$Zn_4O(L3)_{1.5}$	83.7	273/298	1	193/130	—	156
$Zn_4O(L4)_{1.5}$	84.6	273/298	1	225/165	—	156
$[Zn_3(BTC)_2\{Cr_3O(isonic)_6(H_2O)_2(OH)\}]\cdot xDMF$	80	273	1	211	26	157
$\{[Zn_2(\mu_4\text{-}ao_2btc)(\mu\text{-}pbix)_2]\cdot 2DMF\cdot 8H_2O\}_n$	—	273	1	38	—	158
$\{[Ni_3(TBIB)_2(BTC)_2(H_2O)_6]5C_2H_5OH.9H_2O\}_n$	—	273/298	1	106/77	—	159
$\{[Co_2(\mu_8\text{-}abtc)(betib)]\cdot H_2O\}_n$	—	273	1	104	—	174
PCN-26ac	1733	273/298	0.8	210/351	—	161
$Cu_2(DBIP)\text{-}(H_2O)_2$	1773	273	0.95	334	—	163
$(InL5)\cdot(Me_2NH_2)\cdot 3DMA\cdot 2H_2O$	326	195	1	230	—	166
$[Cu_{24}(TPBTM^{6-})8\text{-}(H_2O)_{24}]_3$	3160	298	20	1035	26.3	167
$\{Cu_3(btb^{6-})\}_n$	3288	273	20	1035	—	168
$\{Cu_3(tatb^{6-})\}_n$	3360	273	20	1034	—	168
NJU-Bai3	2690	273	20	973	36.5	147
HNUST-1	1400	273	1	303	31.2	169
(Cu-tdpat)	1938	273	0.1	62	42.2	173
(Cu-tdpat)	—	—	1	445	—	173
$\{[Zn(btz)]\cdot DMF\cdot 0.5H_2O\}n$	1151	273	1	356	31.2	175
SNU- 100′	814	298	1	141	29.3	176
SNU- 100′-Li	924	298	1	153	35.7	176
SNU- 100′-Mg	259	298	1	151	36.3	176
SNU- 100′-Ca	235	298	1	151	37.4	176
SNU- 100′-Co	1000	298	1	168	36.0	176
SNU- 100′-Ni	982	298	1	166	34.5	176
Zn-DDQ	—	273	0.8	128	—	177
Cu-DDQ	—	—	—	261	—	177
SNU-9	—	195	0.6	426	—	178
JLU-Liu3	325	195	0.76	650	—	179
JLU-Liu4	306	195	0.76	645	—	179
FMOF-1	—	298	55	271	—	180
Cu-tdpat	—	298	1	237	—	181
$[Cu(Hoxonic)(bpy)_{0.5}]_n$	—	273	25	110	—	182

[a]L2 = 4,4′-(2,2-bis((4-carboxy-2-methoxyphenoxy)methyl)propane-1,3-diyl)bis(oxy)bis(3-methoxybenzoic acid); L3 = 3,3′-(4,4′-(2,2-bis((4-(2-carboxyvinyl)-2-methoxyphenoxy)methyl) propane-1,3-diyl)bis(oxy)bis(3-methoxy-4,1-phenylene))diacrylic acid; L4 = 6,6′-(2,2-bis((6-carboxynaphthalen-2-yloxy)methyl)propane-1,3-diyl)bis(oxy)di-2-naphthoic acid; pbix = pbix = 1,4-bis(imidazol-1-ylmethyl)benzene; TBIB = 1,3,5-tri(1H-benzo[d]imidazol-1-yl)benzene; abtc = 3,3′,5,5′-azobenzenetetracarboxylate; PCN = porous coordination network; DBIP = 5-(3,5-dicarboxybenzyloxy)isophthalic acid; L5 = 5-(3,5-dicarb oxybenzyloxy)isophthalic acid; TPBTM = N,N′,N″-tris(isophthalyl)-1,3,5- benzenetricarboxamide; HNUST = Hunan University of Science and Technology; btz = 1,5-bis(5-tetrazolo)-3-oxapentane; DDQ = N,N′-dibenzoic acid2,3-diaminoquinoxaline; JLU-Liu = Jilin University Liu-group; Hoxonic = 4,6-dihydroxy-1,3,5-triazine-2-carboxylic acid

3.5.2 Maintaining Porosity in FL-MOFs After the Removal of the Solvent

The BET surface areas and gas uptake capacities of some FL-MOFs have been found to be significantly lower than the estimates of computational studies because of partial channel collapse upon solvent removal. Thus, it is essential to maintain the porosity of FL-MOFs after the complete removal of the solvent molecules for gas adsorption and storage applications. This is a challenging task for FL-MOFs, considering the long flexible ligands used. Apparently, mild activation by solvent removal is an effective strategy to reduce the structural collapse in FL-MOFs.

The synthesis of $Cu_4(H_2O)_4(TDM){\cdot}xS$ (also referred to as PCN-26·*x*S; S = solvent) using a flexible tetrapodal octacarboxylate (H_8TDM) ligand has been reported.[160,161] The as-synthesized PCN-26 was immersed in methanol for 24 h, followed by dichloromethane, allowing the exchange of the DMF molecules. Subsequently, the sample was activated at room temperature overnight under a high vacuum. The mildly activated PCN-26 (PCN-26ac) had two octahedral and cuboctahedral cages, which led to a large solvent accessible volume (64.0%) after the removal of the solvent from the channels. PCN-26ac was found to have a BET surface area of 1733 $m^2\ g^{-1}$, pore volume of 0.84 $cm^3\ g^{-1}$, and CO_2 uptake of 351 mg g^{-1} (at 298 K and 0.8 bar). In contrast, the PCN-26 sample heated directly at 120 °C was found to have a lower surface area (1115 $m^2\ g^{-1}$) and pore volume (0.61 $cm^3\ g^{-1}$).[162] Liang *et al.* synthesized the *nbo* topological copper-based FL-MOF ($Cu_2(DBIP)$-$(H_2O)_2$) using the 5-(3,5-dicarboxybenzyloxy)isophthalic acid (H_4DBIP) ligand.[163] The sample was activated by solvent exchange and heating at 65 °C, and the activated sample had a BET surface area of 1773 $m^2\ g^{-1}$ with a narrow pore size of 1.1 nm, CO_2 uptake of 334 mg g^{-1} (at 0.95 bar and 273 K), and a CO_2–N_2 selectivity of 20.6. However, the directly heated sample (100 °C) had a BET surface area of only 232 $m^2\ g^{-1}$.[162]

Supercritical CO_2 drying was also used to activate FL-MOFs to produce stable MOFs (without collapse) with high surface areas and porosities.[164] For example, Du *et al.* prepared an FL-MOF of $In_2(TDM)[NH_2(CH_3)_2]$ based on a H_8TDM ligand. The framework, after activation using supercritical CO_2 drying, exhibited a BET surface area of 1555 $m^2\ g^{-1}$ and a H_2 uptake of 14.9 mg g^{-1} at 77 K and 1 bar, whereas the surface area was reduced to 752 $m^2\ g^{-1}$ after direct activation at 80 °C.[165]

An alternative strategy to enhance the mechanical and thermal stabilities of the FL-MOFs is the impregnation of tetrabutylammonium cation guests into the framework, which can prevent framework collapse after the activation. Cao *et al.* synthesized two anionic FL-MOFs based on multicarboxylate linkers and indium ions. Whereas the fresh FL-MOFs collapsed completely (even at 100 °C) after the removal of solvent molecules, framework stability was maintained up to 300 °C after the impregnation of tetrabutylammonium cations, and the CO_2 uptake at 195 K and 1 bar was high.[166]

3.5.3 Increasing the Gas Binding Affinity in FL-MOFs

To improve the gas adsorption and separation of FL-MOFs, strategies utilizing OMSs, pore size effects, the introduction of metal ions, the impregnation of amines, and nitrogen-containing organic linkers have been attempted. These strategies aim to increase the gas binding affinity. The addition of nitrogen-containing organic linkers is the most frequently attempted and facile method to improve the gas adsorption and separation properties. In particular, flexible, Lewis basic, nitrogen-containing amine-based ligands are mostly used in FL-MOFs. Bai and co-workers synthesized several amide-based FL-MOFs for CO_2 adsorption.[167–169] The *rht*-type FL-MOF was built from Cu or Co paddlewheel-type clusters and a flexible amide-containing hexacarboxylate ligand, which exhibits the same topology as the PCN-61 series.[170,171] The desolvated FL-MOF has a large free pore volume of *ca.* 74% (calculated using the PLATON program), BET surface area of 3160 $m^2\ g^{-1}$, a CO_2 uptake of 1035 mg g^{-1} at 298 K and 20 bar, and CO_2 selectivity over N_2 of 22/33 at 1/20 bar.[167] To improve the CO_2 adsorption capacity further, $\{Cu_3(btb^{6-})\}_n$ and $\{Cu_3(tatb^{6-})\}_n$ were synthesized using hexacarboxylate linkers of H_6btb and H_6tatb (ligand with a triazine unit), respectively. Both FL-MOFs showed almost the same adsorption (up to 1570 mg g^{-1} at 273 K and 20 bar).[168] The amide-inserted *agw*-type framework (NJU-Bai3: Nanjing University Bai group) was prepared *via* the solvothermal reaction of 5-(4-carboxybenzoyl-amino)-isophthalic acid and $CuCl_2$. NJU-Bai3 contains three types of cages (bowl-like, hexagonal bipyramidal, and trigonal bipyramidal cages) with closely decorated amide units that are directly exposed to each individual cavity. The fully desolvated NJU-Bai3 has a solvent accessible pore volume of *ca.* 76.9% (calculated using the PLATON program). NJU-Bai3 has a CO_2 adsorption capacity of 973 mg g^{-1} at 273 K and 20 bar and a selectivity reaching 60.8 for CO_2–N_2 and 46.6 for CO_2–CH_4.[147] An expanded microporous *nbo*-type Cu_2(BDPT) MOF structure designated as HNUST-1 (Hunan University

Table 3.4 Interpenetrated MOFs for CO_2 capture.

Interpenetrated MOFs[a]	BET area ($m^2\ g^{-1}$)	Temp. (K)	Pressure (bar)	CO_2 adsorption (mg g^{-1})	$-Q_{st}$ (kJ mol^{-1})	Ref.
Zn_2(tcom)(4,4-bpy)	1150	298	1	50	—	184
SIFSIX-2-Cu-i	735	298	1	238	31.9	185
CuTATB-60	3811	298	1	189	35	186
SUMOF-2	1167	273	1	187	—	187
SUMOF-3	1167	273	1	151	—	187
SUMOF-4	1612	273	1	158	—	187
NJU-Bai9	4258	298	40	1070	—	188
PCN-5	225[b]	195	0.7	210	—	189
$[Ni(bpe)_2(N(CN)_2)](N(CN)_2)$	—	195	1	68	—	190
$Ni_2(cyclam)_2$(mtb)	141	195	1	110	—	191

[a]tcom = tetrakis[4-(carboxyphenyl)oxamethyl]methane; bpe = 1,2-bis(4-pyridyl)ethane; cyclam = 1,4,8,11-tetraazacyclotetradecane; mtb = methanetetrabenzoate.
[b]Langmuir surface area.

of Science and Technology) was constructed from bis(3,5-dicarboxyphenyl) terephthalamide (H_4BDPT) and $Cu(NO_3)_2.3H_2O$ in a mixture of DMF, H_2O, ethanol, and HNO_3. This MOF is made of shuttle-shaped and spherical pores of *ca.* 1.1 and 1.6 nm (1:1 ratio), with a solvent accessible pore volume of 71% (calculated by the PLATON program).[172] HNUST-1 exhibits a high CO_2 uptake (303 mg g^{-1} at 1 bar) and CO_2–N_2 selectivity (39.8) at 273 K.[169] Shi *et al.* constructed examples of FL-MOFs containing both Lewis base sites and OMSs (Cu-tdpat) based on the hexacarboxylate linker 2,4,6-tris(3,5-dicarboxylphenylamino)-1,3,5-triazine (H_6tdpat). Cu-tdpat exhibited a high CO_2 adsorption capacity at both low and high pressures (445 mg g^{-1} at 273 K and 1 bar and 310 v/v, 298 K and 48 bar) compared with other leading MOF materials.[173]

3.6 CO_2 Adsorption by MOFs with Interpenetration

The use of long organic linkers frequently results in interpenetration/catenation in the MOF structure, which reduces the solvent accessibility in the framework and produces elongated pores advantageous for flue gas adsorption and the separation of CO_2 over N_2 (Table 3.4). Zhou *et al.* prepared octahedral crystals of an interpenetrated MOF (PCN-6′) by the solvothermal reaction of $Cu(NO_3)_2$ and 4,4′,4″-triazine-2,4,6-triyl-tribenzoic acid in the presence of oxalic acid. The resulting PCN-6′ exhibited a lower Langmuir surface area and pore volume compared with the non-interpenetrated PCN-6. However, PCN-6′ has a solvent accessible volume of 86% (calculated using the PLATON program) and more microporous channels than PCN-6 (74% solvent accessible volume), thus making it more effective for small molecule adsorption.[183] Thallapally *et al.* synthesized interpenetrated Zn-MOF by the self-assembly of tetrakis[4-(carboxyphenyl)oxamethyl]methane with $Zn(NO_3)_2.6H_2O$ and bpy in DMF. The interpenetrated Zn-MOF was made of a pair of identical *PtS* nets that were mutually interpenetrated, and the pores in the interpenetrated Zn-MOF were filled partially with bipyridine. The remaining channels were filled with water and DMF molecules. The CO_2 adsorptions of the interpenetrated Zn-MOF were 50 and 312 mg g^{-1} at 298 K and 1 and 30 bar, respectively. In addition, the breathing behavior of the interpenetrated Zn-MOF at the time of solvent removal and CO_2 insertion was detected.[184] Nugent *et al.* synthesized rod-shaped crystals of $[Cu(dpa)_2(SiF_6)]_n$ (also referred to as SIFSIX-2-Cu) from 4,4′-dipyridylacetylene (dpa) and $CuSiF_6$. The interpenetrated polymorph SIFSIX-2-Cu-i is composed of doubly interpenetrated nets of SIFSIX-2-Cu (see Figure 3.11). The SIFSIX-2-Cu-i surface area and pore size (735 m^2 g^{-1} and 0.5 nm, respectively) are smaller than those (3140 m^2 g^{-1} and 1.3 nm, respectively) of SIFSIX-2-Cu. However, SIFSIX-2-Cu-i exhibited the higher CO_2 uptake and $-Q_{st}$ (238 mg g^{-1} and 31.9 kJ mol^{-1}) compared to the parent SIFSIX-2-Cu (81 mg g^{-1} and 22.0 kJ mol^{-1}) at 1 bar and 298 K. In addition, the CO_2 selectivity over N_2 for SIFSIX-2-Cu and SIFSIX-2-Cu-i were 13.7 and 140, respectively.[185] Some interpenetrated MOFs show an increased

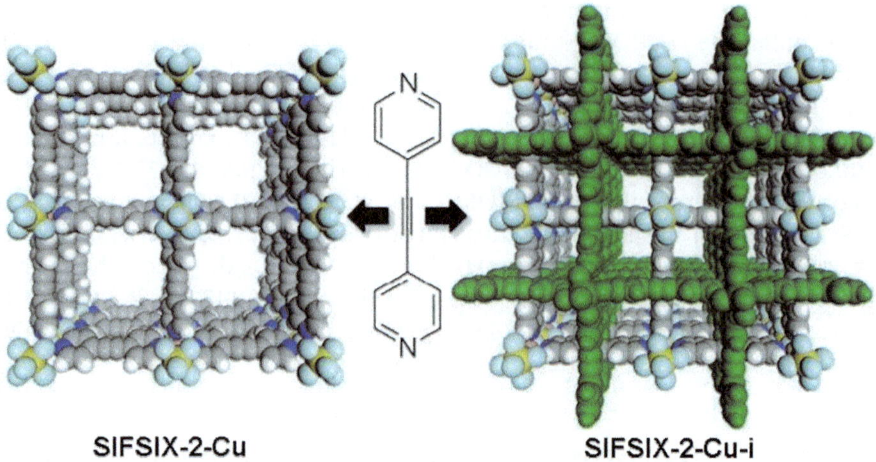

Figure 3.11 Schematic representations of SIFSIX-2-Cu and SIFSIX-2-Cu-i. Reprinted by permission from Macmillan Publishers Ltd: *Nature* (P. Nugent, Y. Belmabkhout, S. D. Burd, A. J. Cairns, R. Luebke, K. Forrest, T. Pham, S. Ma, B. Space, L. Wojtas, M. Eddaoudi and M. J. Zaworotko, *Nature*, 2013, **495**, 80), Copyright 2013.

surface area and pore volume compared to their parent MOFs and have been applied for gas adsorption. For example, interpenetrated and non-interpenetrated CuTATB-n were synthesized *via* the sonochemical reaction of 4,4′,4″-triazine-2,4,6-triyl-tribenzoic acid (TATB) and $Cu(NO_3)_2$. The interpenetrated CuTATB-60 showed CO_2 adsorptions of 189 and 1171 mg g^{-1} at 1 and 30 bar, respectively, and 298 K, which is higher than that of non-interpenetrated CuTATB-30 because of the higher surface area and pore volume of CuTATB-60.[186] Zou and co-workers synthesized SUMOF-2 for CO_2 adsorption. This MOF is composed of Zn clusters and BDC. SUMOF-2 is an interpenetrated analog of MOF-5 and shows a higher CO_2 uptake (*ca.* 193 mg g^{-1}) than MOF-5 (*ca.* 66.1 mg g^{-1}) because of the increasing electric field gradients arising from the interpenetration in the MOF.[187] Bai *et al.* constructed the double interpenetrated [Zn(BCP)·H_2O]·3.5H_2O·DMF (also referred to as NJU-Bai9) from $Zn(NO_3)_2$·6H_2O and 3,5-bis[4-carboxyanilino-carbonyl]pyridine(BCP), which was stabilized *via* the intermolecular π–π interactions of the phenyl-pyridyl rings in the cage.[188] The free pore volume and surface area are 78.1% (calculated by the PLATON program) and 4258 m^2 g^{-1}, respectively, for desolvated NJU-Bai9. The CO_2 uptake of NJU-Bai9 is 1070 mg g^{-1} at 40 bar and 298 K, which is the highest reported for interpenetrated MOFs.[30]

3.7 CO_2 Adsorption by Zeolitic Imidazolate Frameworks (ZIFs)

ZIFs are a subfamily of MOFs and are composed of metal ions, such as Zn or Co, connected by different imidazolate linkers. The topologies of ZIFs resemble those of zeolites. The primary building units of ZIFs have a T–Im–T structure, where T is a tetrahedrally coordinated metal ion and Im is imidazolate or its derivatives. The T–Im–T bonding angle is 145°, which is similar to the Si–O–Si angle in zeolites. The ZIF structure is governed by the Im linkers. ZIFs have several standard zeolite topologies (see Figure 3.12), including *sod* (ZIF-8), *ana* (ZIF-14), *gme* (ZIF-68), *rho* (ZIF-71), *poz* (ZIF-95), and *moz* (ZIF-100), depending on the synthetic method and imidazole linkers used.[192] ZIFs are potential candidates for gas adsorption/separation because of their high nitrogen contents and small pores. ZIFs also exhibit exceptionally high thermal and chemical stabilities and high resistance to water and alkaline conditions compared to other types of MOFs and zeolites.[193] Table 3.5 presents the representative ZIF materials reported to date for CO_2 adsorption. Despite the structural diversity of ZIFs, ZIF-8 is most commonly investigated for adsorption applications because of its commercial availability (Basolite Z1200 from Sigma Aldrich) and easy/reproducible synthesis. ZIF-8 (2-methylimidazole linker) has a *sod* topology formed of four- and six-membered ring ZnN_4 clusters with internal cavities of 1.16 nm in diameter connected by 0.34 nm windows.[193] The CO_2 adsorption capacity of ZIF-8 is only 24 mg g^{-1} at 298 K and 1 bar,[194] but it adsorbs 400 mg g^{-1} CO_2 at 303 K and 45 bar, higher than ZIF-71 (4,5-dichloroimidazole linker) and ZIF-14 (2-ethylimidazole linker).[195] The results of using ZIF-8 for CO_2 adsorption at low pressures indicate that ZIF-8 can adsorb significantly less CO_2 and has moderate selectivity compared to zeolite 13X and other microporous materials.[196] To improve the CO_2 adsorption capacity, a series of functionalized linkers were used in other ZIFs (ZIF-68, -69, −70, −78, −79, −81, and −82) with different pore diameters ranging from 7.1 to 15.9 Å. The functionalized ZIFs have been

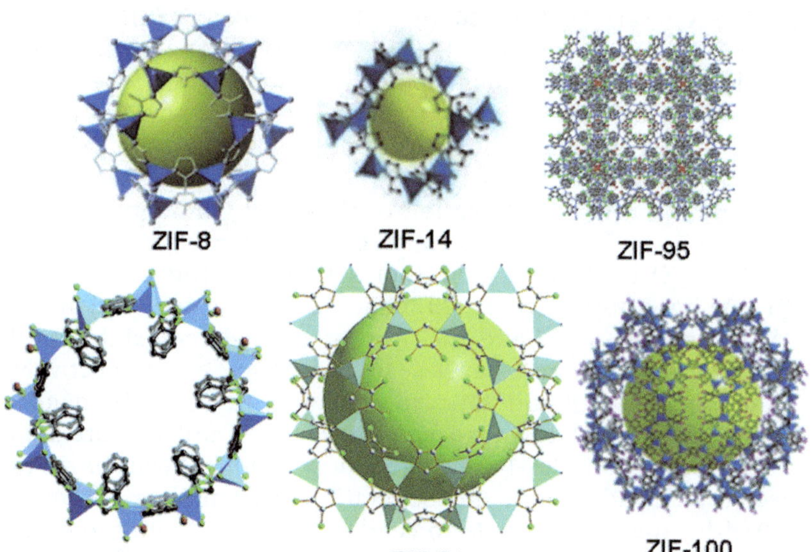

Figure 3.12 Representative ZIF structures.

Table 3.5 Representative ZIFs for CO_2 adsorption.

ZIF-n	CO_2 capture (mg g^{-1})	Selectivity		Conditions		Ref.
		CO_2/N_2	CO_2/N_2	Temp. (K)	Pressure (bar)	
ZIF-8	24	11.0^a		298	1	194
ZIF-8	400	—	5.6^b	303	45	195
ZIF-14	220	—	4.5^b	303	45	195
ZIF-68	73	18.7^c	5.0^d	298	1	197
ZIF-69	79	19.9^c	5.1^d	298	1	197
ZIF-70	106	17.3^c	5.2^d	298	1	197
ZIF-71	356	—	13^b	303	45	195
ZIF-78	100	50.1^c	10.6^d	298	1	197
ZIF-79	65	23.2^c	5.4^d	298	1	197
ZIF-81	74	23.8^c	5.7^d	298	1	197
ZIF-82	102	35.3^c	9.6^d	298	1	197
ZIF-95	38	18.0^c	4.3^d	298	1	197
ZIF-95	40	18.0^e	4.3^f	298	1	199
ZIF-100	63	25.0^c	5.9^d	298	1	197
ZIF-100	44	25.0^e	5.9^f	298	1	199

[a]Feed flow rate of 60 mL min^{-1} of CO_2, 40 mL min^{-1} of N_2, and 30 mL min^{-1} of He.
[b]15 mol% CO_2, 85 mol% CH_4 feed mixture.
[c]CO_2/N_2 (50 : 50 v/v) gas mixtures.
[d]CO_2/CH_4 (50 : 50 v/v) gas mixtures.
[e]50 mol% CO_2, 50 mol% N_2 feed mixture.
[f]50 mol% CO_2, 50 mol% CH_4 feed mixture.

studied for CO_2, CH_4, N_2, and O_2 adsorption and compared with that by BPL carbon, in which the imidazole linker functionality was altered from polar (–NO_2, ZIF-78; –CN, ZIF-82) to nonpolar (–CH_3, ZIF-79). It was found that the CO_2 uptake values varied widely in the following order at 1 bar and 298 K: –NO_2 (ZIF-78) (*ca.* 120 mg g^{-1}) > –CN, -Br, –Cl (ZIF-82, –81, –69) > -C_6H_6, –C_7H_8 (ZIF-68, –79) > –H (ZIF-70) > BPL carbon (*ca.* 41 mg g^{-1}), which indicates the greater quadrupole attraction between the polar functional groups in ZIFs and CO_2. In addition, the CO_2 uptake of ZIF-78 was four times higher than CH_4 and 8–10 times higher than for O_2 and N_2 at 1 bar 298 K.[197] Combined experimental and computational studies have also indicated that the modification of the dipole moment of the linker can lead to a 5–7-fold improvement in CO_2 selectivity for CO_2–CH_4, CO_2–N_2, and CO_2–CO mixtures.[198] The thermally stable (up to 773 K) ZIF-95 and ZIF-100 have huge cavities (24.0 and 35.6 Å, respectively) and highly constricted windows (3.65 and 3.35 Å, respectively) and exhibit moderate CO_2 capture capacities at 1 bar (*ca.* 40 and 44 mg g^{-1}, respectively) and CO_2–N_2 selectivities (18 and 25, respectively) because of the quadrupole interactions of CO_2 with the nitrogen atoms in ZIFs and 3 Å pore diameter. Both ZIFs exhibit completely reversible adsorption.[199] Nair *et al.* further improved the CO_2 adsorption and separation by (i) making a mixed-linker ZIF-8-90 with 2-methylimidazole and carboxaldehyde-2-imidazole (used for ZIF-8 and ZIF-90, respectively) and (ii) PSM of the carbonyl groups in the mixed-linker ZIF-8-90 material using ethylenediamine (ZIF-EN). These two materials, ZIF-8-90 and ZIF-EN, show enhanced adsorption selectivities for

CO_2–CH_4 (5.6 and 13.1, respectively) compared to ZIF-8 (2.4) and showed increased $-Q_{st}$ for CO_2, which indicates that tuning the surface properties of ZIFs by either employing mixed-linkers and/or post-synthetic amine modification can produce adsorbent materials with improved gas adsorption/separation properties.[200] For example, EN was also grafted onto *sod*-ZMOF (EN-*sod*-ZMOF) constructed from 4,5-imidazoledicarboxylic acid and $In(NO_3)_3.2H_2O$ through a condensation reaction with the uncoordinated carboxylic acid groups in the *sod*-ZMOF structure. The ED-*sod*-ZMOF has *ca.* 30% greater adsorption capacity (53 to 69 mg g^{-1} at 298 K and 1 bar), as well as higher capture selectivity towards CO_2 against N_2, compared to the parent *sod*-ZMOF. In addition, ED-*sod*-ZMOF exhibited excellent stability in the cyclic CO_2 adsorption–desorption runs with no amine leaching.[201] Partially alkali-ion-exchanged *sod*-ZMOFs also show further improvement in CO_2 capture. ZIF-11 was predicted to satisfy the industrial feasibility target for H_2–CO_2 and H_2–N_2 separation for pre-combustion carbon capture, and simulation studies have also revealed ZIF-8, −90 and −77 to be candidate materials for natural gas purification (CO_2–CH_4).[202]

3.8 CO_2 Adsorption by Composite MOFs

Composites are multicomponent materials of different (non-gaseous) phase domains in which at least one of the phase domains is continuous. Because of the synergistic effect of the mixed components, as well as the presence of chemical functionality on the individual component, they are widely used in various industrial processes including CO_2 adsorption. Traditional adsorbent materials such as carbons, zeolite, silica, polymers, and activated alumina can be used to make MOF composites (see Figure 3.13), and among them, carbons are extensively used to synthesize MOF composites for CO_2 adsorption and separation applications.[203,204]

3.8.1 MOF–Carbon Composites

Several types of carbon-based MOF composites, such as MOF/CNTs (carbon nanotubes),[205] MOF/AC (activated carbon),[206] MOF/GO (graphene oxide),[207] MOF/CMK-3 (mesoporous CMK-3 carbon prepared from mesoporous

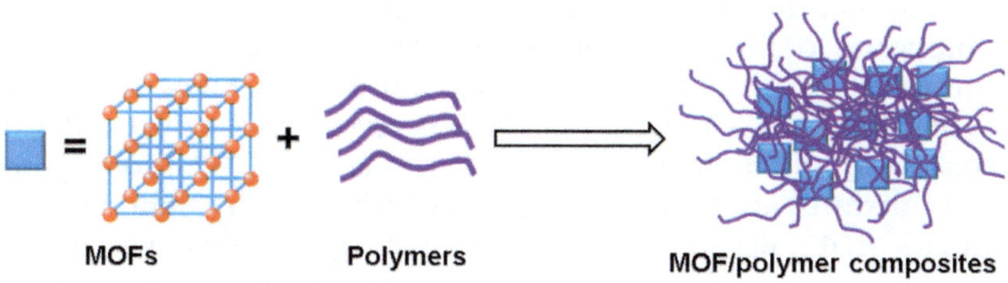

Figure 3.13 Schematic representation of the preparation of the MOF/polymer composites.

SBA-15 silica),[208] MOF/HCM (hierarchical porous carbon monolith),[209] and MOF/VACNT (vertically-aligned carbon nanotube),[210] have been synthesized *via* various methods and applied for CO_2 adsorption. Li and co-workers prepared a MIL-101(Cr)/GO composite and tested it for CO_2–CH_4 separation. MIL-101(Cr)/GO possesses a CO_2 adsorption capacity of 985.6 mg g^{-1} at 25 bar and 298 K, which is higher than that of MIL-101(Cr). The selectivity of MIL-101(Cr)/GO for CO_2–CH_4 (10 : 90) mixture at 1.5 bar is about 32, which is three times higher than that of parent MIL-101(Cr), indicating that CO_2 was more favorably adsorbed on MIL-101(Cr)/GO than CH_4. In addition, the MIL-101(Cr)/GO composite showed significant quasi-reversibility for CO_2 adsorption, showing more than 95% desorption efficiency over five cycles.[207] A water-stable MIL-101(Cr)/CMK-3 was *in situ* synthesized with different ratios of MIL-101(Cr) and CMK-3 using a hydrothermal method. The composite had the same morphology and crystal structure as its parent MIL-101(Cr) and showed an improvement in CO_2 uptake capacity compared to MIL-101(Cr) and CMK-3. The increased CO_2 adsorption was ascribed to the combined effect of the development of extra micropores, increased micropore volume, and the activation of open Cr sites by CMK-3 incorporation.[208] The nano-sized Cu-BTC was incorporated with different proportions of GO, and the composites showed increased CO_2 capture, almost 30% (from 281 to 363 mg g^{-1} at 1 bar and 273 K).[211] Further, Cu-BTC/GO composites with different GO loadings (0.5%, 1%, 3%, and 5%) were prepared by a solvothermal method, and these showed higher surface areas and increased pore volumes compared to the parent Cu-BTC. Among these, the Cu-BTC/GO (1%) composite has a higher adsorption and improved separation of CO_2–CH_4 binary mixtures.[212] Similarly, an interfacially-grown Cu-BTC/GO composite was prepared *via* an emulsion-based strategy and used for CO_2 capture from simulated humid flue gas. The CO_2 adsorption capacity of Cu-BTC/GO increased significantly from 122 to 148 mg g^{-1} and this performance continued for about 120 min under humid flue gas conditions. This performance was possible because most of the H_2O molecules were adsorbed on the highly exfoliated GO sheets in Cu-BTC/GO.[213] Bandosz *et al.* prepared Cu-BTC/urea(U)-modified GO composites and measured the CO_2 adsorption capacity. Cu-BTC/GO-U has a high surface area of 1367 m^2 g^{-1} and high CO_2 uptake of 590 mg g^{-1} at 1.5 bar and 298 K. In addition, the composite material has CO_2–N_2 and CO_2–CH_4 selectivities of 24.77 and 9.57, respectively, and the CO_2 adsorption was completely reversible without any external thermal treatment.[214] Cao *et al.* synthesized the Cu-BTC/CNT composite for the separation of CO_2–CH_4 mixtures, and their efficiency was compared with benchmark materials (MOF-177, UMCM-1, ZIF-8, MIL-53(Al), and Cu-BTC). The IAST calculations revealed that the Cu-BTC/CNT composite exhibits high selectivity in the range of 5.5–7.0 for equimolar CO_2–CH_4 mixtures at 298 K and 1–20 bar, which is higher than those of the benchmark MOFs. Because of the presence of a quadrupole moment in the CO_2 molecules and lack of a quadrupole moment in the CH_4 molecules, any polarized sites in the structure, such as OMSs, or flexible structures, may enhance the separation efficiency of a CO_2–CH_4 mixture.[215] Ding and co-workers reported the *in situ* synthesis of amine- or acid-functionalized CNT-doped Cu-BTC for CO_2 adsorption in both low- and high-pressure conditions. The doped Cu-BTC had a high BET surface area (1180 m^2 g^{-1}) and exhibited a good CO_2 adsorption capacity, which was mainly due to the presence of acid- and amine-functionalized CNTs and large, narrow micropores. Specifically, the Cu-BTC/amine functionalized CNT exhibited a better CO_2 adsorption capacity (around 250 mg g^{-1}) compared to Cu-BTC/acid-functionalized CNT (208 mg g^{-1}) at 5 bar.[216] A Cu-BTC/HCM composite was synthesized by a step-by-step impregnation and crystallization method. The resulting Cu-BTC/HCM structure retained the monolithic shape of the HCM and exhibited a high CO_2 uptake of 121 mg g^{-1} at 1 bar and 298 K, which is almost twice that of the CO_2 uptake of the parent HCM. In addition, the CO_2 separation of a CO_2–N_2 (16 : 84) mixture was found to be *ca.* 100 for the Cu-BTC/HCM composite, indicating highly selective CO_2 adsorption. Furthermore, the composite was regenerated easily by purging with a flow of argon.[209] Cu-BTC/VACNT was prepared by confined growth of Cu-BTC in the VACNT interspace. Compared with a physical mixture of the components, the Cu-BTC/VACNT composite has a higher surface area and porosity, larger CO_2 uptake capacity (154 mg g^{-1} at 273 K and 1.3 bar), and better CO_2–N_2 selectivity, revealing the synergistic effect between Cu-BTC and VACNT.[210] The NH_2-MIL-101(Al)/CNT composite was prepared from the growth of NH_2-MIL-101(Al) on the surface of the CNTs. The BET surface area of the NH_2-MIL-101(Al)/CNT composite decreased (651 m^2 g^{-1}) from that of the original NH_2-MIL-101(Al) (1309 m^2 g^{-1}), but the NH_2-MIL-101(Al)/CNT composite exhibited a greater CO_2 adsorption capacity (*ca.* 60 mg g^{-1}) than CNT-COOH (*ca.* 8 mg g^{-1}) at 1 bar and 298 K. In addition, the composite showed an ideal CO_2–CH_4 selectivity of 4.75, which is higher than those of both CNT (3.13) and NH_2-MIL-101(Al) (4.18). Further, to improve the CO_2–CH_4 separation, 10% 4,4′-hexafluoroisopropylidene diphthalic anhydride (6FDA) was dispersed on the NH_2-MIL-101(Al)/CNT composite. The 6FDA-containing the NH_2-MIL-101(Al)/CNT composite showed improved CO_2 adsorption, and higher CO_2 selectivity in a CO_2–CH_4 (50 : 50) gas mixture than those of free 6FDA-Durene (Durene = 1,2,4,5-tetramethylbenzene) and NH_2-MIL-101(Al)/CNT. These results clearly indicate that the introduction of more active sorption functional groups on the NH_2-MIL-101(Al)/CNT/6FDA-Durene composite increases the specific CO_2 gas adsorption and separation. The performance of NH_2-MIL-101(Al)/CNT/6FDA-Durene clearly exceeds the 1991 upper bound for polymeric membrane performances for CO_2 separation (see Figure 3.14).[217]

A Pd-containing AC-doped MOF, MOF-74(Ni, Co), was synthesized for CO_2 adsorption and CO_2–N_2 separation. The BET surface areas of the material decreased significantly on doping, but the CO_2 adsorption capacities and CO_2–N_2 separation selectivities were improved over the original MOF-74(Ni, Co), which may be due to the interaction between the partial negative charge of the oxygen atoms of the polarized CO_2 and partially positive Pd

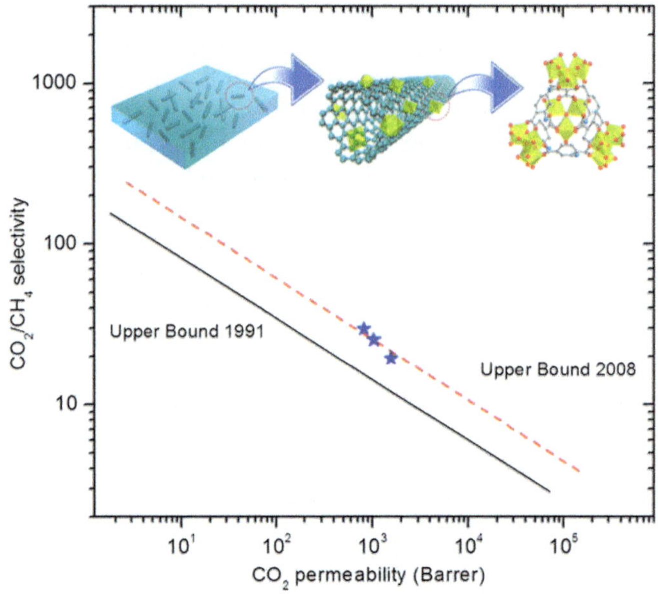

Figure 3.14 CO_2–CH_4 separation performance of NH_2-MIL-101(Al)/CNT/6FDA-Durene. Reprinted with permission from R. Lin, L. Ge, S. Liu, V. Rudolph and Z. Zhu, *ACS Appl. Mater. Interfaces*, 2015, 7, 14750, Copyright 2015 American Chemical Society.

metal species.[218] A series of ZIF-8/CNT composites were prepared *via* a solvothermal method and the resulting composite crystal sizes were found to be smaller with increasing CNT content in the composites (see Figure 3.15). In addition, the BET surface areas increased (ZC1, ZC2, and ZC3) when the CNT content was increased from 0.77 to 15.62 wt%. As the CNT content was further increased from 19.29 to 34.23 wt%, the surface areas and micropore volumes of composites (ZC4, ZC5, and ZC6) decreased gradually. The isotherm results showed that the ZC2 composite is highly porous, having a BET surface area of 1997 $m^2\ g^{-1}$ and micropore volume of 0.747 $cm^3\ g^{-1}$. The ZC2 composite shows slightly higher CO_2 and N_2 uptake capacities at both low (1.3 bar and 273 K) and high (40 bar and 308 K) pressures than the original ZIF-8 because of the larger surface area of ZC2.[219]

A series of core–shell-type UTSA-16/GO composites were synthesized by Xiang *et al.*[220] The BET surface areas of these composites containing 9.5, 19, and 28.5 mg GO were found to be 325, 529 and 487 $m^2\ g^{-1}$, respectively, all less than that of the original UTSA-16 (628 $m^2\ g^{-1}$). However, the thermal stabilities of all the composites increased compared to GO and UTSA-16. UTSA-16-GO19 can adsorb 157 mg g^{-1} of CO_2 and 7.7 mg g^{-1} of CH_4 at 1 bar and 296 K, which is slightly lower than the respective values of 186 mg g^{-1} (CO_2) and 8.4 mg g^{-1} (CH_4) for UTSA-16. The CO_2–CH_4 separation selectivity of the composites for CO_2 adsorption and separation from CO_2–CH_4 mixtures (50 : 50

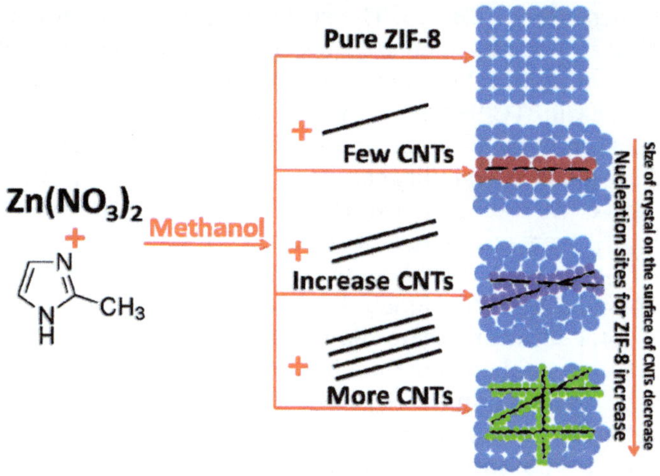

Figure 3.15 Schematic representation of ZIF-8 crystal size reduction with increasing CNT loading.

gas mixtures) is about 114, which is three times higher than that of the original UTSA-16 and higher than that of MOF-74(Mg) (105).[221,222]

3.8.2 Composites of MOFs with Other Support Materials

A limited number of alternative supports have been used in MOF composites for CO_2 adsorption and separation. The MOF-74(Mg)/SBA-15 composite was synthesized *via* the immobilization of MOF-74(Mg) into mesoporous SBA-15 rods. However, the CO_2 adsorption of MOF-74(Mg)/SBA-15 showed an almost 45% decrease in total uptake with respect to pristine MOF-74(Mg) at 293 K and 1 bar, despite having a dual pore structure of micro- and mesopores.[223] Liu *et al.* synthesized four types of composites using mesoporous alumina/carbon-alumina and MIL-53/DUT-5 and the resulting carbon-alumina/MIL-53(Al), alumina/MIL-53(Al), carbon-alumina/DUT-5, and alumina/DUT-5(Al) composites were found to have BET surface areas of 78.8, 88.6, 253.9, and 269.9 $m^2 g^{-1}$, respectively. Further, the CO_2 adsorption isotherm at 273 K and 1 bar showed that the composite derived from alumina/DUT-5 shows a significantly higher adsorption than the others, which may be due to the high surface area of alumina/DUT-5.[224] Flexible MOF/aminoclay nanocomposites have also been synthesized in which two different types of MOFs, {[Cu(2,3-pyridine dicarboxylic acid)(1,3-bis(4-pyridyl)-propane)](5H_2O)}$_n$ (F-MOF1) and {[Cu(2,5-dihydroxybenzoic acid)(4,4′-bpy)](H_2O)}$_n$ (F-MOF2), were used to prepare the composites. Interestingly, the activated F-MOF1/aminoclay exhibited a stepwise adsorption at 195 K with an initial CO_2 uptake of 58 mg g^{-1} at 0.39 bar and then a significant uptake of up to 284 mg g^{-1} at 1 bar, which is 35% higher than those of activated F-MOF1 and F-MOF2.[225] Guan *et al.* constructed Cu-BTC/SBA-15 composites with hierarchical pore structures by the *in situ* self-assembly of

Cu-BTC with mesoporous SBA-15. The surface areas and micropore volumes of the resulting composites increased because the interactions between the surface silanol groups and Cu centers of Cu-BTC induced structural changes in the MOF. Further, Cu-BTC/SBA-15-1 (1 wt% of SBA-15 based on the mass of the metal precursor) composite exhibited an increase of 15.9% in CO_2 adsorption capacity and excellent reversibility compared with that of Cu-BTC.[226]

Mixed matrix membranes (MMMs) containing NH_2-MIL-53(Al) or NH_2-MIL-101(Al) were prepared with a polymer matrix of copolyimides (6FDA:DSDA/4MPD:4,4′-SDA 1:1 (designated as polymer P1) or 6FDA/4MPD:4,4′-SDA 1:1 (designated as polymer P2), where DSDA is diphenylsulfone-3,3′,4,4′-tetracarboxylic dianhydride, SDA is bis(4-aminophenyl) sulfide and MPD is *m*-phenylenediamine), and the resulting materials were applied for the permeation of H_2, CH_4, and CO_2. The P1 matrix MMMs exhibited higher interactions with the fillers used than those with P2, as well as improved separation properties, especially NH_2-MIL-101(Al). These enhancements were explained as a consequence of the controlled pore size of NH_2-MIL-101(Al), flexibility, and the functional groups of sulfone-containing DSDA, and polymer rigidification.[227] An MMM composite was prepared from polyether-*block*-amide (Pebax 1657, which contains 60 wt% polyethylene oxide and 40 wt% polyamide) mixed with nano-sized ZIF-7, which was deposited on a porous polyacrylonitrile thin-layer (less than 1 mm) support. The single gas permeation measurements of ZIF-7/Pebax showed increased permeability (CO_2 up to 145 barrer) and gas selectivity (CO_2–N_2 up to 97 and CO_2–CH_4 up to 30) at 22 wt% ZIF-loading. When the ZIF-7 loading was increased to 34 wt%, the CO_2–CH_4 selectivity further increased to 44 but the permeability decreased compared to the pure Pebax membrane, which was caused by the polymer chain rigidification at a high filler loading.[228] A poly(2,6-dimethyl-1,4-phenylene oxide) (PPO) composite membrane with dispersed Cu-BTC was fabricated *via* a sonication treatment. The sonicated Cu-BTC material had a higher surface area and lower crystal size compared to pristine Cu-BTC. Similarly, compared to MMMs with pristine Cu-BTC particles, sonication-treated Cu-BTC/PPO exhibited good CO_2–N_2 and H_2–CH_4 selectivities because of the enhancement of the interaction between sonicated Cu-BTC and the PPO matrix, facilitating the fabrication of defect-free MMMs.[229] MMM hollow fibers of Cu-BTC and polyimide (PI) were prepared by a dry/wet-spinning method and applied for gas adsorption and separation. The gas permeances of both PI and Cu-BTC/PI hollow fibers achieved high separation levels in the order $H_2 > CO_2 > O_2 > N_2 > CH_4$. In addition, the selectivities for H_2–CH_4, H_2–N_2, H_2–O_2, and H_2–CO_2 were found to be 240, 163, 42, and 28, respectively, for Cu-BTC/PI, which are almost twice those of pure PI.[230] NH_2-MIL-53(Al) particles with the 6FDA-4,4′-oxidianiline (6FDA-ODA) PI were prepared without the addition of any compatibility agent and displayed an ideal CO_2–CH_4 selectivity of up to 77 with increasing pressure because of the breathing nature of NH_2-MIL-53(Al).[231] ZIF-90-based (15 wt%) composite MMMs were fabricated with three PIs (Ultem, Matrimid, and 6FDA-DAM (diaminomesitylene)) as polymer matrices, and used for CO_2–CH_4 separation.

Ultem and Matrimid MMMs displayed considerably enhanced CO_2 permeability without any loss of CO_2–CH_4 selectivity compared to the original ZIF-90. The 6FDA-DAM MMM exhibited a large increase in CO_2 permeability and CO_2 selectivity, indicating that the membrane does not have any defects and the permeabilities of the MOF and the polymer are well matched.[232]

3.9 CO_2 Adsorption by MOFs under Humid Conditions

Industrial flue gas from the burning of fossil fuels contains almost 7% H_2O.[7] H_2O can break the coordination bonding between organic linkers and metal ions, resulting in the collapse of the MOF structure. Therefore, water-stable MOFs are very important for various applications, including CO_2 capture. Recently, significant numbers of water-stable MOFs have been reported for CO_2 capture applications.[58,61,233–238] Llewellyn and co-workers reported that the presence of H_2O molecules in MIL-53(Cr) considerably improves the adsorption and selectivity for CO_2 over CH_4.[233] MIL-53(Cr) showed a breathing effect upon hydration (in the presence of H_2O) and dehydration (activated MIL-53(Cr)), resulting in two forms of MIL-53(Cr) channels (see Figure 3.16). The pores are slightly deformed in the hydrated form because of H-bond interactions between H_2O molecules and carboxylate ligands, whereas regular diamond-shaped channels with high porosity were formed after the activation in MIL-53(Cr).[34] When both hydrated and dehydrated MIL-53(Cr) were tested for CO_2 and CH_4 adsorption at 304 K, the hydrated MIL-53(Cr) did not show any CO_2 adsorption at 1 bar, but showed better CO_2 adsorption than dehydrated MIL-53(Cr) when the pressure was increased to 20 bar owing to the breathing effect of MIL-53(Cr) and the quadrupole interaction of CO_2 and the μ_2-OH groups of MIL-53(Cr). On

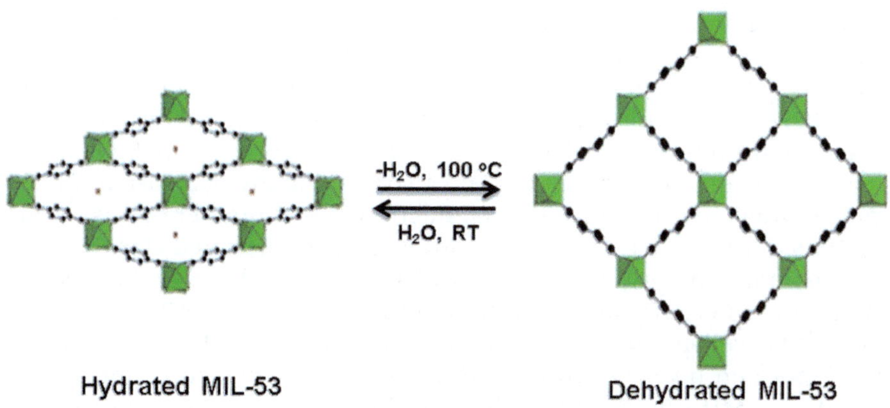

Figure 3.16 Hydration and dehydration process in MIL-53(Cr). Reproduced from ref. 233 with permission from John Wiley and Sons, Copyright © 2006 WILEY-VCH Verlag GmbH & Co. KGaA, Weinheim.

the other hand, the hydrated material showed practically no CH_4 adsorption, even at 20 bar, because of the repulsive effect of polar H_2O molecules and nonpolar CH_4, which resulted in the increased selectivity of CO_2 over CH_4 by the hydrated MIL-53(Cr). The influence of water molecules in the HKUST-1 analogs with different metal ions (Mg, Zn, Co, and Ni) for CO_2 adsorption was also studied. The CO_2 uptake and selectivity over N_2 and CH_4 by Zn-, Cu-, Co-, and Ni-based HKUST-1 frameworks considerably improved in the presence of a small number of water molecules coordinated to the OMSs in the framework because of the quadrupole interaction between CO_2 and the electric field created by H_2O molecules. However, in the case of Mg-HKUST-1, the Mg metal was oxidized easily in the presence of moisture, which diminished the CO_2 adsorption capacity.[62] Snurr *et al.* also presented the CO_2 uptake and selectivity over N_2 and CH_4 in half-hydrated Cu-HKUST-1, which were significantly increased by the presence of H_2O molecules coordinated to the open Cu sites in the framework. The interaction between the quadrupole moment of CO_2 and the electric field created by water molecules was beneficial for the enhanced CO_2 uptake. However, upon further increasing the water loading on Cu-BTC to full hydration, the CO_2 uptake was significantly lower because the pores contained uncoordinated water molecules that blocked the pore spaces.[58] LeVan *et al.* also proved that a small amount of water in Cu-BTC increases the amount of captured CO_2.[61]

Matzger *et al.* studied the effect of humidity on the performance of the MOF-74(M) series (M = Zn, Ni, Co, and Mg) using $N_2/CO_2/H_2O$ breakthrough curves at 9%, 36%, and 70% RHs. MOF-74(Mg) has a CO_2 uptake of 236 mg g^{-1} at 1 bar and 298 K after exposure to H_2O at 70% RH, which is only 16% of the initial uptake capacity under dry conditions.[234] MOF-74(Ni) maintained almost 60% of its initial CO_2 adsorption capacity, whereas MOF-74(Co) maintained almost 85% of its initial capacity under the same experimental conditions. These results indicate that MOF-74(Co) may be a more desirable material for CO_2 capture considering the added costs associated with flue gas dehumidification. In the same manner, LeVan *et al.* studied the stability of MOF-74(Ni) and MOF-74(Mg), which showed that MOF-74(Ni) could maintain a longer adsorption performance after steam conditioning than MOF-74(Mg). The N_2 isotherm of MOF-74(Mg) showed a decreased surface area and a corresponding decrease in the CO_2 uptake, which may be caused by a reduction in the number of unsaturated Mg OMSs in MOF-54(Mg).[50] Eddaoudi *et al.* demonstrated the recyclability and moisture stability of SIFSIX-3-Cu, which shows enhanced CO_2 uptake and selectivity in highly diluted gas streams under 74% RH conditions.[235] Llewellyn and co-workers investigated the MOF stability against H_2O vapor and investigated CO_2 adsorption at low pressures for Cu-BTC, UiO-66, and MIL-100(Fe) under different RHs (3%, 10%, 20%, and 40%).[239] The water adsorption cycle of UiO-66 showed a type-V isotherm for H_2O uptake. Cu-BTC, on the other hand, exhibited a type-I H_2O isotherm and showed more than 50% loss in pore volume after the sixth run because

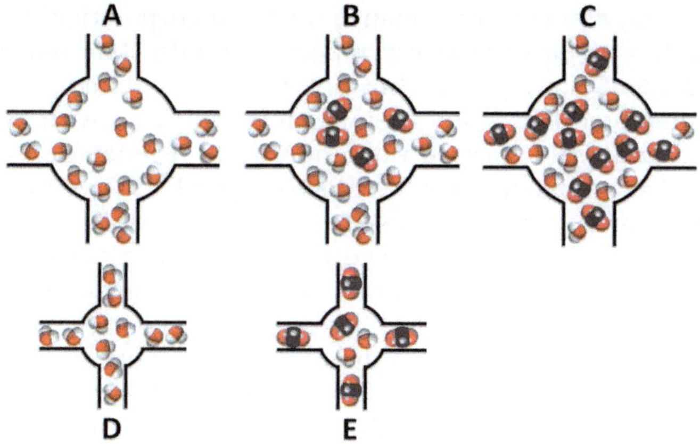

Figure 3.17 Possible mechanisms of CO_2 adsorption by UiO-66 and MIL-100(Fe) in the presence of humidity. Reprinted with permission from E. Soubeyrand-Lenoir, C. Vagner, J. W. Yoon, P. Bazin, F. Ragon, Y. K. Hwang, C. Serre, J.-S. Chang and P. L. Llewellyn, *J. Am. Chem. Soc.*, 2012, **134**, 10174, Copyright 2012 American Chemical Society.

of structural collapse. MIL-100(Fe) exhibited a type-V isotherm for H_2O uptake, and the sixth run isotherm practically overlapped with the first, indicating that no sample degradation took place. The CO_2 adsorption capacities of MIL-100(Fe) showed promising results. A five-fold increase in CO_2 adsorption capacity (105 mg g^{-1}) was observed at 40% RH compared to under anhydrous conditions. These results show that the pre-equilibrated water in the mesoporous MIL-100(Fe) forms microporous pockets that are filled with both CO_2 and H_2O at low pressure (see Figure 3.17A–C). The CO_2 uptakes of UiO-66 showed only small differences under 3%, 10%, 20%, and 40% RH conditions because UiO-66 is microporous and the pre-adsorbed water does not produce any new microporosity accessible for CO_2 adsorption (see Figure 3.17D). The adsorbed CO_2 can also replace some H_2O (see Figure 3.17E). A different adsorption behavior was observed in Cu-BTC; the CO_2 uptake was slightly lower (35 mg g^{-1}) at 3% RH/1 bar/298 K than under anhydrous conditions (40 mg g^{-1}), but the CO_2 uptake (52 mg g^{-1}) increased at 10% RH, and, upon further increasing the RH to 20% and 40%, the CO_2 uptakes decreased to 41 and 20 mg g^{-1}, respectively. This decrease in CO_2 uptake was due to the degradation in the MOF structure, which was confirmed by N_2 isotherm measurements. The overall results prove that, for MIL-100(Fe), CO_2 uptake was increased through the pre-adsorption of water vapor. To increase the stability of UiO-66 in humid conditions, Walton *et al.* prepared the monomethyl-functionalized UiO-66 (UiO-66-MM).[240] UiO-66-MM is less hydrophilic than UiO-66 and adsorbs almost 30% less water than UiO-66. In addition, UiO-66-MM shows a significantly higher CO_2–CH_4 selectivity under dry conditions than UiO-66 because of greater van der Waals interactions with CO_2.

Improved CO_2 capture performance on the incorporation of monodentate hydroxide as the active site was reported for [$Mn^{2+}Mn^{3+}(OH)Cl_2$(1*H*,5*H*-benzo(1,2-d:4,5-d′)bistriazole)]. This material captured up to 134 mg g^{-1} of CO_2 from simulated flue gases, even at high RH (82%), with high CO_2–N_2 selectivity, and quickly released CO_2 under mild regeneration conditions, confirming that monodentate hydroxide is a good additive for combining enhanced CO_2 affinity and reaction reversibility.[236] Hong *et al.*[241] synthesized the water-stable InOF-1 with hydroxo functional groups and its stability was confirmed by powder XRD. InOF-1 shows 54.2 mg g^{-1} CO_2 capture at 30 °C and a two-fold increase (*ca.* 110 mg g^{-1}) in CO_2 capture at 20% RH at 30 °C. This has been attributed to the functional groups (μ_2-OH) acting as directing agents for H_2O in the pores, which allows the more efficient packing of gas molecules.[237] Additionally, a water-stable MOF material named NOTT-400 has also been applied to CO_2 capture in the presence of H_2O.[242] The CO_2 capture capacity of NOTT-400 increased from 42 mg g^{-1} (anhydrous conditions) to 102 mg g^{-1} at 20% RH/30 °C. Similarly, another water-stable MOF material, designated NOTT-401, was reported for CO_2 capture in the presence of H_2O.[238] NOTT-401 has μ_2-OH functional groups inside the channels. The CO_2 capacity of NOTT-401 increased from 12 mg g^{-1} (anhydrous conditions) to 39 mg g^{-1} at 5% RH at 30 °C, representing a 3.2-fold improvement. This was explained as a consequence of the bridging hydroxo functional groups formed at low water loadings, which interact strongly with H_2O molecules *via* hydrogen bonding, thus reducing the microporosity (confinement effect). No enhancement in the CO_2 capture was observed when the pre-adsorption of H_2O was increased because of the increased H_2O molecule occupation within the micropores of NOTT-401. Similarly, the effect of the μ_2-OH functional group on MIL-53(Al)[243] and NH_2-MIL-53(Al)[244] for CO_2 capture was also studied. The CO_2 captured by MIL-53(Al) increased from 30 to 52 and 48 mg g^{-1} at 20% and 40% RHs/30 °C, respectively; the high H_2O loading slightly reduced the CO_2 capture because of an increase in H_2O disorder caused by thermal motion.[243] The CO_2 capture by NH_2-MIL-53(Al) was less than that by MIL-53(Al) under humid conditions because the former exhibited a considerably stronger affinity for water than the latter.[244] Apart from the hydroxo functional group, alkylamine is another effective active site for modifying the CO_2 adsorption affinity. Mg_2(dobpdc) with MMEN was synthesized and applied for CO_2 capture at a high H_2O partial pressure (partial pressure = 19 mbar) and showed a CO_2 capacity of 185 mg g^{-1} at 0.1 bar and 40 °C.[245] Indeed, numerous benchmark materials for CO_2 capture in humid conditions have been prepared by Yaghi and co-workers.[127,246] ZIF-300, ZIF-301, and ZIF-302 show effective dynamic separation of CO_2 from N_2 under both dry and humid conditions without any loss of performance over three cycles.[246] In addition, the amine-functionalized IRMOF-74-III has been used for the selective capture of CO_2 in 65% RH. Compared with the practically constant breakthrough time of IRMOF-74-III-CH_2NH_2 in wet and dry conditions, the breakthrough time of IRMOF-74-III-CH_3 is decreased by 80% under wet conditions. This different behavior is because the CO_2

uptake in IRMOF-74-III-CH_3 mainly occurs on the open Mg sites, which are occupied by water molecules under humid conditions. In contrast, the CO_2 uptake in IRMOF-74-III-CH_2NH_2 takes place at the linker amine sites, and the open Mg sites are not accessible under dry or humid conditions; thus, the effect of H_2O on the CO_2 uptake is negligible.[127]

3.10 Conclusion and Perspectives

Flue gas from the combustion of fossil fuels results in increasing CO_2 emissions in the atmosphere and causes global warming. To alleviate this problem, the development of environmentally benign and cost-effective CO_2 capture and separation techniques *via* adsorption using efficient solid adsorbents is urgently required. In this regard, MOFs are one of the most promising platforms for post-combustion CO_2 capture because of their high surface areas, large pore volumes, high adsorption affinities, varied pore topologies, diverse structures, tunable functionalization, and easily controllable composition. Based on the research activities reported so far, this chapter has summarized the current progress in MOF implementation for CO_2 capture, which we have discussed in seven sections: (i) metal ions with open sites, (ii) functional group attachments, (iii) mixed-ligand-based MOFs, (iv) increased MOF flexibility, (v) interpenetrated MOFs, (vi) ZIF-based MOFs, and (vii) MOF composites.

MOFs with OMSs or amine-functionalized MOFs are flourishing, and materials with new features have enabled high working CO_2 capture capacities under experimental conditions because of the strong interactions between CO_2 molecules and the MOFs. However, some amine-functionalized MOFs are disadvantageous concerning regeneration because of their CO_2 chemisorption with high enthalpies. On the other hand, a few frameworks with weakly polar functionalities show limited CO_2 adsorption capacities and often exhibit low enthalpies. In addition, the synthesis of amine-functionalized MOFs can be challenging because of the thermally labile functional groups in the linkers. Mixed-linker-based MOFs are also promising materials for CO_2 capture because of their increased structural stability and increased pore volume. Further, the use of flexible ligands may provide unique opportunities to prepare novel framework materials with desirable attributes that increase the pore volume and gas binding affinity. Considering the pore size effects, cages with small windows, as well as interpenetration with micropores, can result in high adsorption and CO_2 selectivity, but further research is required in this area. The development of MOF composites is also promising to extend the scope of MOF applications in gas absorption and separation, which enhances the gas adsorption capability and selectivity because of the synergistic effects between the MOF and diverse material supports. A uniform dispersion of MOF particles in a composite matrix also improves the gas separation capability of the membranes. However, several problems remain to be solved concerning the flexible, interpenetrated, and composite-based

MOFs; for example, (i) the weak stability and low porosity of the flexible and interpenetrated MOFs arising from the longer length of the linkers and (ii) the reduction in surface area and pore volume after composite synthesis because of polymer immobilization in the MOFs. In addition, industrial flue gas contains almost 7% water, which causes framework deterioration and may enhance adsorption or compete with CO_2 for the active sites of MOFs. Some water-stable MOFs have been constructed with oxidatively stable metal nodes and hydrophobic functional groups, and these show significantly improved stability. Extensive efforts to prepare these compounds are being made.

Based on the CO_2 capture work by MOFs considered in this chapter, it can be said that continued research efforts for the construction of novel amine-functionalized MOFs with OMSs, weak polarity, high porosity, and water stability are still necessary for the practical application of MOFs for CO_2 adsorption.

Acknowledgements

This work was supported by the Basic Science Research Program through the National Research Foundation of Korea (NRF) funded by the Ministry of Education (Grant number: NRF-2015R1A4A1042434). PP also acknowledges the support by the Korea Research Fellowship (Grant number: 2017H1D3A1A02013620) provided by the Ministry of Science and Information and Communication Technology (ICT).

References

1. J. M. Huck, L.-C. Lin, A. H. Berger, M. N. Shahrak, R. L. Martin, A. S. Bhown, M. Haranczyk, K. Reuter and B. Smit, *Energy Environ. Sci.*, 2014, **7**, 4132.
2. Q. Wang, J. Bai, Z. Lu, Y. Pan and X. You, *Chem. Commun.*, 2016, **52**, 443.
3. International Energy Agency, *CCS Technical and Economic Issues*, http://www.iea.org.
4. *The Global Status of CCS: 2011*, The Global CCS Institute, Canberra, Australia, 2011.
5. B. Seoane, J. Coronas, I. Gascon, M. E. Benavides, O. Karvan, J. Caro, F. Kapteijn and J. Gascon, *Chem. Soc. Rev.*, 2015, **44**, 2421.
6. E. F. da Silva and H. F. Svendsen, *Int. J. Greenhouse Gas Control*, 2007, **1**, 151.
7. K. Sumida, D. L. Rogow, J. A. Mason, T. M. McDonald, E. D. Bloch, Z. R. Herm, T.-H. Bae and J. R. Long, *Chem. Rev.*, 2012, **112**, 724.
8. M.-M. Titirici, R. J. White, N. Brun, V. L. Budarin, D. S. Su, F. del Monte, J. H. Clark and M. J. MacLachlan, *Chem. Soc. Rev.*, 2015, **44**, 250.
9. Y. Liu, Z. U. Wang and H.-C. Zhou, *Greenhouse Gases: Sci. Technol.*, 2012, **2**, 239.

10. D. M. D'Alessandro, B. Smit and J. R. Long, *Angew. Chem., Int. Ed.*, 2010, **49**, 6058.
11. Y. Ide, N. Kagawa, M. Sadakane and T. Sano, *Chem. Commun.*, 2013, **49**, 9027.
12. P. Puthiaraj, S.-M. Cho, Y.-R. Lee and W.-S. Ahn, *J. Mater. Chem. A*, 2015, **3**, 6792.
13. Z. Zhang, Y. Zhao, Q. Gong, Z. Li and J. Li, *Chem. Commun.*, 2013, **49**, 653.
14. B. F. Hoskins and R. Robson, *J. Am. Chem. Soc.*, 1990, **112**, 1546.
15. D. J. Tranchemontagne, J. L. Mendoza-Cortes, M. O'Keeffe and O. M. Yaghi, *Chem. Soc. Rev.*, 2009, **38**, 1257.
16. P. Puthiaraj, P. Suresh and K. Pitchumani, *Green Chem.*, 2014, **16**, 2865.
17. Z. Wang and S. M. Cohen, *Chem. Soc. Rev.*, 2009, **38**, 1315.
18. J.-R. Li, J. Sculley and H.-C. Zhou, *Chem. Rev.*, 2012, **112**, 869.
19. B. Chen, M. Eddaoudi, T. M. Reineke, J. W. Kampf, M. O'Keeffe and O. M. Yaghi, *J. Am. Chem. Soc.*, 2000, **122**, 11559.
20. R. Kitaura, G. Onoyama, H. Sakamoto, R. Matsuda, S.-i. Noro and S. Kitagawa, *Angew. Chem., Int. Ed.*, 2004, **43**, 2684.
21. S. S.-Y. Chui, S. M.-F. Lo, J. P. H. Charmant, A. G. Orpen and I. D. Williams, *Science*, 1999, **283**, 1148.
22. P. Horcajada, S. Surble, C. Serre, D.-Y. Hong, Y.-K. Seo, J.-S. Chang, J.-M. Greneche, I. Margiolaki and G. Ferey, *Chem. Commun.*, 2007, 2820.
23. G. Férey, C. Serre, C. Mellot-Draznieks, F. Millange, S. Surblé, J. Dutour and I. Margiolaki, *Angew. Chem., Int. Ed.*, 2004, **43**, 6296.
24. G. Férey, C. Mellot-Draznieks, C. Serre, F. Millange, J. Dutour, S. Surblé and I. Margiolaki, *Science*, 2005, **309**, 2040.
25. K. M. L. Taylor-Pashow, J. D. Rocca, Z. Xie, S. Tran and W. Lin, *J. Am. Chem. Soc.*, 2009, **131**, 14261.
26. A. G. Wong-Foy, O. Lebel and A. J. Matzger, *J. Am. Chem. Soc.*, 2007, **129**, 15740.
27. B. Chen, N. W. Ockwig, A. R. Millward, D. S. Contreras and O. M. Yaghi, *Angew. Chem., Int. Ed.*, 2005, **44**, 4745.
28. J. H. Cavka, S. Jakobsen, U. Olsbye, N. Guillou, C. Lamberti, S. Bordiga and K. P. Lillerud, *J. Am. Chem. Soc.*, 2008, **130**, 13850.
29. H. Furukawa, N. Ko, Y. B. Go, N. Aratani, S. B. Choi, E. Choi, A. Ö. Yazaydin, R. Q. Snurr, M. O'Keeffe, J. Kim and O. M. Yaghi, *Science*, 2010, **329**, 424.
30. A. R. Millward and O. M. Yaghi, *J. Am. Chem. Soc.*, 2005, **127**, 17998.
31. M. Eddaoudi, J. Kim, N. Rosi, D. Vodak, J. Wachter, M. O'Keeffe and O. M. Yaghi, *Science*, 2002, **295**, 469.
32. L. Grajciar, A. D. Wiersum, P. L. Llewellyn, J.-S. Chang and P. Nachtigall, *J. Phys. Chem. C*, 2011, **115**, 17925.
33. T. Devic, P. Horcajada, C. Serre, F. Salles, G. Maurin, B. Moulin, D. Heurtaux, G. Clet, A. Vimont, J.-M. Grenèche, B. L. Ouay, F. Moreau, E. Magnier, Y. Filinchuk, J. Marrot, J.-C. Lavalley, M. Daturi and G. Férey, *J. Am. Chem. Soc.*, 2010, **132**, 1127.

34. T. Loiseau, C. Serre, C. Huguenard, G. Fink, F. Taulelle, M. Henry, T. Bataille and G. Férey, *Chem.–Eur. J.*, 2004, **10**, 1373.
35. P. L. Llewellyn, S. Bourrelly, C. Serre, A. Vimont, M. Daturi, L. Hamon, G. De Weireld, J.-S. Chang, D.-Y. Hong, Y. Kyu Hwang, S. Hwa Jhung and G. Férey, *Langmuir*, 2008, **24**, 7245.
36. Q. Liu, L. Ning, S. Zheng, M. Tao, Y. Shi and Y. He, *Sci. Rep.*, 2013, **3**, 2916.
37. N. L. Rosi, J. Kim, M. Eddaoudi, B. Chen, M. O'Keeffe and O. M. Yaghi, *J. Am. Chem. Soc.*, 2005, **127**, 1504.
38. P. D. C. Dietzel, Y. Morita, R. Blom and H. Fjellvåg, *Angew. Chem., Int. Ed.*, 2005, **44**, 6354.
39. P. D. C. Dietzel, B. Panella, M. Hirscher, R. Blom and H. Fjellvag, *Chem. Commun.*, 2006, 959.
40. P. D. C. Dietzel, R. Blom and H. Fjellvåg, *Eur. J. Inorg. Chem.*, 2008, **2008**, 3624.
41. W. Zhou, H. Wu and T. Yildirim, *J. Am. Chem. Soc.*, 2008, **130**, 15268.
42. S. R. Caskey, A. G. Wong-Foy and A. J. Matzger, *J. Am. Chem. Soc.*, 2008, **130**, 10870.
43. E. D. Bloch, L. J. Murray, W. L. Queen, S. Chavan, S. N. Maximoff, J. P. Bigi, R. Krishna, V. K. Peterson, F. Grandjean, G. J. Long, B. Smit, S. Bordiga, C. M. Brown and J. R. Long, *J. Am. Chem. Soc.*, 2011, **133**, 14814.
44. H.-Y. Cho, D.-A. Yang, J. Kim, S.-Y. Jeong and W.-S. Ahn, *Catal. Today*, 2012, **185**, 35.
45. J. P. Turley, F. Romer, M. L. Trudeau, M. L. Dias, M. E. Smith, J. V. Hanna and D. M. Antonelli, *Microporous Mesoporous Mater.*, 2014, **190**, 284.
46. D. Yu, A. O. Yazaydin, J. R. Lane, P. D. C. Dietzel and R. Q. Snurr, *Chem. Sci.*, 2013, **4**, 3544.
47. P. D. C. Dietzel, V. Besikiotis and R. Blom, *J. Mater. Chem.*, 2009, **19**, 7362.
48. J. Park, H. Kim, S. S. Han and Y. Jung, *J. Phys. Chem. Lett.*, 2012, **3**, 826.
49. R. Poloni, K. Lee, R. F. Berger, B. Smit and J. B. Neaton, *J. Phys. Chem. Lett.*, 2014, **5**, 861.
50. J. Liu, A. I. Benin, A. M. B. Furtado, P. Jakubczak, R. R. Willis and M. D. LeVan, *Langmuir*, 2011, **27**, 11451.
51. D. Britt, H. Furukawa, B. Wang, T. G. Glover and O. M. Yaghi, *Proc. Natl. Acad. Sci. U. S. A.*, 2009, **106**, 20637.
52. B. Liu, Y.-H. Jiang, Z.-S. Li, L. Hou and Y.-Y. Wang, *Inorg. Chem. Front.*, 2015, **2**, 550.
53. O. K. Farha, A. M. Spokoyny, K. L. Mulfort, M. F. Hawthorne, C. A. Mirkin and J. T. Hupp, *J. Am. Chem. Soc.*, 2007, **129**, 12680.
54. Y.-S. Bae, O. K. Farha, A. M. Spokoyny, C. A. Mirkin, J. T. Hupp and R. Q. Snurr, *Chem. Commun.*, 2008, 4135.
55. Y.-S. Bae, A. M. Spokoyny, O. K. Farha, R. Q. Snurr, J. T. Hupp and C. A. Mirkin, *Chem. Commun.*, 2010, **46**, 3478.

56. A. Ö. Yazaydın, R. Q. Snurr, T.-H. Park, K. Koh, J. Liu, M. D. LeVan, A. I. Benin, P. Jakubczak, M. Lanuza, D. B. Galloway, J. J. Low and R. R. Willis, *J. Am. Chem. Soc.*, 2009, **131**, 18198.
57. P. Aprea, D. Caputo, N. Gargiulo, F. Iucolano and F. Pepe, *J. Chem. Eng. Data*, 2010, **55**, 3655.
58. A. Ö. Yazaydın, A. I. Benin, S. A. Faheem, P. Jakubczak, J. J. Low, R. R. Willis and R. Q. Snurr, *Chem. Mater.*, 2009, **21**, 1425.
59. P. Chowdhury, C. Bikkina, D. Meister, F. Dreisbach and S. Gumma, *Microporous Mesoporous Mater.*, 2009, **117**, 406.
60. D. Farrusseng, C. Daniel, C. Gaudillère, U. Ravon, Y. Schuurman, C. Mirodatos, D. Dubbeldam, H. Frost and R. Q. Snurr, *Langmuir*, 2009, **25**, 7383.
61. J. Liu, Y. Wang, A. I. Benin, P. Jakubczak, R. R. Willis and M. D. LeVan, *Langmuir*, 2010, **26**, 14301.
62. J. Yu, Y. Wu and P. B. Balbuena, *ACS Sustainable Chem. Eng.*, 2016, **4**, 2387.
63. A. Demessence, D. M. D'Alessandro, M. L. Foo and J. R. Long, *J. Am. Chem. Soc.*, 2009, **131**, 8784.
64. G. E. Cmarik, M. Kim, S. M. Cohen and K. S. Walton, *Langmuir*, 2012, **28**, 15606.
65. P. Chowdhury, C. Bikkina and S. Gumma, *J. Phys. Chem. C*, 2009, **113**, 6616.
66. Z. Bao, L. Yu, Q. Ren, X. Lu and S. Deng, *J. Colloid Interface Sci.*, 2011, **353**, 549.
67. P. D. C. Dietzel, R. E. Johnsen, H. Fjellvag, S. Bordiga, E. Groppo, S. Chavan and R. Blom, *Chem. Commun.*, 2008, 5125.
68. C. Tan, S. Yang, N. R. Champness, X. Lin, A. J. Blake, W. Lewis and M. Schroder, *Chem. Commun.*, 2011, **47**, 4487.
69. Z. Zhang, S. Xiang, X. Rao, Q. Zheng, F. R. Fronczek, G. Qian and B. Chen, *Chem. Commun.*, 2010, **46**, 7205.
70. B. Zheng, R. Yun, J. Bai, Z. Lu, L. Du and Y. Li, *Inorg. Chem.*, 2013, **52**, 2823.
71. Y. Lin, C. Kong and L. Chen, *RSC Adv.*, 2016, **6**, 32598.
72. M. E. Braun, C. D. Steffek, J. Kim, P. G. Rasmussen and O. M. Yaghi, *Chem. Commun.*, 2001, 2532.
73. S. Wu, X. Ma, J. Ran, Y. Zhang, F. Qin and Y. Liu, *RSC Adv.*, 2015, **5**, 14221.
74. N. T. S. Phan, T. T. Nguyen, Q. H. Luu and L. T. L. Nguyen, *J. Mol. Catal. A: Chem.*, 2012, **363**, 178.
75. S. Rostamnia, H. Xin and N. Nouruzi, *Microporous Mesoporous Mater.*, 2013, **179**, 99.
76. J. R. Karra and K. S. Walton, *J. Phys. Chem. C*, 2010, **114**, 15735.
77. Y. Lin, C. Kong and L. Chen, *RSC Adv.*, 2012, **2**, 6417.
78. K. Zhang, Y. Chen, A. Nalaparaju and J. Jiang, *CrystEngComm*, 2013, **15**, 10358.

79. S. Couck, J. F. M. Denayer, G. V. Baron, T. Rémy, J. Gascon and F. Kapteijn, *J. Am. Chem. Soc.*, 2009, **131**, 6326.
80. J. Kim, W. Y. Kim and W.-S. Ahn, *Fuel*, 2012, **102**, 574.
81. E. Stavitski, E. A. Pidko, S. Couck, T. Remy, E. J. M. Hensen, B. M. Weckhuysen, J. Denayer, J. Gascon and F. Kapteijn, *Langmuir*, 2011, **27**, 3970.
82. M. Mihaylov, K. Chakarova, S. Andonova, N. Drenchev, E. Ivanova, A. Sabetghadam, B. Seoane, J. Gascon, F. Kapteijn and K. Hadjiivanov, *J. Phys. Chem. C*, 2016, **120**, 23584.
83. X. Si, C. Jiao, F. Li, J. Zhang, S. Wang, S. Liu, Z. Li, L. Sun, F. Xu, Z. Gabelica and C. Schick, *Energy Environ. Sci.*, 2011, **4**, 4522.
84. P. Serra-Crespo, E. V. Ramos-Fernandez, J. Gascon and F. Kapteijn, *Chem. Mater.*, 2011, **23**, 2565.
85. Y. Fu, D. Sun, Y. Chen, R. Huang, Z. Ding, X. Fu and Z. Li, *Angew. Chem., Int. Ed.*, 2012, **51**, 3364.
86. S.-N. Kim, J. Kim, H.-Y. Kim, H.-Y. Cho and W.-S. Ahn, *Catal. Today*, 2013, **204**, 85.
87. S. J. Garibay and S. M. Cohen, *Chem. Commun.*, 2010, **46**, 7700.
88. Q. Yang, A. D. Wiersum, P. L. Llewellyn, V. Guillerm, C. Serre and G. Maurin, *Chem. Commun.*, 2011, **47**, 9603.
89. Y. Huang, W. Qin, Z. Li and Y. Li, *Dalton Trans.*, 2012, **41**, 9283.
90. L. Liu, J. Zhang, H. Fang, L. Chen and C.-Y. Su, *Chem.–Asian J.*, 2016, **11**, 2278.
91. K. Peikert, F. Hoffmann and M. Froba, *Chem. Commun.*, 2012, **48**, 11196.
92. J. Jiao, L. Dou, H. Liu, F. Chen, D. Bai, Y. Feng, S. Xiong, D.-L. Chen and Y. He, *Dalton Trans.*, 2016, **45**, 13373.
93. C. Song, Y. Ling, L. Jin, M. Zhang, D.-L. Chen and Y. He, *Dalton Trans.*, 2016, **45**, 190.
94. C. Song, J. Hu, Y. Ling, Y. Feng, R. Krishna, D.-L. Chen and Y. He, *J. Mater. Chem. A*, 2015, **3**, 19417.
95. J. An, S. J. Geib and N. L. Rosi, *J. Am. Chem. Soc.*, 2010, **132**, 38.
96. T. Panda, P. Pachfule, Y. Chen, J. Jiang and R. Banerjee, *Chem. Commun.*, 2011, **47**, 2011.
97. R.-B. Lin, D. Chen, Y.-Y. Lin, J.-P. Zhang and X.-M. Chen, *Inorg. Chem.*, 2012, **51**, 9950.
98. S. Xiong, Y. Gong, H. Wang, H. Wang, Q. Liu, M. Gu, X. Wang, B. Chen and Z. Wang, *Chem. Commun.*, 2014, **50**, 12101.
99. Q. Yan, Y. Lin, P. Wu, L. Zhao, L. Cao, L. Peng, C. Kong and L. Chen, *ChemPlusChem*, 2013, **78**, 86.
100. P.-Z. Li, X.-J. Wang, K. Zhang, A. Nalaparaju, R. Zou, R. Zou, J. Jiang and Y. Zhao, *Chem. Commun.*, 2014, **50**, 4683.
101. S. Bernt, V. Guillerm, C. Serre and N. Stock, *Chem. Commun.*, 2011, **47**, 2838.
102. A. Khutia and C. Janiak, *Dalton Trans.*, 2014, **43**, 1338.
103. C. Volkringer and S. M. Cohen, *Angew. Chem., Int. Ed.*, 2010, **49**, 4644.
104. Z. Wang and S. M. Cohen, *J. Am. Chem. Soc.*, 2009, **131**, 16675.

105. Z. Xiang, S. Leng and D. Cao, *J. Phys. Chem. C*, 2012, **116**, 10573.
106. Y. K. Hwang, D.-Y. Hong, J.-S. Chang, S. H. Jhung, Y.-K. Seo, J. Kim, A. Vimont, M. Daturi, C. Serre and G. Férey, *Angew. Chem., Int. Ed.*, 2008, **47**, 4144.
107. X. Wang, H. Li and X.-J. Hou, *J. Phys. Chem. C*, 2012, **116**, 19814.
108. S.-N. Kim, S.-T. Yang, J. Kim, J.-E. Park and W.-S. Ahn, *CrystEngComm*, 2012, **14**, 4142.
109. C. P. Cabello, G. Berlier, G. Magnacca, P. Rumori and G. T. Palomino, *CrystEngComm*, 2015, **17**, 430.
110. T. M. McDonald, D. M. D'Alessandro, R. Krishna and J. R. Long, *Chem. Sci.*, 2011, **2**, 2022.
111. S. Choi, T. Watanabe, T.-H. Bae, D. S. Sholl and C. W. Jones, *J. Phys. Chem. Lett.*, 2012, **3**, 1136.
112. W. R. Lee, H. Jo, L.-M. Yang, H. Lee, D. W. Ryu, K. S. Lim, J. H. Song, D. Y. Min, S. S. Han, J. G. Seo, Y. K. Park, D. Moon and C. S. Hong, *Chem. Sci.*, 2015, **6**, 3697.
113. J. S. Yeon, W. R. Lee, N. W. Kim, H. Jo, H. Lee, J. H. Song, K. S. Lim, D. W. Kang, J. G. Seo, D. Moon, B. Wiers and C. S. Hong, *J. Mater. Chem. A*, 2015, **3**, 19177.
114. T. M. McDonald, W. R. Lee, J. A. Mason, B. M. Wiers, C. S. Hong and J. R. Long, *J. Am. Chem. Soc.*, 2012, **134**, 7056.
115. T. M. McDonald, J. A. Mason, X. Kong, E. D. Bloch, D. Gygi, A. Dani, V. Crocella, F. Giordanino, S. O. Odoh, W. S. Drisdell, B. Vlaisavljevich, A. L. Dzubak, R. Poloni, S. K. Schnell, N. Planas, K. Lee, T. Pascal, L. F. Wan, D. Prendergast, J. B. Neaton, B. Smit, J. B. Kortright, L. Gagliardi, S. Bordiga, J. A. Reimer and J. R. Long, *Nature*, 2015, **519**, 303.
116. W. S. Drisdell, R. Poloni, T. M. McDonald, T. A. Pascal, L. F. Wan, C. D. Pemmaraju, B. Vlaisavljevich, S. O. Odoh, J. B. Neaton, J. R. Long, D. Prendergast and J. B. Kortright, *Phys. Chem. Chem. Phys.*, 2015, **17**, 21448.
117. P.-Q. Liao, X.-W. Chen, S.-Y. Liu, X.-Y. Li, Y.-T. Xu, M. Tang, Z. Rui, H. Ji, J.-P. Zhang and X.-M. Chen, *Chem. Sci.*, 2016, **7**, 6528.
118. X. Su, L. Bromberg, V. Martis, F. Simeon, A. Huq and T. A. Hatton, *ACS Appl. Mater. Interfaces*, 2017, **9**, 11299.
119. A. Das, P. D. Southon, M. Zhao, C. J. Kepert, A. T. Harris and D. M. D'Alessandro, *Dalton Trans.*, 2012, **41**, 11739.
120. H. R. Abid, H. M. Ang and S. Wang, *Nanoscale*, 2012, **4**, 3089.
121. Y. C. Lin, Q. J. Yan, C. L. Kong and L. Chen, *Sci. Rep.*, 2013, **3**, 1859.
122. Q. Yan, Y. Lin, C. Kong and L. Chen, *Chem. Commun.*, 2013, **49**, 6873.
123. Y. Lin, H. Lin, H. Wang, Y. Suo, B. Li, C. Kong and L. Chen, *J. Mater. Chem. A*, 2014, **2**, 14658.
124. F. Zhang, X. Zou, X. Gao, S. Fan, F. Sun, H. Ren and G. Zhu, *Adv. Funct. Mater.*, 2012, **22**, 3583.
125. B. Arstad, H. Fjellvåg, K. O. Kongshaug, O. Swang and R. Blom, *Adsorption*, 2008, **14**, 755.

126. R. Haldar, S. K. Reddy, V. M. Suresh, S. Mohapatra, S. Balasubramanian and T. K. Maji, *Chem.–Eur. J.*, 2014, **20**, 4347.
127. A. M. Fracaroli, H. Furukawa, M. Suzuki, M. Dodd, S. Okajima, F. Gándara, J. A. Reimer and O. M. Yaghi, *J. Am. Chem. Soc.*, 2014, **136**, 8863.
128. Q.-G. Zhai, Q. Lin, T. Wu, L. Wang, S.-T. Zheng, X. Bu and P. Feng, *Chem. Mater.*, 2012, **24**, 2624.
129. R. Vaidhyanathan, S. S. Iremonger, K. W. Dawson and G. K. H. Shimizu, *Chem. Commun.*, 2009, 5230.
130. G. Beobide, W.-G. Wang, O. Castillo, A. Luque, P. Román, G. Tagliabue, S. Galli and J. A. R. Navarro, *Inorg. Chem.*, 2008, **47**, 5267.
131. J. A. R. Navarro, E. Barea, J. M. Salas, N. Masciocchi, S. Galli, A. Sironi, C. O. Ania and J. B. Parra, *Inorg. Chem.*, 2006, **45**, 2397.
132. E. D. Bloch, D. Britt, C. Lee, C. J. Doonan, F. J. Uribe-Romo, H. Furukawa, J. R. Long and O. M. Yaghi, *J. Am. Chem. Soc.*, 2010, **132**, 14382.
133. C. A. Fernandez, P. K. Thallapally, R. K. Motkuri, S. K. Nune, J. C. Sumrak, J. Tian and J. Liu, *Cryst. Growth Des.*, 2010, **10**, 1037.
134. X.-J. Wang, P.-Z. Li, Y. Chen, Q. Zhang, H. Zhang, X. X. Chan, R. Ganguly, Y. Li, J. Jiang and Y. Zhao, *Sci. Rep.*, 2013, **3**, 1149.
135. T. K. Prasad, D. H. Hong and M. P. Suh, *Chem.–Eur. J.*, 2010, **16**, 14043.
136. J.-S. Qin, D.-Y. Du, W.-L. Li, J.-P. Zhang, S.-L. Li, Z.-M. Su, X.-L. Wang, Q. Xu, K.-Z. Shao and Y.-Q. Lan, *Chem. Sci.*, 2012, **3**, 2114.
137. C. Montoro, E. Garcia, S. Calero, M. A. Perez-Fernandez, A. L. Lopez, E. Barea and J. A. R. Navarro, *J. Mater. Chem.*, 2012, **22**, 10155.
138. Y. Lin, Q. Yan, C. Kong and L. Chen, *Sci. Rep.*, 2013, **3**, 1859.
139. J. Sahu, A. Aijaz, Q. Xu and P. K. Bharadwaj, *Inorg. Chim. Acta*, 2015, **430**, 193.
140. H. Chen, K. Riascos-Rodríguez, M. E. Marcano-González and A. J. Hernández-Maldonado, *Chem. Eng. J.*, 2016, **283**, 806.
141. S. A. A. Razavi, M. Y. Masoomi, T. Islamoglu, A. Morsali, Y. Xu, J. T. Hupp, O. K. Farha, J. Wang and P. C. Junk, *Inorg. Chem.*, 2017, **56**, 2581.
142. S. Horike, S. Bureekaew and S. Kitagawa, *Chem. Commun.*, 2008, 471.
143. Y. Hijikata, S. Horike, M. Sugimoto, M. Inukai, T. Fukushima and S. Kitagawa, *Inorg. Chem.*, 2013, **52**, 3634.
144. K. Nakagawa, D. Tanaka, S. Horike, S. Shimomura, M. Higuchi and S. Kitagawa, *Chem. Commun.*, 2010, **46**, 4258.
145. Z.-H. Xuan, D.-S. Zhang, Z. Chang, T.-L. Hu and X.-H. Bu, *Inorg. Chem.*, 2014, **53**, 8985.
146. Q. Yang and C. Zhong, *ChemPhysChem*, 2006, **7**, 1417.
147. J. Duan, Z. Yang, J. Bai, B. Zheng, Y. Li and S. Li, *Chem. Commun.*, 2012, **48**, 3058.
148. Z. Liang, M. Marshall and A. L. Chaffee, *Microporous Mesoporous Mater.*, 2010, **132**, 305.
149. J. Ethiraj, E. Albanese, B. Civalleri, J. G. Vitillo, F. Bonino, S. Chavan, G. C. Shearer, K. P. Lillerud and S. Bordiga, *ChemSusChem*, 2014, **7**, 3382.

150. Z. H. Rada, H. R. Abid, H. Sun and S. Wang, *J. Chem. Eng. Data*, 2015, **60**, 2152.
151. D. Tanaka, K. Nakagawa, M. Higuchi, S. Horike, Y. Kubota, T. C. Kobayashi, M. Takata and S. Kitagawa, *Angew. Chem., Int. Ed.*, 2008, **47**, 3914.
152. C. Gücüyener, J. van den Bergh, J. Gascon and F. Kapteijn, *J. Am. Chem. Soc.*, 2010, **132**, 17704.
153. Z.-J. Lin, J. Lu, M. Hong and R. Cao, *Chem. Soc. Rev.*, 2014, **43**, 5867.
154. C.-L. Chen, J.-Y. Zhang and C.-Y. Su, *Eur. J. Inorg. Chem.*, 2007, **2007**, 2997.
155. Q. Yang, D. Liu, C. Zhong and J.-R. Li, *Chem. Rev.*, 2013, **113**, 8261.
156. Y.-Q. Lan, H.-L. Jiang, S.-L. Li and Q. Xu, *Adv. Mater.*, 2011, **23**, 5015.
157. A. Schoedel, A. J. Cairns, Y. Belmabkhout, L. Wojtas, M. Mohamed, Z. Zhang, D. M. Proserpio, M. Eddaoudi and M. J. Zaworotko, *Angew. Chem., Int. Ed.*, 2013, **52**, 2902.
158. M. Arıcı, O. Z. Yeşilel, M. Taş and H. Demiral, *Inorg. Chem.*, 2015, **54**, 11283.
159. R. A. Agarwal and S. Mukherjee, *Polyhedron*, 2016, **105**, 228.
160. Z.-J. Lin, T.-F. Liu, X.-L. Zhao, J. Lü and R. Cao, *Cryst. Growth Des.*, 2011, **11**, 4284.
161. W. Zhuang, D. Yuan, D. Liu, C. Zhong, J.-R. Li and H.-C. Zhou, *Chem. Mater.*, 2012, **24**, 18.
162. R. Patra, H. M. Titi and I. Goldberg, *CrystEngComm*, 2013, **15**, 2853.
163. Z. Liang, J. Du, L. Sun, J. Xu, Y. Mu, Y. Li, J. Yu and R. Xu, *Inorg. Chem.*, 2013, **52**, 10720.
164. O. K. Farha and J. T. Hupp, *Acc. Chem. Res.*, 2010, **43**, 1166.
165. Y.-S. Xue, Y. He, L. Zhou, F.-J. Chen, Y. Xu, H.-B. Du, X.-Z. You and B. Chen, *J. Mater. Chem. A*, 2013, **1**, 4525.
166. Z.-J. Lin, T.-F. Liu, Y.-B. Huang, J. Lü and R. Cao, *Chem.–Eur. J.*, 2012, **18**, 7896.
167. B. Zheng, J. Bai, J. Duan, L. Wojtas and M. J. Zaworotko, *J. Am. Chem. Soc.*, 2011, **133**, 748.
168. B. Zheng, Z. Yang, J. Bai, Y. Li and S. Li, *Chem. Commun.*, 2012, **48**, 7025.
169. B. Zheng, H. Liu, Z. Wang, X. Yu, P. Yi and J. Bai, *CrystEngComm*, 2013, **15**, 3517.
170. D. Yuan, D. Zhao, D. Sun and H.-C. Zhou, *Angew. Chem., Int. Ed.*, 2010, **49**, 5357.
171. D. Zhao, D. Yuan, D. Sun and H.-C. Zhou, *J. Am. Chem. Soc.*, 2009, **131**, 9186.
172. A. Spek, *J. Appl. Crystallogr.*, 2003, **36**, 7.
173. B. Li, Z. Zhang, Y. Li, K. Yao, Y. Zhu, Z. Deng, F. Yang, X. Zhou, G. Li, H. Wu, N. Nijem, Y. J. Chabal, Z. Lai, Y. Han, Z. Shi, S. Feng and J. Li, *Angew. Chem., Int. Ed.*, 2012, **51**, 1412.
174. M. Arici, O. Z. Yeşilel, M. Taş and H. Demiral, *Cryst. Growth Des.*, 2017, **17**, 2654.

175. P. Cui, Y.-G. Ma, H.-H. Li, B. Zhao, J.-R. Li, P. Cheng, P. B. Balbuena and H.-C. Zhou, *J. Am. Chem. Soc.*, 2012, **134**, 18892.
176. H. J. Park and M. P. Suh, *Chem. Sci.*, 2013, **4**, 685.
177. Y. Zhu, Y.-M. Wang, S.-Y. Zhao, P. Liu, C. Wei, Y.-L. Wu, C.-K. Xia and J.-M. Xie, *Inorg. Chem.*, 2014, **53**, 7692.
178. V. Bon, I. Senkovska, D. Wallacher, D. M. Többens, I. Zizak, R. Feyerherm, U. Mueller and S. Kaskel, *Inorg. Chem.*, 2014, **53**, 1513.
179. J. Wang, J. Luo, J. Zhao, D.-S. Li, G. Li, Q. Huo and Y. Liu, *Cryst. Growth Des.*, 2014, **14**, 2375.
180. P. Z. Moghadam, J. F. Ivy, R. K. Arvapally, A. M. dos Santos, J. C. Pearson, L. Zhang, E. Tylianakis, P. Ghosh, I. W. H. Oswald, U. Kaipa, X. Wang, A. K. Wilson, R. Q. Snurr and M. A. Omary, *Chem. Sci.*, 2017, **8**, 3989.
181. K. Zhang, A. Nalaparaju and J. Jiang, *J. Mater. Chem. A*, 2015, **3**, 16327.
182. E. Barea, G. Tagliabue, W.-G. Wang, M. Pérez-Mendoza, L. Mendez-Liñan, F. J. López-Garzon, S. Galli, N. Masciocchi and J. A. R. Navarro, *Chem. –Eur. J.*, 2010, **16**, 931.
183. S. Ma, D. Sun, M. Ambrogio, J. A. Fillinger, S. Parkin and H.-C. Zhou, *J. Am. Chem. Soc.*, 2007, **129**, 1858.
184. P. K. Thallapally, J. Tian, M. Radha Kishan, C. A. Fernandez, S. J. Dalgarno, P. B. McGrail, J. E. Warren and J. L. Atwood, *J. Am. Chem. Soc.*, 2008, **130**, 16842.
185. P. Nugent, Y. Belmabkhout, S. D. Burd, A. J. Cairns, R. Luebke, K. Forrest, T. Pham, S. Ma, B. Space, L. Wojtas, M. Eddaoudi and M. J. Zaworotko, *Nature*, 2013, **495**, 80.
186. J. Kim, S.-T. Yang, S. B. Choi, J. Sim, J. Kim and W.-S. Ahn, *J. Mater. Chem.*, 2011, **21**, 3070.
187. Q. Yao, J. Su, O. Cheung, Q. Liu, N. Hedin and X. Zou, *J. Mater. Chem.*, 2012, **22**, 10345.
188. R. Yun, Z. Lu, Y. Pan, X. You and J. Bai, *Angew. Chem., Int. Ed.*, 2013, **52**, 11282.
189. S. Ma, X.-S. Wang, E. S. Manis, C. D. Collier and H.-C. Zhou, *Inorg. Chem.*, 2007, **46**, 3432.
190. T. K. Maji, R. Matsuda and S. Kitagawa, *Nat. Mater.*, 2007, **6**, 142.
191. Y. E. Cheon and M. P. Suh, *Chem.–Eur. J.*, 2008, **14**, 3961.
192. A. Phan, C. J. Doonan, F. J. Uribe-Romo, C. B. Knobler, M. O'Keeffe and O. M. Yaghi, *Acc. Chem. Res.*, 2010, **43**, 58.
193. K. S. Park, Z. Ni, A. P. Côté, J. Y. Choi, R. Huang, F. J. Uribe-Romo, H. K. Chae, M. O'Keeffe and O. M. Yaghi, *Proc. Natl. Acad. Sci. U. S. A.*, 2006, **103**, 10186.
194. Z. Zhang, S. Xian, Q. Xia, H. Wang, Z. Li and J. Li, *AIChE J.*, 2013, **59**, 2195.
195. D. Danaci, R. Singh, P. Xiao and P. A. Webley, *Chem. Eng. J.*, 2015, **280**, 486.
196. J. McEwen, J.-D. Hayman and A. Ozgur Yazaydin, *Chem. Phys.*, 2013, **412**, 72.

197. R. Banerjee, H. Furukawa, D. Britt, C. Knobler, M. O'Keeffe and O. M. Yaghi, *J. Am. Chem. Soc.*, 2009, **131**, 3875.
198. H. Amrouche, S. Aguado, J. Pérez-Pellitero, C. Chizallet, F. Siperstein, D. Farrusseng, N. Bats and C. Nieto-Draghi, *J. Phys. Chem. C*, 2011, **115**, 16425.
199. B. Wang, A. P. Cote, H. Furukawa, M. O'Keeffe and O. M. Yaghi, *Nature*, 2008, **453**, 207.
200. J. A. Thompson, N. A. Brunelli, R. P. Lively, J. R. Johnson, C. W. Jones and S. Nair, *J. Phys. Chem. C*, 2013, **117**, 8198.
201. C. Chen, J. Kim, D.-W. Park and W.-S. Ahn, *Mater. Lett.*, 2013, **106**, 344.
202. A. W. Thornton, D. Dubbeldam, M. S. Liu, B. P. Ladewig, A. J. Hill and M. R. Hill, *Energy Environ. Sci.*, 2012, **5**, 7637.
203. H. R. Moon, D.-W. Lim and M. P. Suh, *Chem. Soc. Rev.*, 2013, **42**, 1807.
204. A. J. Brown, J. R. Johnson, M. E. Lydon, W. J. Koros, C. W. Jones and S. Nair, *Angew. Chem., Int. Ed.*, 2012, **51**, 10615.
205. S. J. Yang, J. Y. Choi, H. K. Chae, J. H. Cho, K. S. Nahm and C. R. Park, *Chem. Mater.*, 2009, **21**, 1893.
206. P. B. Somayajulu Rallapalli, M. C. Raj, D. V. Patil, K. P. Prasanth, R. S. Somani and H. C. Bajaj, *Int. J. Energy Res.*, 2013, **37**, 746.
207. X. Zhou, W. Huang, J. Miao, Q. Xia, Z. Zhang, H. Wang and Z. Li, *Chem. Eng. J.*, 2015, **266**, 339.
208. Z. Zhang, H. Wang, X. Chen, C. Zhu, W. Wei and Y. Sun, *Adsorption*, 2015, **21**, 77.
209. D. Qian, C. Lei, G.-P. Hao, W.-C. Li and A.-H. Lu, *ACS Appl. Mater. Interfaces*, 2012, **4**, 6125.
210. L. Ge, L. Wang, V. Rudolph and Z. Zhu, *RSC Adv.*, 2013, **3**, 25360.
211. S. Liu, L. Sun, F. Xu, J. Zhang, C. Jiao, F. Li, Z. Li, S. Wang, Z. Wang, X. Jiang, H. Zhou, L. Yang and C. Schick, *Energy Environ. Sci.*, 2013, **6**, 818.
212. W. Huang, X. Zhou, Q. Xia, J. Peng, H. Wang and Z. Li, *Ind. Eng. Chem. Res.*, 2014, **53**, 11176.
213. M. N. Martin, J. I. Basham, P. Chando and S.-K. Eah, *Langmuir*, 2010, **26**, 7410.
214. A. Policicchio, Y. Zhao, Q. Zhong, R. G. Agostino and T. J. Bandosz, *ACS Appl. Mater. Interfaces*, 2014, **6**, 101.
215. Z. Xiang, X. Peng, X. Cheng, X. Li and D. Cao, *J. Phys. Chem. C*, 2011, **115**, 19864.
216. N. Iqbal, X. Wang, J. Yu, N. Jabeen, H. Ullah and B. Ding, *RSC Adv.*, 2016, **6**, 4382.
217. R. Lin, L. Ge, S. Liu, V. Rudolph and Z. Zhu, *ACS Appl. Mater. Interfaces*, 2015, **7**, 14750.
218. A. K. Adhikari and K.-S. Lin, *Chem. Eng. J.*, 2016, **284**, 1348.
219. Y. Yang, L. Ge, V. Rudolph and Z. Zhu, *Dalton Trans.*, 2014, **43**, 7028.
220. Y. Shen, Z. Li, L. Wang, Y. Ye, Q. Liu, X. Ma, Q. Chen, Z. Zhang and S. Xiang, *J. Mater. Chem. A*, 2015, **3**, 593.

221. S. Xiang, Y. He, Z. Zhang, H. Wu, W. Zhou, R. Krishna and B. Chen, *Nat. Commun.*, 2012, **3**, 954.
222. P. Chowdhury, S. Mekala, F. Dreisbach and S. Gumma, *Microporous Mesoporous Mater.*, 2012, **152**, 246.
223. A. Chakraborty and T. K. Maji, *APL Mater.*, 2014, **2**, 124107.
224. Y. Liu, S. Liu and Z. Yue, *RSC Adv.*, 2015, **5**, 31742.
225. A. Chakraborty, S. Roy, M. Eswaramoorthy and T. K. Maji, *J. Mater. Chem. A*, 2017, **5**, 8423.
226. C. Chen, B. Li, L. Zhou, Z. Xia, N. Feng, J. Ding, L. Wang, H. Wan and G. Guan, *ACS Appl. Mater. Interfaces*, 2017, **9**, 23060.
227. B. Seoane, C. Téllez, J. Coronas and C. Staudt, *Sep. Purif. Technol.*, 2013, **111**, 72.
228. T. Li, Y. Pan, K.-V. Peinemann and Z. Lai, *J. Membr. Sci.*, 2013, **425**, 235.
229. L. Ge, W. Zhou, V. Rudolph and Z. Zhu, *J. Mater. Chem. A*, 2013, **1**, 6350.
230. J. Hu, H. Cai, H. Ren, Y. Wei, Z. Xu, H. Liu and Y. Hu, *Ind. Eng. Chem. Res.*, 2010, **49**, 12605.
231. X. Y. Chen, H. Vinh-Thang, D. Rodrigue and S. Kaliaguine, *Ind. Eng. Chem. Res.*, 2012, **51**, 6895.
232. T.-H. Bae, J. S. Lee, W. Qiu, W. J. Koros, C. W. Jones and S. Nair, *Angew. Chem., Int. Ed.*, 2010, **49**, 9863.
233. P. L. Llewellyn, S. Bourrelly, C. Serre, Y. Filinchuk and G. Férey, *Angew. Chem., Int. Ed.*, 2006, **45**, 7751.
234. A. C. Kizzie, A. G. Wong-Foy and A. J. Matzger, *Langmuir*, 2011, **27**, 6368.
235. O. Shekhah, Y. Belmabkhout, Z. Chen, V. Guillerm, A. Cairns, K. Adil and M. Eddaoudi, *Nat. Commun.*, 2014, **5**, 4228.
236. P.-Q. Liao, H. Chen, D.-D. Zhou, S.-Y. Liu, C.-T. He, Z. Rui, H. Ji, J.-P. Zhang and X.-M. Chen, *Energy Environ. Sci.*, 2015, **8**, 1011.
237. R. A. Peralta, B. Alcantar-Vazquez, M. Sanchez-Serratos, E. Gonzalez-Zamora and I. A. Ibarra, *Inorg. Chem. Front.*, 2015, **2**, 898.
238. E. Sánchez-González, J. R. Álvarez, R. A. Peralta, A. Campos-Reales-Pineda, A. Tejeda-Cruz, E. Lima, J. Balmaseda, E. González-Zamora and I. A. Ibarra, *ACS Omega*, 2016, **1**, 305.
239. E. Soubeyrand-Lenoir, C. Vagner, J. W. Yoon, P. Bazin, F. Ragon, Y. K. Hwang, C. Serre, J.-S. Chang and P. L. Llewellyn, *J. Am. Chem. Soc.*, 2012, **134**, 10174.
240. H. Jasuja, J. Zang, D. S. Sholl and K. S. Walton, *J. Phys. Chem. C*, 2012, **116**, 23526.
241. J. Qian, F. Jiang, D. Yuan, M. Wu, S. Zhang, L. Zhang and M. Hong, *Chem. Commun.*, 2012, **48**, 9696.
242. I. A. Ibarra, S. Yang, X. Lin, A. J. Blake, P. J. Rizkallah, H. Nowell, D. R. Allan, N. R. Champness, P. Hubberstey and M. Schroder, *Chem. Commun.*, 2011, **47**, 8304.
243. M. Sanchez-Serratos, P. A. Bayliss, R. A. Peralta, E. Gonzalez-Zamora, E. Lima and I. A. Ibarra, *New J. Chem.*, 2016, **40**, 68.

244. A. Zarate, R. A. Peralta, P. A. Bayliss, R. Howie, M. Sanchez-Serratos, P. Carmona-Monroy, D. Solis-Ibarra, E. Gonzalez-Zamora and I. A. Ibarra, *RSC Adv.*, 2016, **6**, 9978.
245. J. A. Mason, T. M. McDonald, T.-H. Bae, J. E. Bachman, K. Sumida, J. J. Dutton, S. S. Kaye and J. R. Long, *J. Am. Chem. Soc.*, 2015, **137**, 4787.
246. N. T. T. Nguyen, H. Furukawa, F. Gándara, H. T. Nguyen, K. E. Cordova and O. M. Yaghi, *Angew. Chem., Int. Ed.*, 2014, **53**, 10645.

CHAPTER 4

Alkali-metal-carbonate-based CO_2 Adsorbents

HIROFUMI KANOH*[a] AND HONGCHAO LUO[b]

[a]Graduate School of Science, Chiba University, 1-33 Yayoi-cho, Inage-ku, Chiba, 263-8522, Japan; [b]Chemistry and Material Engineering, Liupanshui Normal University, No. 19, YUCAI Lane, Minghu Road, Zhongshan District, Liupanshui City, Guizhou Province, China
*E-mail: kanoh@faculty.chiba-u.jp

4.1 Introduction

Anthropogenic greenhouse gas emissions have increased since the industrial revolution, leading to atmospheric concentrations of carbon dioxide, methane and nitrous oxide that are unprecedented, and are now higher than ever. The effects of increased atmospheric concentrations of greenhouse gases caused by fossil fuel burning, industrialization and deforestation together with other anthropogenic activities are extremely likely to have been the dominant reason for the observed warming. Since CO_2 is the major anthropogenic greenhouse gas in the atmosphere contributing to global warming, in recent years, worldwide efforts have been devoted to developing the mitigation options. There are four effective ways to reduce CO_2 emissions into the atmosphere: (I) improving energy utilization efficiency, (II) increasing the use of less carbon-intensive fuels, biomass-based fuels, nuclear power and renewable energy sources such as solar energy, (III) enhancing biological

Inorganic Materials Series No. 2
Post-combustion Carbon Dioxide Capture Materials
Edited by Qiang Wang

Published by the Royal Society of Chemistry, www.rsc.org

sinks and reducing non-CO_2 greenhouse gas emissions, and (IV) carbon dioxide capture and storage (CCS). However, fossil fuels are still the dominant energy source at present. Furthermore, other energy sources, such as nuclear power and solar energy, still cannot replace fossil fuels on a large scale. As a result, CCS is widely used for the reduction of the CO_2 concentration. CCS as a main strategy for CO_2 management is a process consisting of several stages such as the separation and capture of CO_2 from large emission sources, *i.e.*, industrial and energy-related sources, transportation to the storage site and storage in a suitably deep geological formation for long-term isolation from the atmosphere.[1–4] Amine-, Zeolite13X-, and activated-carbon-based CO_2 capture has been the main focus in this field.[5–7] However, these technologies have not been established because of several issues, such as high energy consumption during sorbent regeneration, amine loss during regeneration, and low amine utilization efficiency. Furthermore, Zeolite13X- and activated-carbon-based CO_2 sorbents under ambient conditions have CO_2 efficiency problems because of a decrease in the amount of CO_2 captured by coexisting water vapors, as well as the high cost of gas removal. Thus, these sorbents are not appropriate for CO_2 capture applications under moist conditions. Usable CO_2 sorbents must cause effective enhancement in the sorption rate, saturated amount, and stability for long-term CO_2 capture.

For several years, more attention has been paid to alkali-metal-carbonate-based sorbents as typical low-temperature sorbents for CO_2 capture applications due to both their high sorption capacities and low costs.[8–14] Many studies on the CO_2 capture of alkali-metal-carbonate-based sorbents have focused on both experimental fuel gas conditions and porous supports. Particularly, the experimental conditions of CO_2 capture include reaction or regeneration temperature, CO_2 concentration, H_2O concentration, *etc.*[8]

Most previous studies on the CO_2 capture of alkali-metal-carbonate-based sorbents such as K_2CO_3 and Na_2CO_3 showed that they took place within a relatively high temperature range (323–393 K), such as the following reaction (4.1):

$$M_2CO_3(s) + CO_2(g) + H_2O(g) \leftrightarrows 2MHCO_3(s) \tag{4.1}$$

where the forward, exothermic reaction is the bicarbonate formation of M_2CO_3 (M: Na, K) while the reverse, endothermic reaction is the decomposition of $MHCO_3$.

4.1.1 Sodium Carbonate (Na_2CO_3)

For much of the research into the CO_2 occlusion of Na_2CO_3 under different CO_2 and H_2O concentrations, CO_2 capture did not always occur *via* reaction (4.1). Several reactions of CO_2 occlusion of Na_2CO_3 under moist conditions were as follows:[15]

$$\begin{gathered} Na_2CO_3(s) + H_2O(g) \leftrightarrows Na_2CO_3{\cdot}H_2O(s) \\ H_2O \text{ capture capacity, } 9.43 \text{ mmol g}^{-1} \end{gathered} \tag{4.2}$$

$$5Na_2CO_3(s) + 3CO_2(g) + 3H_2O(g) \leftrightarrows 2Na_5H_3(CO_3)_4(s) \quad \text{CO}_2 \text{ capture capacity, 5.66 mmol g}^{-1} \tag{4.3}$$

$$Na_2CO_3(s) + CO_2(g) + H_2O(g) \leftrightarrows 2NaHCO_3(s) \quad \text{CO}_2 \text{ capture capacity, 9.43 mmol g}^{-1} \tag{4.4}$$

Different precursors, such as Na_2CO_3, $NaHCO_3$, and $Na_2CO_3{\cdot}NaHCO_3{\cdot}2H_2O$, were also applied to CO_2 capture in moist conditions.[15] It is found that Na_2CO_3 decomposed from $NaHCO_3$ has a better CO_2 occlusion performance because of its good pore structure. Thus, in this study, $NaHCO_3$ is used to prepare Na_2CO_3 under a N_2 atmosphere for CO_2 capture.

4.1.2 Potassium Carbonate (K_2CO_3)

The reactions of CO_2 occlusion of K_2CO_3 under moist conditions are shown in reactions (4.5)–(4.8):[12]

$$K_2CO_3(s) + 1.5H_2O(g) \leftrightarrows K_2CO_3{\cdot}1.5H_2O(s) \quad \text{H}_2\text{O capture capacity, 10.8 mmol g}^{-1} \tag{4.5}$$

$$2K_2CO_3(s) + CO_2(g) + 2.5H_2O(g) \leftrightarrows K_4H_2(CO_3)_3{\cdot}1.5H_2O(s) + 0.5H_2O(g) \quad \text{CO}_2 \text{ capture capacity, 3.62 mmol g}^{-1} \tag{4.6}$$

$$K_4H_2(CO_3)_3{\cdot}1.5H_2O(s) + CO_2(g) \leftrightarrows 4KHCO_3(s) + 0.5H_2O(g) \quad \text{CO}_2 \text{ capture capacity, 2.73 mmol g}^{-1} \tag{4.7}$$

$$K_2CO_3(s) + CO_2(g) + H_2O(g) \leftrightarrows 2KHCO_3(s) \quad \text{CO}_2 \text{ capture capacity, 7.24 mmol g}^{-1} \tag{4.8}$$

Furthermore, different precursors of K_2CO_3 also influence the CO_2 capture behavior. Zhao *et al.* reported that K_2CO_3 from different precursors, such as K_2CO_3, $K_2CO_3{\cdot}1.5H_2O$, and $KHCO_3$, was used to occlude CO_2 in moist conditions.[16] It is found that K_2CO_3 decomposed from $KHCO_3$ has a hexagonal crystal structure and a better pore structure, favorable for CO_2 capture, while K_2CO_3 from other precursors has, in general, a monoclinic crystal structure.[17] Thus, in this study, $KHCO_3$ can be used to prepare K_2CO_3 under a N_2 atmosphere for CO_2 capture.

Thus, the kinetics mechanism of pure Na_2CO_3 or K_2CO_3 for CO_2 capture in moist conditions is not clearly understood. Hence, studying the fundamental kinetics of CO_2 occlusion of pure Na_2CO_3 or K_2CO_3 is important. We systematically studied CO_2 occlusion of Na_2CO_3 or K_2CO_3 by changing various reaction conditions, such as temperature, CO_2 concentration, and humidity, based on the CO_2 capture characteristics of pure Na_2CO_3 or K_2CO_3 under moist conditions.

In this chapter, we focus on the capacity and kinetics in the CO_2 capture of Na_2CO_3 [18] and K_2CO_3 [19,20] under different temperatures, and CO_2 and H_2O concentrations. Detailed information on the CO_2 capture process mainly involves crystal structure changes and morphology variations in the CO_2 capture of Na_2CO_3 and K_2CO_3. Therefore, a series of techniques will be powerful methods to investigate the crystal structure changes and morphology variations, such as thermogravimetric-differential thermal analysis (TG-DTA), X-ray diffraction (XRD) and scanning electron microscopy (SEM).

Also, some results for the improvement of the regeneration temperature from bicarbonates and kinetics of the CO_2 occlusion reaction will be described.

4.2 CO_2 Capture of Na_2CO_3 and K_2CO_3 Under Moist Conditions

4.2.1 CO_2 Capture of Na_2CO_3 Under Moist Conditions

With a high occlusion capacity and a low cost, Na_2CO_3 as a CO_2 sorbent under moist conditions at 0.1 MPa and above 313 K has been studied extensively[8,10,21–24] as well as K_2CO_3. Humidity plays a crucial role in the CO_2 occlusion amount and the overall reaction rate. By increasing the H_2O concentration, the CO_2 occlusion amount and rate increased significantly.[10,22–24] Furthermore, it was reported that the hydrate of Na_2CO_3 could be formed slowly through reaction (4.9) in the presence of excess water vapor at below 333 K:[15]

$$Na_2CO_3(s) + H_2O(g) \leftrightarrows Na_2CO_3{\cdot}H_2O(s) \tag{4.9}$$

The temperature also plays an important role in CO_2 capture. Most previous studies regarding the CO_2 capture of Na_2CO_3-based sorbents were performed using the thermogravimetric analyzer, several fixed-bed, and fluidized-bed, where the flue gas was simulated with 5%–10% CO_2 with 10%–20% H_2O at relatively high temperatures (323–353 K) under atmospheric pressure.[8,15,25,26] In numerous studies on the CO_2 sorption of Na_2CO_3 under moist conditions, it was found that Na_2CO_3 decomposed from $NaHCO_3$ had a better CO_2 capture performance, whereas Na_2CO_3 and $Na_2CO_3{\cdot}H_2O$ hardly reacted with CO_2.[22] When the flue gas conditions were 8% CO_2 with 16% H_2O in helium, Na_2CO_3 decomposed from $NaHCO_3$ could be converted to $Na_5H_3(CO_3)_4$ through reaction (4.11) at 343 K and higher, whereas $NaHCO_3$ was formed *via* reaction (4.10) at below 343 K.[8] Dong *et al.* also reported that the formation of $Na_5H_3(CO_3)_4$ could be transformed from Na_2CO_3 supported on Al_2O_3 through reaction (4.11) under moist flue gas (15% CO_2 with 15% H_2O) in nitrogen at around 333 K.[25,26] The relevant reactions are:[8]

$$Na_2CO_3(s) + CO_2(g) + H_2O(g) \leftrightarrows 2NaHCO_3(s) \quad (4.10)$$

$$5Na_2CO_3(s) + 3CO_2(g) + 3H_2O(g) \leftrightarrows 2Na_5H_3(CO_3)_4(s) \quad (4.11)$$

where the forward, exothermic reactions (4.10) and (4.11) are the bicarbonate and Wegscheider's salt ($Na_5H_3(CO_3)_4$) formation of Na_2CO_3, respectively, corresponding to theoretical CO_2 capture amounts of 9.43 and 5.66 mmol g^{-1}, while the reverse, endothermic reactions are the decomposition of $NaHCO_3$ and $Na_5H_3(CO_3)_4$, respectively. These results indicate that lower temperatures favor bicarbonate formation.

However, detailed crystal structure changes during the reaction have not been shown so far. The morphologies of Na_2CO_3 particles also change during the reaction processes because the crystal grows depending on the reaction conditions. The experimental results of structural and morphological changes during the reaction enable us to understand the kinetic behavior of Na_2CO_3 particles. But since details of the morphology changes are not well-known, CO_2 capture mechanisms under different CO_2 and H_2O conditions are not understood clearly.

In this section, the CO_2 sorptivities of Na_2CO_3 at different temperatures under various CO_2 and H_2O concentrations were investigated by thermogravimetric-differential thermal analysis (TG-DTA) and structural changes accompanying the reaction were shown by X-ray diffraction (XRD) *in vacuo*. Morphology changes were observed with a scanning electron microscope (SEM). According to crystal structure and kinetics analyses of the bicarbonate formation of Na_2CO_3, the dependence of the CO_2 capture of Na_2CO_3 on CO_2 concentration and humidity can be explained in detail.

4.2.1.1 *Experimental*

4.2.1.1.1 Sample Preparation. Analytical reagent grade sodium bicarbonate ($NaHCO_3$) was used during experiments of decomposition of $NaHCO_3$ and bicarbonate formation of Na_2CO_3. The decomposition of $NaHCO_3$ was done by treatment with the thermogravimetric-differential thermal analysis (TG-DTA). $NaHCO_3$ (30–50 mg) was heated at 5 K min^{-1} from 298 to 473 K to form Na_2CO_3 in pure N_2 gas at 100 cm^3 min^{-1} and kept at these conditions for 5 min until complete decomposition. Then, the temperature was cooled to 303 K at 5 K min^{-1} and kept for 30 min under pure N_2 gas.

4.2.1.1.2 Bicarbonate Formation Measurements. The obtained samples were processed with the TG-DTA apparatus using a gas composition of CO_2 and H_2O (v/v of 98.2 : 1.8) at a flow rate of 100 cm^3 min^{-1} at temperatures between 303 and 363 K for 18 h.

The bicarbonate formation of Na_2CO_3 was examined with the TG-DTA apparatus with several rates of CO_2 and N_2 containing saturated water vapor at 303 K for 300 min. CO_2 and N_2 gases were supplied as high-purity gases from cylinders with mass flow controllers to control the flow rate. Water vapor was introduced by flowing gases of N_2 and CO_2 into distilled water in a bubbler for more than 5 h to reach saturation, where N_2 gas was used as a balance gas.

On the other hand, under different humidity conditions, measurements of the bicarbonate formation of Na_2CO_3 were also performed with the TG-DTA instrument. Water vapor was supplied by introducing N_2 gas into distilled water inside a bubbler and N_2 gas was also used as a dry gas inside a bubbler to adjust the H_2O partial pressure.

For the bicarbonate formation measurements, the mixed gas of CO_2, H_2O and N_2 was put into a moisture detector with a thermometer and hygrometer at 313 K. The relative humidity of the mixed gas was determined with the hygrometer. The total gas flow rate was maintained at 100 $cm^3\ min^{-1}$ for all experiments.

4.2.1.1.3 Crystal Structure and Morphology Measurements. The crystal structures of the products after CO_2 occlusion reactions with 10, 20, 40, and 300 min were measured by using an X-ray diffractometer (XRD) *in vacuo* to avoid reactions with CO_2 or H_2O under the ambient atmosphere. The powder XRD patterns were obtained in the 2θ range of 25°–45° using Cu Kα radiation (40 kV, 25 mA, and λ = 0.15406 nm) at room temperature. The morphology of Na_2CO_3 before and after CO_2 sorption for various reaction times was observed with a scanning electron microscope (SEM) after applying a metallic-osmium coating onto the samples.

4.2.1.2 Results and Discussion

4.2.1.2.1 CO_2 Capture of Na_2CO_3 at Different Temperatures. The decomposition of $NaHCO_3$ occurred with TG-DTA under pure N_2 gas *via* the reverse reaction of (4.10) to form Na_2CO_3, CO_2, and H_2O at 473 K, as shown in Figure 4.1. The weight for the decomposition of $NaHCO_3$ decreased to 63.0% (±0.2%), which is in good agreement with the theoretical value of 63.1% according to the reverse reaction of (4.10). This confirms that the decomposition reaction proceeds completely. After the decomposition, a mixture of CO_2 and H_2O was supplied to Na_2CO_3 at different temperatures. As shown in Figure 4.1, the sample weight and bicarbonate formation rate proceeded slowly in the temperature range of 303 K to 343 K. This indicates that Na_2CO_3 should be a promising sorbent under moist conditions at temperatures near room temperature, although the capture rate is much slower than that of K_2CO_3.[20] When the temperature was 303 K, there was a rapid increase in sample weight over 3 h and the CO_2 occlusion amount was 8.23 mmol g^{-1}, which is 87.2% of the theoretical amount (*i.e.*, 9.43 mol g^{-1}). Compared with the result at 303 K, both CO_2 occlusion rate and amount decreased slightly at 323 K. As the temperature increased from 323 K to 343 K, the CO_2 occlusion

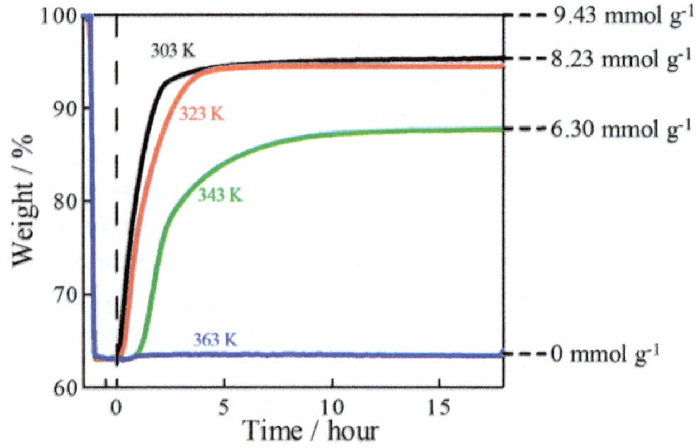

Figure 4.1 Weight change by the decomposition of $NaHCO_3$ (<0 hour) at 473 K and the bicarbonate formation of Na_2CO_3 with CO_2 and H_2O (>0 hour) at different temperatures.

amount only increased to 6.30 mmol g^{-1} along with the initial formation proceeding slowly. When the temperature was increased to 363 K, the CO_2 occlusion amount was 0 mmol g^{-1}. These CO_2 occlusion processes at different temperatures reached a plateau after 5 h. In general, since bicarbonate formation requires a minimum activation energy for the occurrence of the chemical reaction, a reaction rate would become more rapid at higher temperatures. However, the slow increase in the weight change in Figure 4.1 at higher temperatures indicated that CO_2 occlusion of Na_2CO_3 did not simply depend upon the temperature.

Because the temperature dependence of the CO_2 occlusion amount and rate was not understood clearly, the crystal structure after CO_2 occlusion at different temperatures was examined to obtain XRD patterns, as shown in Figure 4.2. The XRD patterns showed peaks consisting of two phases, *i.e.*, $NaHCO_3$ and $Na_5H_3(CO_3)_4$, at temperatures from 303 K to 343 K. Furthermore, the peak intensity of the $Na_5H_3(CO_3)_4$ phase increased with increasing temperature. This indicated that reaction (4.11) occurred favorably for relatively high temperatures, whereas reaction (4.10) proceeded at lower temperatures. Only the pure Na_2CO_3 phase existed in the XRD patterns of the CO_2 occluded sample at 363 K. These results were in good agreement with the decline in CO_2 occlusion amount with increasing temperature. Figure 4.3 shows DTA curves and derivative thermogravimetric (DTG) curves (obtained by differentiating the weight change in Figure 4.1 with respect to time) obtained at temperatures of 303–363 K. The DTA curves indicated exothermic reactions for the CO_2 occlusion of Na_2CO_3 under moist conditions while both DTA and DTG curves showed a striking similarity. With increasing temperature, the peak intensity diminished in the DTG and DTA curves, which corresponds to exothermic reactions of the formation of $NaHCO_3$ and $Na_5H_3(CO_3)_4$ through the reaction of Na_2CO_3 with CO_2 and H_2O. It can be confirmed that the CO_2 occlusion rate of reaction (4.11) is slower than that of

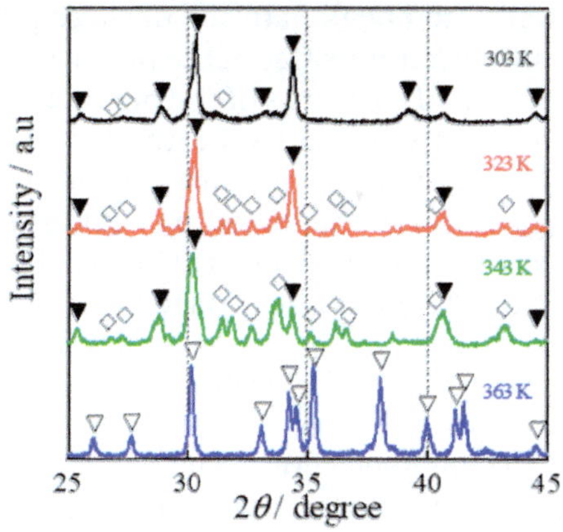

Figure 4.2 XRD patterns of CO_2 occluded samples at different temperatures. ▼: $NaHCO_3$, ◇: $Na_5H_3(CO_3)_4$, ▽: Na_2CO_3. Reproduced from ref. 18 with permission from Elsevier, Copyright 2017.

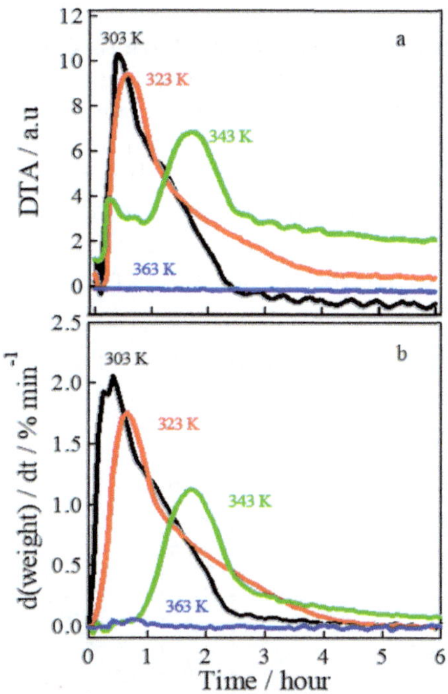

Figure 4.3 Time courses of DTA (a) and DTG (b) curves at different temperatures. Reproduced from ref. 18 with permission from Elsevier, Copyright 2017.

reaction (4.10) on account of the formation of much more $Na_5H_3(CO_3)_4$ at higher temperatures. Consequently, it can also be identified that the released heat of reaction (4.11) is lower than that of reaction (4.10) based on the peaks observed in the DTA curves.

According to the well-fitted peak patterns of these pure phases regarding the relative ratios of the peak intensities of the pure $NaHCO_3$ and $Na_5H_3(CO_3)_4$ phases, the experimental XRD patterns of the pure $NaHCO_3$ and $Na_5H_3(CO_3)_4$ phases were obtained at 303 K and 323 K. Furthermore, the ratio between the peak area of $NaHCO_3$ and the total peak area of $NaHCO_3$ and $Na_5H_3(CO_3)_4$, $I_{NaHCO_3}/(I_{NaHCO_3} + I_{Na_5H_3(CO_3)_4})$, was calculated by peak fitting each peak of the pure $NaHCO_3$ and $Na_5H_3(CO_3)_4$ phases at 303–343 K. Then, it was found that the ratio of the $NaHCO_3$ phase decreased with increasing temperature. This result further demonstrates that a low temperature contributes to the bicarbonate formation of Na_2CO_3 by occluding CO_2 and H_2O *via* the highly exothermic reaction (4.10).

The TG-DTA and XRD analyses show that the CO_2 occlusion of Na_2CO_3 at different temperatures involves two exothermic reactions of the formation of $NaHCO_3$ and $Na_5H_3(CO_3)_4$ from Na_2CO_3. The highly exothermic reaction (4.10) proceeds more easily at low temperatures, whereas reaction (4.11) proceeds at higher temperatures. Moreover, the released heat of reaction (4.11) is lower than that of reaction (4.10). Thus, Na_2CO_3 can be suitable for CO_2 capture at low temperatures.

4.2.1.2.2 CO_2 Capture of Na_2CO_3 Under Various CO_2 Concentrations. The dependence of the sorptivity of Na_2CO_3 on CO_2 concentration was shown with TG-DTA under various CO_2 flow rates, as shown for reaction times of 0–300 min in Figure 4.4. Under a sufficiently high humidity (RH of 74% at 303 K), a mixed gas (total flow rate of 100 cm^3 min^{-1}) of CO_2 and N_2 was supplied to Na_2CO_3. When the CO_2 flow rate was only 1 cm^3 min^{-1} (1%), the occlusion amount was 2.84 mmol g^{-1} at 300 min. As the CO_2 flow rate increased from 1 to 30 cm^3 min^{-1}, the CO_2 occlusion amount of Na_2CO_3 increased from 2.84 to

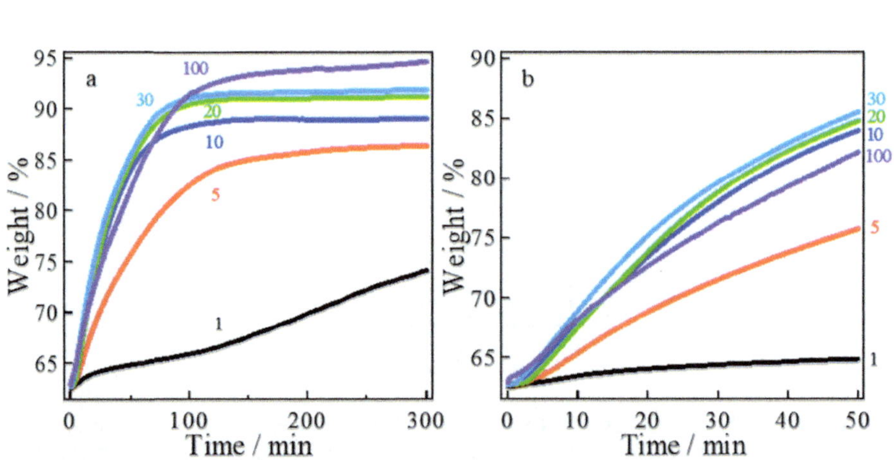

Figure 4.4 TG curves. (a) Bicarbonate formation at various CO_2 flow rates (cm^3 min^{-1}), which are indicated with numbers, with N_2 and H_2O. (b) Magnification for 0–50 min. Reproduced from ref. 18 with permission from Elsevier, Copyright 2017.

7.39 mmol g^{-1} at 300 min, which is 74.2%–92.0% of the theoretical amount of 9.43 mmol g^{-1}. Especially, at the initial 50 min, the CO_2 occlusion amount had already increased from 0.43 to 5.75 mmol g^{-1}, which is 64.8%–85.6% of the theoretical amount. When the CO_2 flow rate increased from 30 to 100 cm^3 min^{-1}, the occlusion amount of Na_2CO_3 increased to 8.10 mmol g^{-1} at 300 min, which is 94.8% of the theoretical amount of 9.43 mmol g^{-1}. These experimental results were consistent with higher CO_2 flow rates favorable for higher CO_2 occlusion amounts according to reactions (4.10) and (4.11). A higher CO_2 occlusion amount could be achieved by increasing the CO_2 flow rate from 30 to 100 cm^3 min^{-1}, but the CO_2 occlusion rate was much slower, as can be seen in Figure 4.4(b). When the CO_2 flow rate was 100 cm^3 min^{-1}, the CO_2 occlusion amount was only 4.91 mmol g^{-1} at 50 min, which is 82.3%, lower than 85.6% at 30 cm^3 min^{-1}. In contrast, when the CO_2 flow rate was 20 and 30 cm^3 min^{-1}, the CO_2 occlusion rate was much faster than other CO_2 flow rates.

Figure 4.5 shows the XRD patterns after the CO_2 occlusion of Na_2CO_3 (300 min in Figure 4.4) at various CO_2 flow rates. When the CO_2 flow rate was only 1 and 5 cm^3 min^{-1}, the $Na_2CO_3{\cdot}H_2O$ phase was present in the samples because of an unfavorable shift toward the forward reactions of (4.10) and (4.11) under lower CO_2 concentrations. While increasing the CO_2 concentration between 5–100 cm^3 min^{-1}, the peaks of the $Na_5H_3(CO_3)_4$ and $Na_2CO_3{\cdot}H_2O$ phases appeared and decreased while those of the $NaHCO_3$ phase increased. This result in the XRD patterns was consistent with the TG data.

The CO_2 occlusion of Na_2CO_3 at high CO_2 flow rates was limited during 50 min based on the TG results in Figure 4.4(b). Since there are different exothermic behaviors along with the formation of $NaHCO_3$ and $Na_5H_3(CO_3)_4$, the

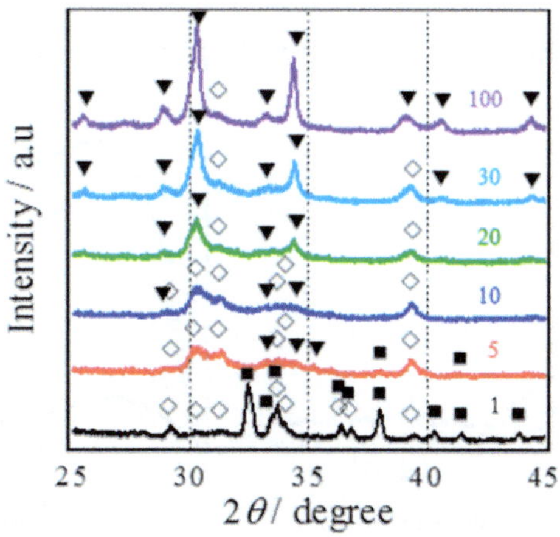

Figure 4.5 XRD patterns after CO_2 occlusion of Na_2CO_3 at various CO_2 flow rates (cm^3 min^{-1}), which are indicated with numbers. ▼: $NaHCO_3$, ◇: $Na_5H_3(CO_3)_4$, ■: $Na_2CO_3{\cdot}H_2O$. Reproduced from ref. 18 with permission from Elsevier, Copyright 2017.

transformation process of Na_2CO_3 leads to a temperature change of the reaction system. Hence, examining the transformation process of Na_2CO_3 and the associated changes in temperature during the formation of $NaHCO_3$ and $Na_5H_3(CO_3)_4$ is important for a better understanding of the CO_2 sorptivity kinetics of Na_2CO_3.

Temperature variation curves (a), DTA curves (b), and DTG curves (c) (obtained from Figure 4.4) for different CO_2 concentrations, which exhibited a series of peaks, can be seen in Figure 4.6.

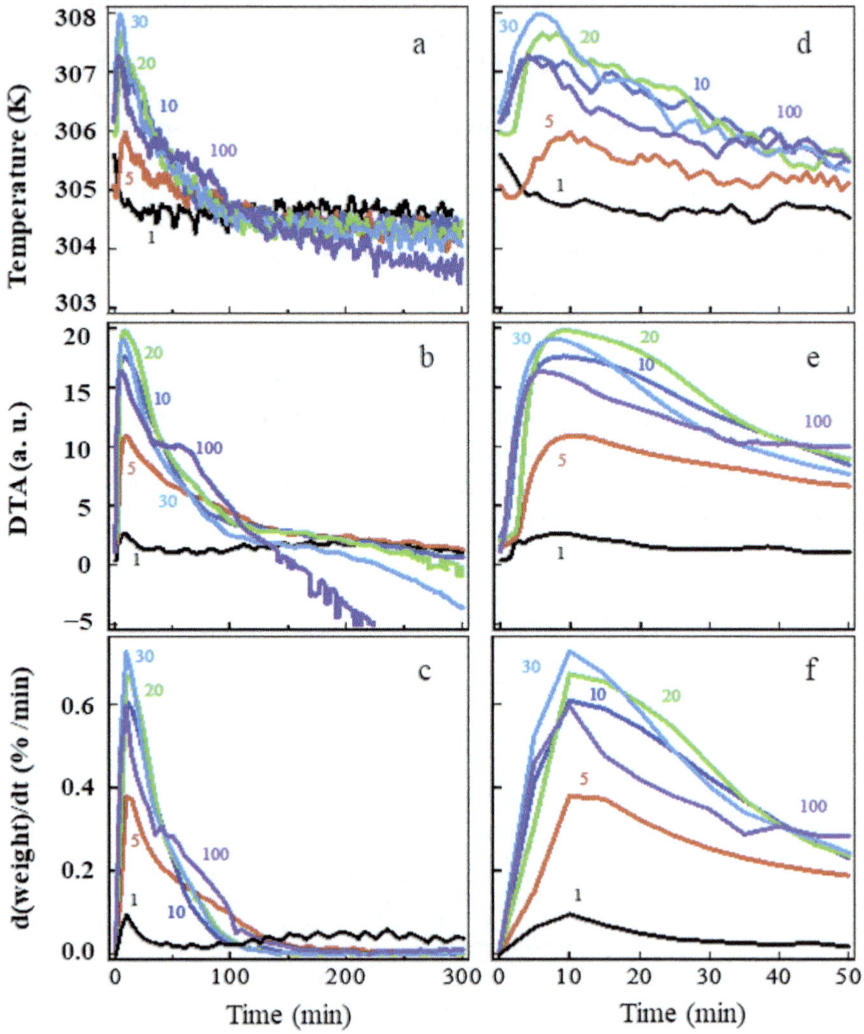

Figure 4.6 Temperature change (a), DTA (b), and DTG (c) curves at various CO_2 flow rates ($cm^3\ min^{-1}$), which are indicated with numbers. These panels are magnified for reaction times of 0–50 min in (d), (e), and (f), respectively. Reproduced from ref. 18 with permission from Elsevier, Copyright 2017.

The maximum rate of this initial stage increased at approximately 10 min with the CO_2 flow rate from 1 to 30 cm^3 min^{-1}, but decreased with a higher CO_2 flow rate at this stage, which were consistent with the TG results in Figure 4.4. After the initial increase in CO_2 occlusion rate, the reaction rate decreased in all curves until the completed CO_2 occlusion (300 min). With initiating the different rates of CO_2 occlusion under various CO_2 concentrations, the exothermic signals, which were similar to the variations of CO_2 occlusion rate, were observed in Figure 4.6(b) and (e). Also, associated temperature changes were obtained with a remarkable similarity to the DTA and DTG curves, which were caused by a highly exothermic reaction for the CO_2 occlusion, as shown in Figure 4.6(a) and (b). These results indicate that the overall reaction, depending strongly upon the initial stage under various CO_2 flow rates, proceeds slowly, which can be understood together with the analysis of the reaction mechanism of the system, as discussed below.

XRD patterns of CO_2-occluded samples of Na_2CO_3 in the presence of water vapor with various reaction times were examined, as shown in Figure 4.7. When the reaction proceeded to 10 min, major Na_2CO_3 and minor $Na_2CO_3{\cdot}H_2O$ phases were present in the XRD patterns for various CO_2 flow rates. Liang reported that $Na_2CO_3{\cdot}H_2O$ was formed slowly when the temperature was lower than 333 K under excess water vapor.[15] $Na_2CO_3{\cdot}H_2O$ was also transformed under lower CO_2 concentrations (1 and 5 cm^3 min^{-1}) at room temperature, as can be seen in Figure 4.5. During this transformation period, it can be claimed that the hydration of Na_2CO_3 occurs *via* reaction (4.9) in the presence of saturated water vapor at room temperature. In addition, the XRD patterns showed the presence of the $Na_5H_3(CO_3)_4$ phase for CO_2 flow rates of 10 cm^3 min^{-1}, whereas few $NaHCO_3$ phases were present in the XRD patterns for high CO_2 concentrations (30 and 100 cm^3 min^{-1}). The formation of $Na_5H_3(CO_3)_4$ and $NaHCO_3$ phases could be transformed from either Na_2CO_3 or $Na_2CO_3{\cdot}H_2O$. Thus, it can be concluded that the temperature changes are a result of the exothermic reactions at this stage. With the reaction occurring over 20 min, the Na_2CO_3 phase remained in the XRD patterns for all CO_2 flow rates. For a low CO_2 flow rate (10 cm^3 min^{-1}), the peak intensity of the $Na_5H_3(CO_3)_4$ phase increased significantly with the $Na_2CO_3{\cdot}H_2O$ phase diminishing. This result indicated that $Na_5H_3(CO_3)_4$ was transformed from $Na_2CO_3{\cdot}H_2O$ rather than Na_2CO_3. Conversely, for high CO_2 flow rates (30 and 100 cm^3 min^{-1}), the $NaHCO_3$ and $Na_5H_3(CO_3)_4$ phases increased slightly. Furthermore, the peaks of the Na_2CO_3 phase in the XRD patterns for 100 cm^3 min^{-1} remained longer than those for 10 and 30 cm^3 min^{-1}.

These results indicated that the transformation process of Na_2CO_3 was inhibited between 10–20 min for high CO_2 flow rates. Thus, it can be identified that the inhibited transformation at high CO_2 flow rates caused the peak intensity to decrease sharply in the DTG and DTA curves between 10–20 min in Figure 4.6(d)–(f).

While increasing the reaction time to 40 min, the peaks of the $NaHCO_3$ phases increased significantly while the peak intensity of the Na_2CO_3 phase

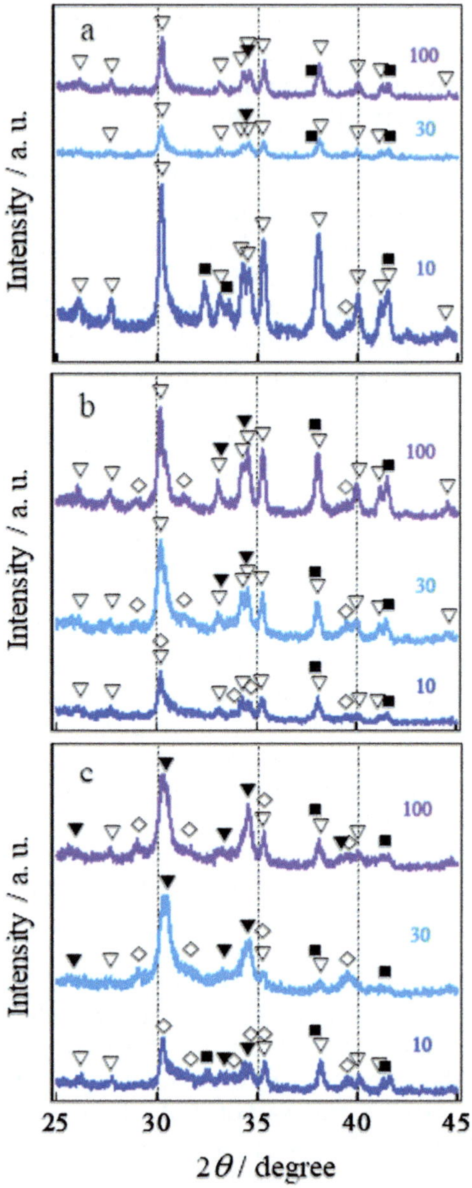

Figure 4.7 XRD patterns after CO_2 occlusion of Na_2CO_3 at various CO_2 flow rates ($cm^3\ min^{-1}$), which are indicated with numbers, with reaction times of: (a) 10 min, (b) 20 min, (c) 40 min. ▼: $NaHCO_3$, ◇: $Na_5H_3(CO_3)_4$, ■: $Na_2CO_3{\cdot}H_2O$, ▽: Na_2CO_3. Reproduced from ref. 18 with permission from Elsevier, Copyright 2017.

decreased for 30 and 100 cm^3 min^{-1}. During this transformation process, the peak intensity of $Na_5H_3(CO_3)_4$ nearly increased between 20 and 40 min, corresponding to the $Na_2CO_3{\cdot}H_2O$ phase barely decreasing (Figure 4.7(c)). It can be deduced that the formation of $NaHCO_3$ was derived from Na_2CO_3. During the transformation process, it was found that there was a decline in the DTG and DTA curves in Figure 4.6(d)–(f). Consequently, the temperature continued to decrease along with the formation of $NaHCO_3$. Compared to the XRD patterns for 30 and 100 cm^3 min^{-1}, the XRD patterns for 10 cm^3 min^{-1} between 20–40 min varied, which were assigned to the phases of Na_2CO_3, $Na_2CO_3{\cdot}H_2O$, and $Na_5H_3(CO_3)_4$.

Considering the XRD patterns at 300 min for the CO_2 flow rates of 30 and 100 cm^3 min^{-1}, with the $Na_5H_3(CO_3)_4$ phase decreasing between 40 and 300 min (Figures 4.5 and 4.7(c)), the XRD patterns showed that the peak intensity of the $NaHCO_3$ phase increased slightly at 300 min. Thus, it can be confirmed that $Na_5H_3(CO_3)_4$ is converted slowly to $NaHCO_3$. These results were consistent with the DTG and DTA curves.

On the basis of the transformation process discussed, the following reactions, instead of reaction (4.11), could be summarized under various CO_2 flow rates in the presence of saturated water vapor, with the exception of reaction (4.10).

$$Na_2CO_3(s) + H_2O(g) \leftrightarrows Na_2CO_3{\cdot}H_2O(s) \tag{4.12}$$

$$5Na_2CO_3{\cdot}H_2O(s) + 3CO_2(g) \leftrightarrows 2Na_5H_3(CO_3)_4(s) + 2H_2O(g) \tag{4.13}$$

$$Na_5H_3(CO_3)_4(s) + CO_2(g) + H_2O(g) \leftrightarrows 5NaHCO_3(s) \tag{4.14}$$

We found that the transformation process, consisting of the formation of $Na_2CO_3{\cdot}H_2O$ from Na_2CO_3, $Na_5H_3(CO_3)_4$ transformed from $Na_2CO_3{\cdot}H_2O$ and Na_2CO_3, and the bicarbonate formation of $Na_5H_3(CO_3)_4$, should be the main pathway of CO_2 occlusion for low CO_2 concentrations (below 10 cm^3 min^{-1}) rather than for high CO_2 concentrations (30 and 100 cm^3 min^{-1}). Conversely, for high CO_2 concentrations, the bicarbonate formation was mostly derived from Na_2CO_3 directly through reaction (4.10), although the CO_2 capture accompanied the formation of few $Na_5H_3(CO_3)_4$ and $Na_2CO_3{\cdot}H_2O$. Thus, the bicarbonate formation through reaction (4.10) accounts for a higher proportion of these two pathways with increasing the CO_2 concentration. From these results, it can be summarized that, for the changes in Figure 4.6, it is mainly reactions (4.10) and (4.12) that occur during the initial 10 min, reactions (4.10), (4.13) and (4.14) from 10 min to 20 min, and reactions (4.10) and (4.14) from 20 min to 40 min that proceed. These reaction processes depend on the CO_2 and H_2O concentrations. The effect of H_2O concentration will be described below.

As mentioned in Section 4.2.1.2.1, reaction (4.10) is more exothermic than reaction (4.11), but Figure 4.6 displays the opposite results regarding the CO_2 occlusion rate and exothermic properties for 100 cm^3 min^{-1}. Hence, the initial exothermic properties and the CO_2 sorption rate of Na_2CO_3 under various CO_2 flow rates cannot be obtained from the XRD data alone according to the transformation processes and kinetics of the CO_2 sorptivity of Na_2CO_3 under various CO_2 flow rates and with the coexistence of saturated water vapor. To understand these properties, the morphological variation of Na_2CO_3 was examined *via* SEM before and after CO_2 sorption for different reaction times. The SEM images for the 10 and 30 cm^3 min^{-1} CO_2 atmospheres are shown in Figures 4.8 and 4.9, respectively.

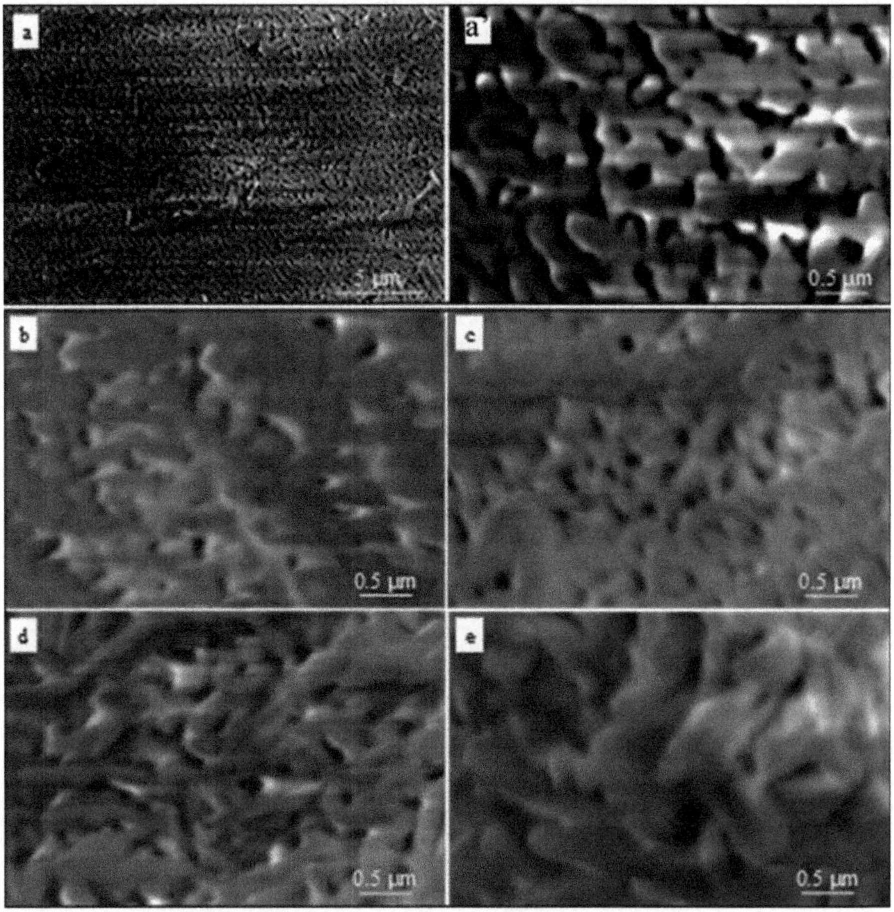

Figure 4.8 SEM images of Na_2CO_3 before and after CO_2 occlusion at a CO_2 flow rate of 10 cm^3 min^{-1} (a)–(e) ((a): 0 min, (a′): magnification of (a), (b): 10 min, (c): 20 min, (d): 40 min, and (e): 300 min). Reproduced from ref. 18 with permission from Elsevier, Copyright 2017.

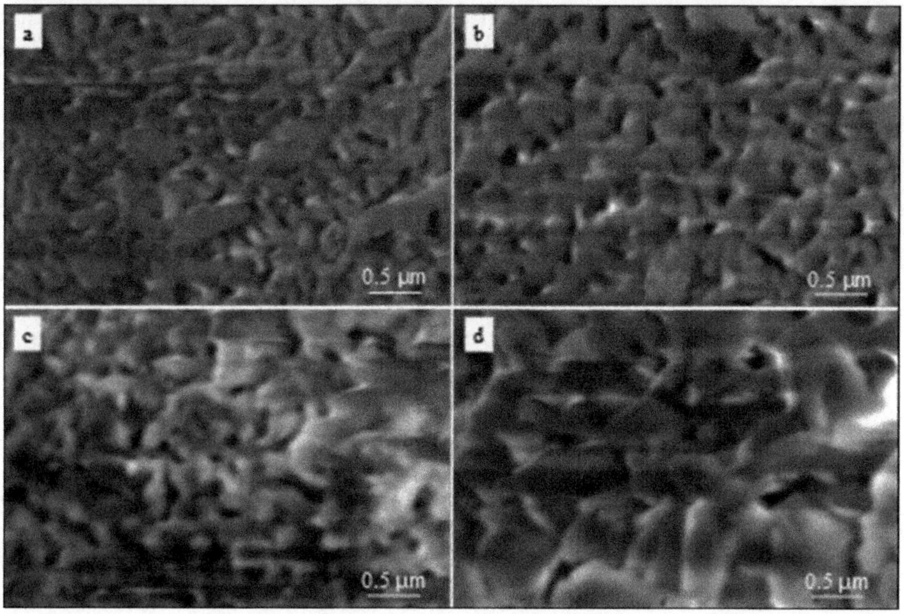

Figure 4.9 SEM images of Na_2CO_3 after CO_2 occlusion at a CO_2 flow rate of 30 cm^3 min^{-1} (a)–(d) ((a): 10 min, (b): 20 min, (c): 40 min, and (d): 300 min). Reproduced from ref. 18 with permission from Elsevier, Copyright 2017.

Na_2CO_3 particles formed by the decomposition of $NaHCO_3$ under an N_2 atmosphere (Figure 4.8(a)) possess a macroporous surface. For a CO_2 flow rate of 10 cm^3 min^{-1}, the diameter of the macropores on the surface diminished slightly after the initial 10 min (Figure 4.8(b)) with the formation of major $Na_2CO_3{\cdot}H_2O$ and minor $Na_5H_3(CO_3)_4$, at which the maximum in CO_2 occlusion rate was obtained. After 20 min, the pores on the surface developed (Figure 4.8(c)), corresponding to the transformation of $Na_2CO_3{\cdot}H_2O$ to $Na_5H_3(CO_3)_4$. Large granules of dominant $Na_5H_3(CO_3)_4$ developed on the external surface (Figure 4.8(d) and (e)) along with $Na_2CO_3{\cdot}H_2O$ reacting with CO_2 and H_2O for 40–300 min (Figures 4.5 and 4.7(c)).

For a CO_2 flow rate of 30 cm^3 min^{-1}, the diameter of the macropores on the surface decreased significantly and few macropores remained on the surface after the initial 10 min (Figure 4.9(a)), whereas more macropores were present on the surface after 20 min. The initial formation of large granules proceeded on the external surface after 40 min (Figure 4.9(c)) along with the formation of major $NaHCO_3$ and minor $Na_5H_3(CO_3)_4$ between 20 and 40 min (Figure 4.7(b) and (c)). After 300 min, few macropores appeared on the surface. This result demonstrates that the evolution of water vapor that stems from reaction (4.13) leads to macropores on the surface between 20–300 min. Furthermore, the formation of $NaHCO_3$ covering the Na_2CO_3 particles results in an increase in surface resistance to inhibit the CO_2 occlusion. Thus, we can observe that there is an obvious decline in the DTG and DTA curves (Figure 4.6(e) and (f)).

For a CO_2 flow rate of 100 cm^3 min^{-1}, the macropores on the surface almost disappeared after the initial 10 min and then formed a wrinkled surface after 20 min. This result leads to the restricted transformation of Na_2CO_3 for 100 cm^3 min^{-1} (Figure 4.7(a) and (b)). Thus, it was observed that the peak intensity at the initial stage decreased in the DTG and DTA curves (Figure 4.6(e) and (f)). The initial formation of granules on the surface continued to increase the surface resistance after 40 min. Granules of $NaHCO_3$ developed on the external surface along with the bicarbonate formation of Na_2CO_3 and $Na_2CO_3{\cdot}H_2O$ reacting with CO_2 and H_2O between 40–300 min (Figures 4.5 and 4.7(c)).

On the basis of the discussion on structural and morphological changes with reaction times under various CO_2 concentrations, it is found that the bicarbonate formation of Na_2CO_3 has two different pathways, such as reactions (4.12)–(4.14) and reaction (4.10). With increasing the CO_2 concentrations, bicarbonate formation through reaction (4.10) accounts for the higher proportion. The formation of $Na_5H_3(CO_3)_4$ and $Na_2CO_3{\cdot}H_2O$ favorably proceeds under low CO_2 concentrations, although their CO_2 occlusion rates are slow. Furthermore, the macropores on the surface decrease or almost disappear within 10 min. This result leads to increasing the surface resistance to limit the transformation of Na_2CO_3 reacting with CO_2 and H_2O at the initial stage for higher CO_2 concentrations.

4.2.1.2.3 CO_2 Capture of Na_2CO_3 Under Various H_2O Concentrations. As mentioned in Section 4.2.1.2.2, the bicarbonate formation of Na_2CO_3 proceeds *via* two pathways under various CO_2 concentrations. The proportion of the two pathways influences the overall CO_2 occlusion rate of Na_2CO_3 along with the morphology variation. To understand the effect of H_2O concentration on the CO_2 occlusion of Na_2CO_3 at room temperature, studies on CO_2 capture under different humidities is important. The kinetics of CO_2 occlusion and crystal structure changes after CO_2 occlusion were examined.

Figure 4.10 shows the CO_2 occlusion of Na_2CO_3 under various H_2O concentrations at a CO_2 flow rate of 30 cm^3 min^{-1}. For a H_2O concentration of RH38%, the CO_2 sorption amount was 7.21 mmol g^{-1}, which is 76.5% of the theoretical value of 9.43 at 300 min. At a H_2O concentration of RH74%, the amount of CO_2 occlusion increased to 7.39 mmol g^{-1}. Furthermore, with increasing the H_2O concentration from RH38% to RH74%, the equilibrium time was shortened from 300 to 100 min. These results indicated that the CO_2 occlusion reaction rate is significantly accelerated with increasing the H_2O concentration, whereas the CO_2 occlusion amount barely increased.

Figure 4.11 shows XRD patterns after the CO_2 occlusion of Na_2CO_3 under H_2O concentrations between RH38% and RH74%. The XRD patterns showed that the major $NaHCO_3$ phase was present for a H_2O concentration above RH38%. With increasing the H_2O concentration, the peak intensity of

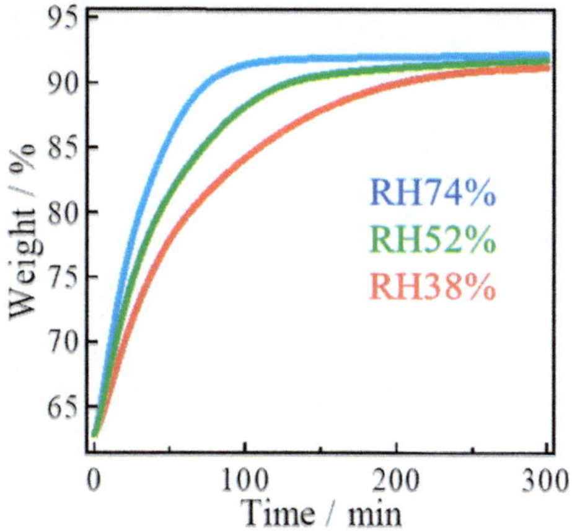

Figure 4.10 TG curves for the CO_2 occlusion of Na_2CO_3 under various H_2O concentrations at a CO_2 flow rate of 30 cm^3 min^{-1} and 303 K. Reproduced from ref. 18 with permission from Elsevier, Copyright 2017.

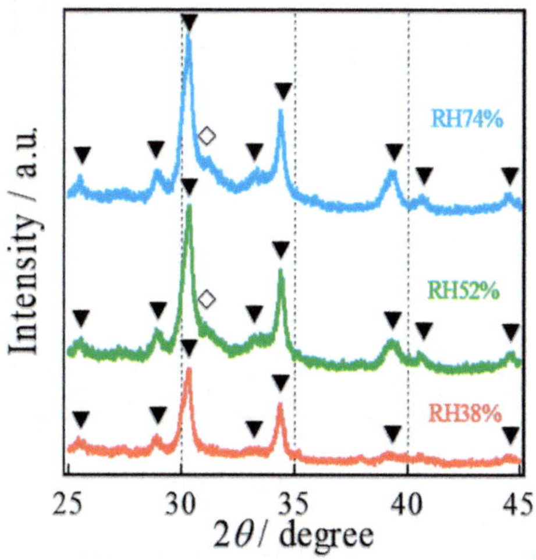

Figure 4.11 XRD patterns after CO_2 occlusion for various H_2O concentrations. ▼: $NaHCO_3$, ◇ : $Na_5H_3(CO_3)_4$. Reproduced from ref. 18 with permission from Elsevier, Copyright 2017.

$Na_5H_3(CO_3)_4$ increased in the XRD patterns. This result demonstrates that higher H_2O concentrations are favorable for the formation of $Na_5H_3(CO_3)_4$ from $Na_2CO_3{\cdot}H_2O$ through reactions (4.12) and (4.13).

The CO_2 sorptivity of Na_2CO_3 can be tracked by monitoring the temperature under various H_2O concentrations with the bicarbonate formation of Na_2CO_3 leading to a temperature change by the exothermic reactions. Figure 4.12 shows that both DTG and DTA curves obviously exhibited a series

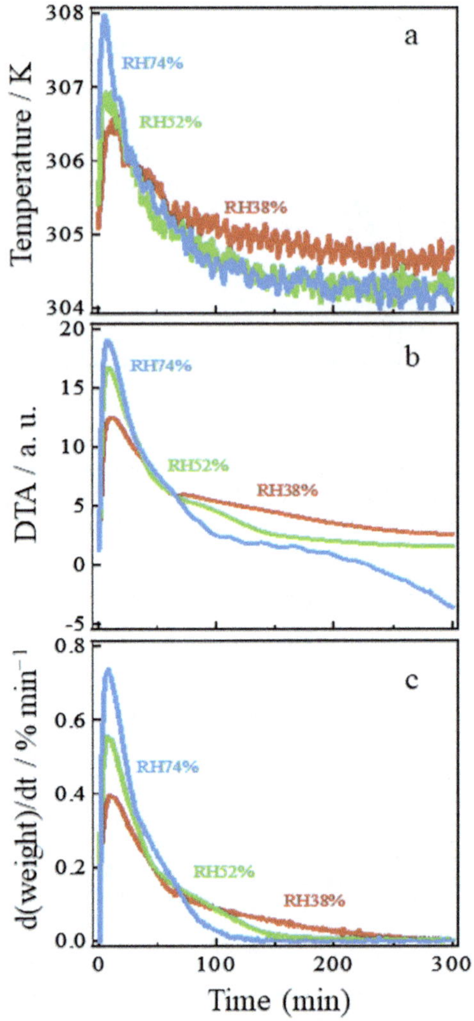

Figure 4.12 Temperature changes (a), DTA (b) and DTG (c) curves for various H_2O concentrations at a CO_2 flow rate of 30 cm^3 min^{-1}. Reproduced from ref. 18 with permission from Elsevier, Copyright 2017.

of peaks accompanied by the exothermic reactions of the CO_2 occlusion of Na_2CO_3. In particular, DTG curves showed that the CO_2 occlusion rate had a sharper rise at the initial stage (within 50 min) for higher H_2O concentrations. This implied that the higher H_2O concentration accelerated the CO_2 occlusion process at the initial stage. Furthermore, the DTA and DTG curves showed that the exothermic signals increased strongly with increasing the H_2O concentration at the initial stage. Consequently, it was observed that there was a significant increase in temperature at the initial stage of the reaction with increasing the H_2O concentration between RH38% and

RH74%. After the initial temperature increase, the temperature decreased in all curves. These results demonstrated that the heat was obtained from the exothermic reactions at the initial stage. Moreover, these results indicated that the overall reaction rate depended upon the CO_2 occlusion rate at the initial stage. These peaks observed in the DTG and DTA curves for various H_2O concentrations can be understood together with analyses of the reactions at the initial stage on the basis of the crystalline structural change, which were discussed below.

Figure 4.13 shows the crystal structure changes at different times when the exothermic reaction proceeds within 40 min under various H_2O concentrations at a CO_2 flow rate of 30 $cm^3\ min^{-1}$. As the reaction occurred beyond 10 min, as shown in Figure 4.13(a), the XRD patterns showed that all peaks were assigned to the Na_2CO_3 phase for a H_2O concentration of RH52%, whereas the peaks for RH74% consisted of three phases such as Na_2CO_3, $Na_2CO_3{\cdot}H_2O$, and $NaHCO_3$. It was confirmed that a higher H_2O concentration was favorable for the occurrence of the exothermic reactions (4.12) and (4.13).

With the formation of $Na_2CO_3{\cdot}H_2O$ and $NaHCO_3$, it was observed that the peak intensity in the DTG and DTA curves at the initial stage increased with increasing the H_2O concentration (Figure 4.12).

When the reaction proceeded to 20 min, the XRD patterns showed that the peaks of the $Na_2CO_3{\cdot}H_2O$, $Na_5H_3(CO_3)_4$, and $NaHCO_3$ phases were present for H_2O concentrations of RH52% and RH74%, although the dominant Na_2CO_3 phase remained after 20 min. With the reaction time increasing to 40 min, for the H_2O concentration of RH74%, the peak intensity of $NaHCO_3$ increased significantly while the Na_2CO_3 phase decreased significantly. Moreover, the $Na_2CO_3{\cdot}H_2O$ and $Na_5H_3(CO_3)_4$ phases increased or decreased. This result confirms that Na_2CO_3 is converted mostly to $NaHCO_3$ through reaction (4.10). For a low H_2O concentration (RH52%), the peak intensity of $Na_5H_3(CO_3)_4$ phase increased along with the formation of $Na_2CO_3{\cdot}H_2O$.

The morphological variation of Na_2CO_3 after the CO_2 occlusion for different reaction times was examined with SEM. For a H_2O concentration of RH52%, the diameters of the macropores on the surface decreased slightly after the initial 10 min, when the macropores significantly diminished for RH74% (Figure 4.9(a)). After 20 min, much macropores remained on the surface, corresponding to the formation of few $Na_2CO_3{\cdot}H_2O$ and $Na_5H_3(CO_3)_4$. The macroporous surface almost disappeared at 40 min along with the formation of dominant $NaHCO_3$. The winkled surface was formed at 300 min, when the product was totally $NaHCO_3$ phase, as observed in the XRD patterns (Figure 4.13). In comparison to the morphological variation between 40 and 300 min for RH74%, it was found that granules on the surface did not develop obviously for RH52%. These results indicate that a higher H_2O concentration is favorable for the formation of granules on the surface of Na_2CO_3 particles.

Regarding the analyses of XRD and morphological variations with the reaction times under various H_2O concentrations, we find that the crystal structure transformation from Na_2CO_3 limited the overall CO_2 occlusion rate.

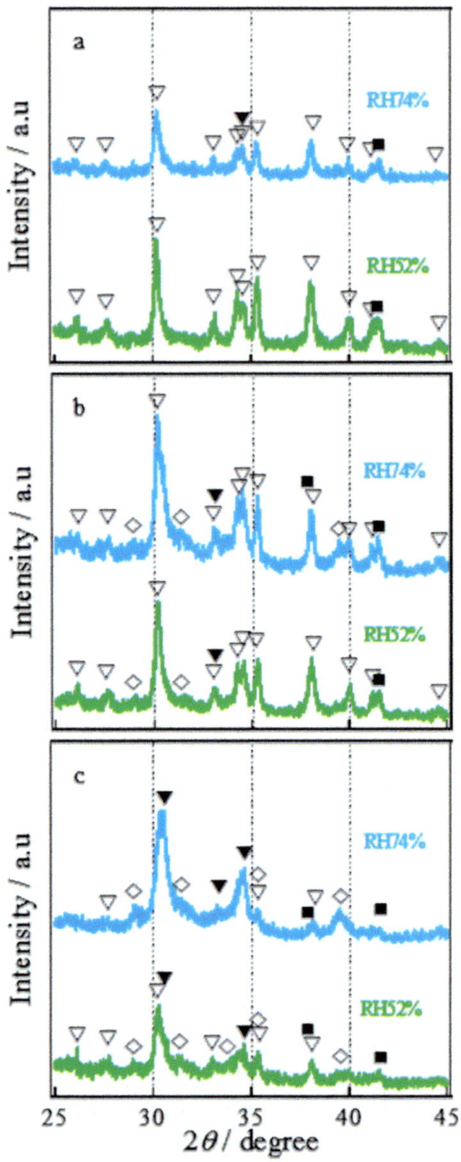

Figure 4.13 XRD patterns after the CO_2 occlusion of Na_2CO_3 for various H_2O concentrations at a CO_2 flow rate of 30 cm^3 min^{-1} with reaction times: (a) 10 min, (b) 20 min, and (c) 40 min. ▼: $NaHCO_3$, ◇: $Na_5H_3(CO_3)_4$, ■ : $Na_2CO_3 \cdot H_2O$, ▽: Na_2CO_3. Reproduced from ref. 18 with permission from Elsevier, Copyright 2017.

In particular, the transformation of Na_2CO_3 at the initial stage occurs slowly because of the weak hydration of Na_2CO_3. During this transformation process, the morphological variations do not play a crucial role in the CO_2 occlusion rate.

4.2.1.3 Conclusions

The analyses of TG and XRD showed the dependence of the CO_2 sorptivity of Na_2CO_3 on temperature, CO_2 concentration, and H_2O concentration. The formation of $NaHCO_3$ favors a low temperature with a highly exothermic reaction, whereas the formation of $Na_5H_3(CO_3)_4$ proceeds readily at higher temperatures. On the basis of the structure changes and morphological variations with reaction times under various CO_2 and H_2O concentrations, it is found that the bicarbonate formation of Na_2CO_3 has two different pathways, such as reactions (4.12)–(4.14) and reaction (4.10). For higher CO_2 and H_2O concentrations, bicarbonate formation through reaction (4.10) accounts for the higher proportion. The formation of $Na_5H_3(CO_3)_4$ and $Na_2CO_3{\cdot}H_2O$ favorably occurs under low CO_2 and high H_2O concentrations, although bicarbonate formation proceeds slowly through reactions (4.13) and (4.14). Furthermore, the macropores on the surface almost disappear within 10 min depending mainly upon higher CO_2 and H_2O concentrations, inducing the bicarbonate formation of Na_2CO_3. This result leads to increasing the surface resistance to limit the transformation of Na_2CO_3 reacting with CO_2 and H_2O at the initial stage for higher CO_2 and H_2O concentrations, although the morphology does not play a crucial role in the transformation of Na_2CO_3. In addition, Na_2CO_3 is an abundant and cheap sorbent and its CO_2 occlusion amount is 5.75 mmol g^{-1} within 50 min for 30 cm^3 min^{-1}, which is 85.6% of theoretical CO_2 capture capacity, thus, Na_2CO_3 as the CO_2 capture sorbent has some application value for CO_2 capture.

4.2.2 Capture of CO_2 of K_2CO_3 Under Moist Conditions

Feasible CO_2 sorbents must indicate a significant effectiveness in reaction rate, saturation amount, and long-term stability for CO_2 capture. Zhao *et al.* found that K_2CO_3 obtained from $KHCO_3$ showed an excellent CO_2 sorption capacity. This can be used to solve the issue of an overall low CO_2 sorption rate.[27] The practical application of K_2CO_3 as a CO_2 sorbent has already been exhibited in a pilot-scale plant for an actual coal power plant.[28,29] K_2CO_3 under moist conditions at atmospheric pressure has been studied extensively in the following reaction:[16,27,29–36]

$$K_2CO_3(s) + CO_2(g) + H_2O(g) \leftrightarrows 2KHCO_3(s) \tag{4.15}$$

whereas the forward reaction, the formation of $KHCO_3$, is an exothermic reaction (bicarbonate formation; the theoretical CO_2 capture amount of K_2CO_3 is 7.24 mmol g^{-1}), whereas the reverse reaction, the decomposition of $KHCO_3$, is an endothermic reaction. The CO_2 sorption amount of K_2CO_3 increases according to the forward reaction of (4.15) at a lower temperature; thus, K_2CO_3 can be used at room temperature for a high saturation amount. Moreover, Chioyama *et al.* reported that the sorption amount (6.48 mmol g^{-1}, 285 mg g^{-1}) at 0.1 MPa and 313 K was higher than that at a higher temperature and showed that K_2CO_3 can be effective at room temperature.[19]

Furthermore, the desirable CO_2 sorption ability of K_2CO_3 supported on activated carbon at a low temperature has also been exhibited by increasing the surface area.[37] As a result, K_2CO_3 is considered to be promising as a CO_2 sorbent under moist conditions at room temperature.

However, most previous studies concerning the CO_2 sorption of K_2CO_3 were performed at relatively high temperatures (>333 K) under ambient pressure. CO_2 was not always captured directly *via* reaction (4.15) in a number of studies on the CO_2 capture of K_2CO_3 under moist conditions. An improved CO_2 capture capacity using $K_2CO_3{\cdot}1.5H_2O$ as the active species under a flow gas of 13.8% CO_2 with 10% H_2O in helium at 373 K was reported by Hayashi *et al.*[33] Shigemoto *et al.* proposed that K_2CO_3 supported on activated carbon could change to $KHCO_3$ through the reaction of $K_2CO_3{\cdot}1.5H_2O$ with CO_2 under a humid gas (10% H_2O, 11.8% CO_2) at 363 K. The relevant reactions are:[34]

$$K_2CO_3(s) + 1.5H_2O(g) \rightleftarrows K_2CO_3{\cdot}1.5H_2O(s) \tag{4.16}$$

$$K_2CO_3{\cdot}1.5H_2O(s) + CO_2(g) \rightleftarrows 2KHCO_3(s) + 0.5H_2O(g) \tag{4.17}$$

The CO_2 sorption of K_2CO_3 supported on Al_2O_3 proceeded *via* reactions (4.16) and (4.17), not *via* reaction (4.15), under a 10:12.2:77.8 vol% mixed gas of CO_2–H_2O–N_2 at 323–343 K, as reported by Seo *et al.*[35] Additionally, a similar CO_2 capture of K_2CO_3 on Al_2O_3 under a 1:9:90 vol% mixed gas of CO_2–H_2O–N_2 at 323 K was reported by Lee *et al.*[36] However, the formation of $K_4H_2(CO_3)_3{\cdot}1.5H_2O$ may also promote the bicarbonate formation of K_2CO_3[38] and proceed *via* reactions (4.18) and (4.19) rather than reaction (4.15) in a mixed gas of 15:15:70 mol% of CO_2–H_2O–N_2 at 333 K,[39] as shown by Zhao *et al.*

$$2K_2CO_3(s) + 2.5H_2O(g) + CO_2(g) \rightleftarrows K_4H_2(CO_3)_3{\cdot}1.5H_2O(s) \tag{4.18}$$

$$K_4H_2(CO_3)_3{\cdot}1.5H_2O(s) + CO_2(g) \rightleftarrows 4KHCO_3(s) + 0.5H_2O(g) \tag{4.19}$$

These reactions depended on the experimental conditions, in which dependence of the crystal growth causes the morphology change of the K_2CO_3 particle. This is important to understand the kinetic property of K_2CO_3. However, the morphological changes with reaction time have not been well known. Thus, the CO_2 sorption mechanism of K_2CO_3 under various CO_2 and H_2O conditions is still unclear.

In this section, the CO_2 sorption ability of K_2CO_3 under various CO_2 concentrations with saturated water vapor was examined with TG-DTA apparatus and structural changes during the reaction were studied by XRD under a vacuum. Also, morphological changes were observed using SEM. From the experimental data of the crystal structure and kinetic analysis of the bicarbonate formation of K_2CO_3, three proper reactions are proposed for $KHCO_3$ formation from K_2CO_3 at ambient temperatures.

4.2.2.1 *Experimental*

4.2.2.1.1 Sample Preparation. $KHCO_3$ (99.5% chemical purity) was used during the experiments of decomposition of $KHCO_3$ and bicarbonate formation of K_2CO_3. The decomposition of $KHCO_3$ was carried out with TG-DTA. $KHCO_3$ (44–50 mg) was heated at 5 K min^{-1} from 298 to 473 K in pure N_2 gas at 100 cm^3 min^{-1} to form K_2CO_3 and was kept at these conditions for 5 min until complete decomposition. Then, the material was cooled to 313 K at −5 K min^{-1} and maintained for 30 min under pure N_2 gas.

4.2.2.1.2 Bicarbonate Formation Measurements. K_2CO_3 bicarbonate formation was measured with the TG-DTA apparatus, as shown in Section 4.2.1.1.2, with different compositions of CO_2 and N_2 containing saturated water vapor at 313 K.

4.2.2.1.3 Crystal Structure and Morphology Measurements. The crystal structures of the products after reaction times of 1, 5, 20, 40, 60, and 120 min were examined by XRD *in vacuo* without reactions with CO_2 or H_2O under an ambient atmosphere, as shown in Section 4.2.1.1.3. The morphology of the materials before and after CO_2 sorption for the different reaction times was observed using SEM after osmium coating.

4.2.2.2 *Results and Discussion*

4.2.2.2.1 Bicarbonate Formation of K_2CO_3. $KHCO_3$ decomposed to K_2CO_3, CO_2, and H_2O as per the reverse of reaction (4.15). This was confirmed by a weight decrease between 20 and 35 min upon heating to 473 K, as shown in Figure 4.14. The sample weight decreased to 69.1% (±0.1%) after heat treatment. Since the theoretical value based on the reverse reaction of (4.15) is 69.0%, the decomposition reaction proceeded completely. After the decomposition, a mixed gas of CO_2 and H_2O was supplied to K_2CO_3 at different temperatures. The increase in sample weight depended on temperature, as shown in Figure 4.14. This could be attributed to the bicarbonate formation of K_2CO_3 with CO_2 and H_2O, as shown below. The change increased more rapidly at higher temperatures because bicarbonate formation requires a minimum activation energy for the chemical reaction. The initial formation rate is much faster at 383 K than at lower temperatures. However, the saturated occlusion amount at 383 K was lower than half of the theoretical value of 7.24 mmol g^{-1}. In contrast, the initial formation rate is slower at 313 K, but the saturated occlusion amount was 6.48 mmol g^{-1}, which is 89.3% of the theoretical amount, *i.e.*, 7.24 mol g^{-1}. The occlusion amount of 6.48 mmol g^{-1} (285 mg g^{-1}) is exceedingly high at 1 bar and 313 K under a moist atmosphere in comparison with the CO_2 amount from the

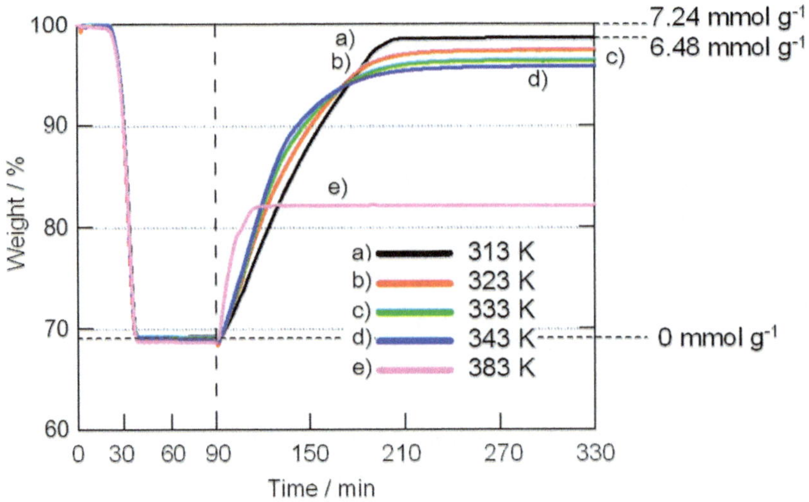

Figure 4.14 Weight change by the decomposition of $KHCO_3$ (<40 min) at 473 K and the bicarbonate formation of K_2CO_3 with CO_2 and H_2O (>90 min) at different temperatures: (a) 313 K, (b) 323 K, (c) 333 K, (d) 343 K, (e) 383 K.

physical adsorption of activated carbon fibers (120–200 mg g^{-1} at 273 K).[40] Thus, K_2CO_3 is a promising material as a CO_2 sorbent under ambient conditions. CO_2 absorption or occlusions by K_2CO_3 under moist conditions were examined at relatively high temperatures, *i.e.*, >333 K, in most of the previous reports, but K_2CO_3 should be used at room temperature for the highest saturated occlusion amount.

Since the temperature dependence of the saturated occlusion amount is not yet understood, XRD patterns were obtained for the samples treated at different temperatures to examine the structural changes of K_2CO_3 on CO_2 occlusion with H_2O. The patterns of the CO_2 occluded samples at different temperatures are shown in Figure 4.15. All peaks of the sample formed at 383 K were assigned to the $K_4H_2(CO_3)_3{\cdot}1.5H_2O$ phase. Thus, no formation of $KHCO_3$ was observed at 383 K. The XRD patterns showed peaks consistent with two phases, *i.e.*, $KHCO_3$ and $K_4H_2(CO_3)_3{\cdot}1.5H_2O$, at temperatures from 313 K to 343 K. The peaks attributed to the $KHCO_3$ phase appeared, and the peak intensity increased with decreasing temperature, but the peaks of $K_4H_2(CO_3)_3{\cdot}1.5H_2O$ remained at 323 K and 313 K. The formation of $KHCO_3$ from $K_4H_2(CO_3)_3{\cdot}1.5H_2O$ was reported and proceeded through reactions (4.16) and (4.17).[41] These two reactions, however, have not intensively been considered for the CO_2 occlusion of K_2CO_3.

The XRD results showed that only reaction (4.16) proceeded at 383 K, but reaction (4.17) did proceed as well at lower temperatures. Figure 4.16 shows derivative thermogravimetric (DTG) curves (upper) and DTA curves (lower) obtained at 313, 323, 333, and 343 K, which indicate exothermic reactions. DTG curves were obtained by differentiating the weight change

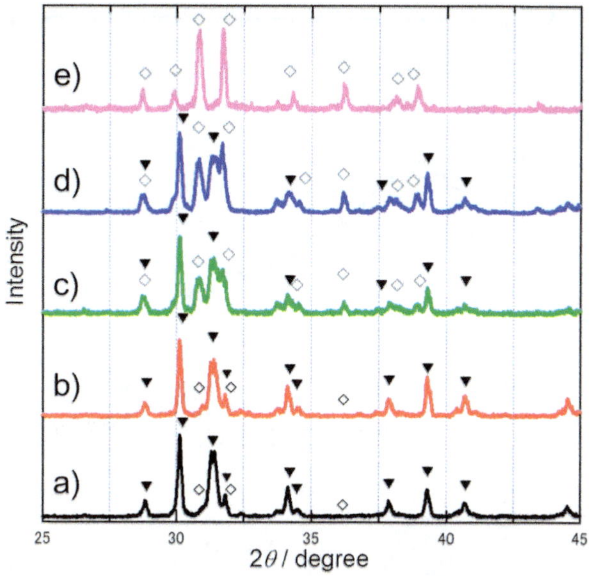

Figure 4.15 XRD patterns of CO_2 occluded samples at different temperatures: (a) 313 K, (b) 323 K, (c) 333 K, (d) 343 K, (e) 383 K. ▼: $KHCO_3$, ◇: $K_4H_2(CO_3)_3{\cdot}1.5H_2O$.

in Figure 4.14 over time. Evidently, the patterns of both curves exhibit a similar shape: (i) the curves consist of two components, and (ii) the 1st peak intensity increases and 2nd peak intensity decreases as the temperature increases. Thus, the bicarbonate formation of K_2CO_3 is likely to involve a double-step reaction.

XRD patterns of pure $KHCO_3$ and $K_4H_2(CO_3)_3{\cdot}1.5H_2O$ phases, where peak intensities for each phase are kept as relative ratios, were obtained. Peak fitting for the experimental patterns obtained at 323 K and 313 K using the peak patterns of these pure phases gave us well-fitted patterns. This indicates that these crystals consist of only the two phases. The peak areas were obtained for the two phases by the peak fitting. Calculation of the peak area for each peak provided a ratio of the peak area of $KHCO_3$ to the total peak area of $KHCO_3$ and $K_4H_2(CO_3)_3{\cdot}1.5H_2O$, $I_{KHCO_3}/(I_{KHCO_3} + I_{K_4H_2(CO_3)_3{\cdot}1.5H_2O})$, as shown in Figure 4.17. It was found that the ratio of the $KHCO_3$ phase increased with decreasing temperature. Since the $KHCO_3$ phase was not observed at 383 K, it was concluded that reaction (4.17) does not proceed at 383 K but reaction (4.16) does. Since the two reactions are exothermic, as observed in the two DTA peaks in Figure 4.16 (lower), reactions (4.16) and (4.17) should be unfavorable at higher temperatures as per Le Châtelier's principle. At temperatures from 313 K to 383 K, reaction (4.16) can proceed to some extent. However, reaction (4.17) probably shows a stronger temperature dependence, and little formation of $KHCO_3$ proceeds at 383 K or higher. This should be considered in developing an efficient CO_2 occlusion of K_2CO_3 under moist conditions.

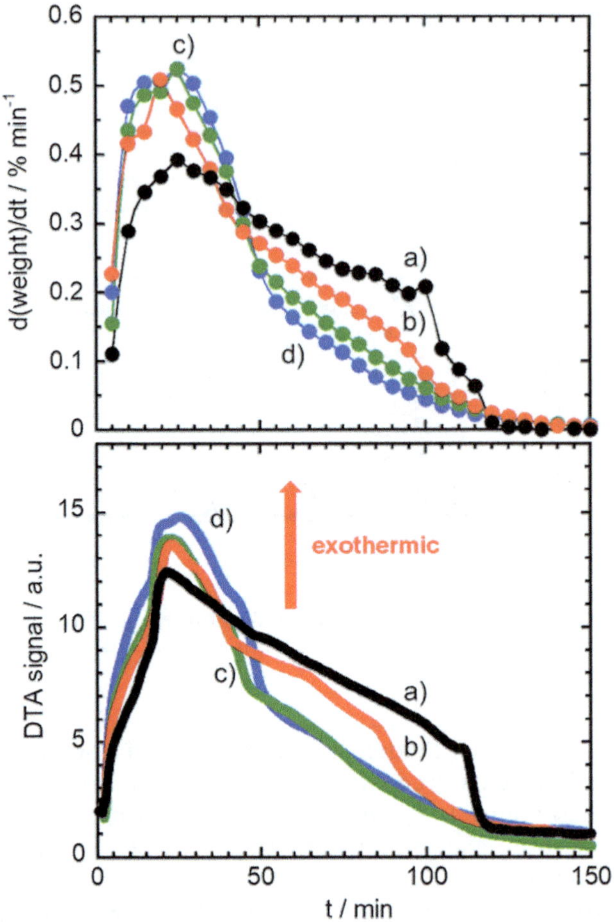

Figure 4.16 DTG (upper) and DTA curves (lower) over time at different temperatures: (a) 313 K, (b) 323 K, (c) 333 K, (d) 343 K.

TG measurements for the CO_2 sorption of K_2CO_3 were carried out under several partial pressures of CO_2 containing water vapor to examine the influence of the CO_2 concentration. The results are shown for reaction times of 0–120 min in Figure 4.18. Sufficient moisture (RH of 68% at 313 K) was given to K_2CO_3 with a mixed gas of CO_2 and N_2 containing saturated water vapor. For a flow rate of 1 cm^3 min^{-1} CO_2 + 99 cm^3 min^{-1} N_2 (1% v/v CO_2), the sorbed amount was 6.24 mmol g^{-1} (86.2% of the theoretical amount of 7.24 mmol g^{-1}), and it increased from 6.24 to 6.85 mmol g^{-1} when the CO_2 flow rate increased from 1 to 30 cm^3 min^{-1}. Equilibrium shifted toward bicarbonate formation of K_2CO_3 according to reaction (4.15) with increasing CO_2 concentration. The sorbed amount was 6.92 mmol g^{-1} for a CO_2 flow rate of 100 cm^3 min^{-1} (95.6% of the theoretical value). From the experimental results in comparison with those at a CO_2 flow rate of 30 cm^3 min^{-1}, increase of the CO_2 concentration led to a higher reaction efficiency of K_2CO_3 to

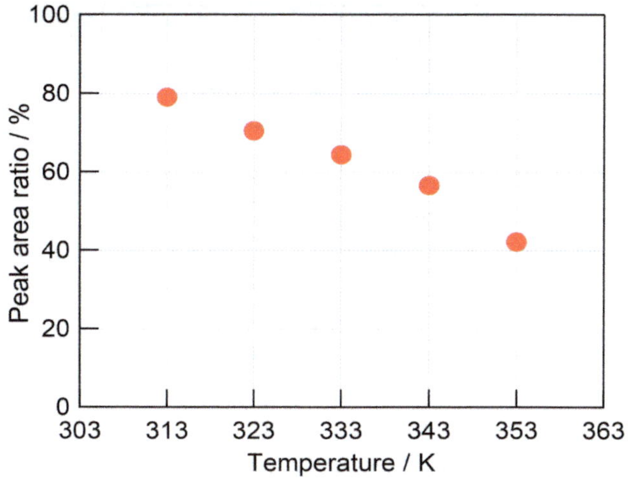

Figure 4.17 Temperature dependence of the peak area ratio of $KHCO_3$ to the total peak area of $KHCO_3$ and $K_4H_2(CO_3)_3{\cdot}1.5H_2O$.

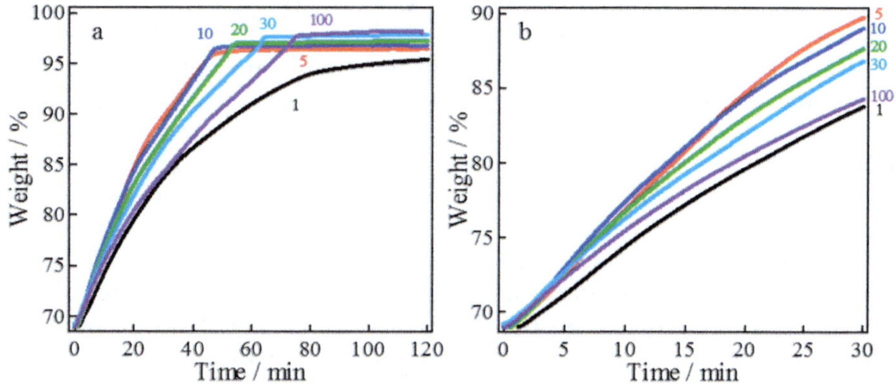

Figure 4.18 TG curves (a) for bicarbonate formation under various CO_2 flow rates ($cm^3\ min^{-1}$), which are indicated with numbers, with N_2 and H_2O and magnified for 0–30 min (b). Reproduced from ref. 20 with permission from the American Chemical Society, Copyright 2015.

$KHCO_3$ under a saturated humidity at 313 K. Figure 4.19 indicates the XRD patterns after the completion of the bicarbonate formation (after 120 min in Figure 4.18) at several CO_2 flow rates. It is obvious that the peaks of both the $K_4H_2(CO_3)_3{\cdot}1.5H_2O$ and $KHCO_3$ phases were observed in all samples, while the peaks of the $K_4H_2(CO_3)_3{\cdot}1.5H_2O$ phase decreased with an increase of CO_2 concentration. This corresponded to the TGA results.

However, the CO_2 sorption was slow at CO_2 flow rates of 20–100 $cm^3\ min^{-1}$ under moist conditions. In contrast, the reactions at CO_2 flow rates of 5 and 10 $cm^3\ min^{-1}$ became much faster than other CO_2 flow rates, as shown in Figure 4.18(b). The layer of transformed $KHCO_3$ on the interface

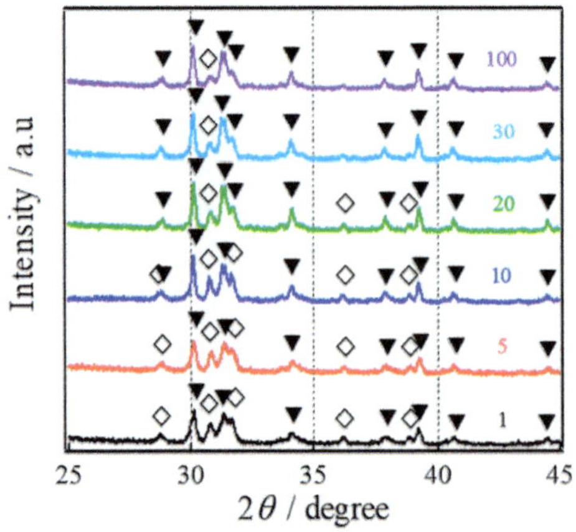

Figure 4.19 XRD patterns after CO_2 occlusion of K_2CO_3 at various CO_2 flow rates ($cm^3\ min^{-1}$), which are indicated with numbers. ▼: $KHCO_3$; ◇: $K_4H_2(CO_3)_3{\cdot}1.5H_2O$. Reproduced from ref. 20 with permission from the American Chemical Society, Copyright 2015.

of K_2CO_3 particles seemed to inhibit the CO_2 and H_2O transfer process because of increasing surface resistance.[39] By contrast, the initial sorption rate was restricted at CO_2 flow rates above 20 $cm^3\ min^{-1}$ (Figure 4.18(b)). The formation of $K_2CO_3{\cdot}1.5H_2O$ could enhance the CO_2 sorption ability as an active species, with its formation being the limiting step, as reported before.[33,34] However, Zhao *et al.* found that the initial product after the reaction of K_2CO_3 with CO_2 and H_2O was not $K_2CO_3{\cdot}1.5H_2O$ but rather $K_4H_2(CO_3)_3{\cdot}1.5H_2O$. The bicarbonate formation rate of K_2CO_3 was much faster than that of $K_2CO_3{\cdot}1.5H_2O$.[38,39] The formation of the initial two products, the transformation processes, and the morphological changes on the surface of K_2CO_3 particles can affect the reaction rate of CO_2 sorption under various CO_2 concentrations. Examination of the transformation process of K_2CO_3 and the accompanying changes in different temperatures during the reactions is very important for a better understanding of the CO_2 sorption kinetics of K_2CO_3.

4.2.2.2.2 Exothermic Properties and Temperature Variation. Since the process of bicarbonate formation of K_2CO_3 leads to a temperature elevation of the reaction system, the kinetic properties were examined by measuring the temperature during the reaction under several CO_2 concentrations. Figure 4.20 shows the temperature curves (a), DTA curves (b), and derivative thermogravimetric (DTG) curves (c), which were obtained by differentiating the weight change in Figure 4.18 over time, for various CO_2 flow rates. A marked similarity was found in both DTA and DTG curves, showing a series of peaks at the initial and final stages (close to equilibrium) corresponding to exothermic reactions of the initial product formation from K_2CO_3 and the initial product transformation to $KHCO_3$, respectively. Especially, the temperature elevation in the initial stage exhibited a marked similarity to the DTA and DTG curves at this stage (panels (d)–(f) of Figure 4.20).

A rapid elevation in the temperature in the initial stage of the reaction (reaction time of <5 min) is observed in Figure 4.20(d). The maximum temperature of the stage increased with the CO_2 flow rate from 1 to 10 cm^3 min^{-1},

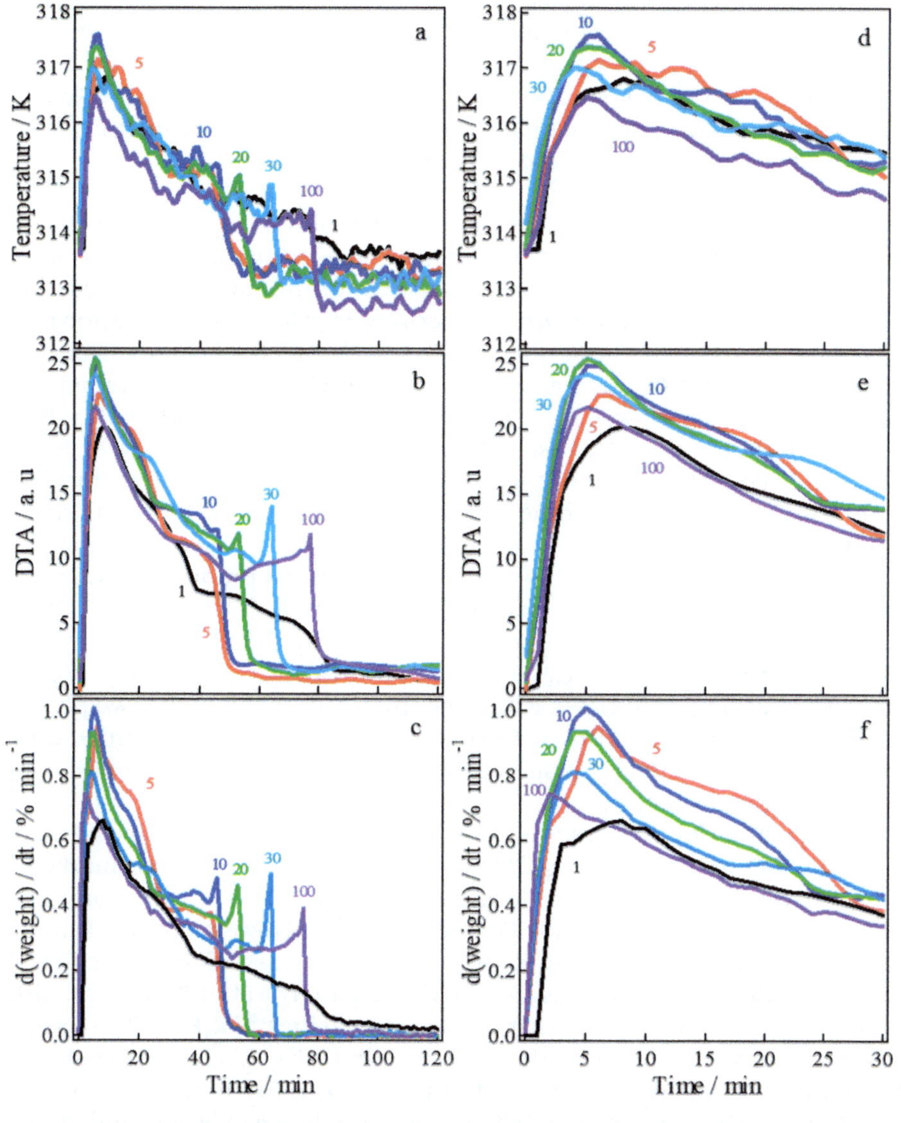

Figure 4.20 Temperature change (a), DTA (b), and DTG (c) curves at various CO_2 flow rates (cm^3 min^{-1}), which are indicated with numbers. These figures are magnified for reaction times of 0–30 min in (d), (e), and (f), respectively. Reproduced from ref. 20 with permission from the American Chemical Society, Copyright 2015.

but decreased with a higher CO_2 flow rate. After the temperature elevation, the temperature moderately decreased in all curves (middle stage). Then, there is a second steep elevation for the CO_2 flow rates between 20 and 100 $cm^3\ min^{-1}$ near equilibrium of the CO_2 sorption, because of a subsequent exothermic reaction, although any clear peak was not observed for CO_2 flow rates of 1 and 5 $cm^3\ min^{-1}$. The second peak exhibited an analogous peak in the DTA and DTG curves (panels (b) and (c) of Figure 4.20). A temperature drop at this stage (final stage) was also observed with the increase of the CO_2 flow rate from 20 to 100 $cm^3\ min^{-1}$. It can be said that the overall reaction and the initial process of CO_2 sorption under a high CO_2 flow rate (above 20 $cm^3\ min^{-1}$) proceed slowly under moist conditions from these results together with considering the reaction mechanism discussed later.

4.2.2.2.3 Structural Changes in the CO_2 Sorption of K_2CO_3. The bicarbonate formation of K_2CO_3 was examined by measuring the XRD patterns of CO_2-sorbed samples of K_2CO_3 under moist conditions at different reaction times. There have been few reports concerning an initial transformation process in the bicarbonate formation of K_2CO_3, or detailed mechanisms. Lee *et al.* reported that K_2CO_3 was transformed to $K_4H_2(CO_3)_3{\cdot}1.5H_2O$ by treatment with water vapor, followed by a further reaction from $K_4H_2(CO_3)_3{\cdot}1.5H_2O$ to $K_2CO_3{\cdot}1.5H_2O$ by dry N_2 gas.[9,42] It was also reported that $K_2CO_3{\cdot}1.5H_2O$ was readily converted to K_2CO_3 at a low temperature under low moisture conditions. This was determined by the heat released in the exothermic reaction.[9,42]

However, from our results in Figure 4.21, the initial product of CO_2 sorption of K_2CO_3 is not a simple transformation from K_2CO_3 to $K_2CO_3{\cdot}1.5H_2O$ and $K_4H_2(CO_3)_3{\cdot}1.5H_2O$. For the reaction over 1 min, the XRD patterns show the peaks of two phases – $K_2CO_3{\cdot}1.5H_2O$ and $K_4H_2(CO_3)_3{\cdot}1.5H_2O$ – for all CO_2 flow rates (Figure 4.21(a)). K_2CO_3 completely changed to the two phases under moist conditions (RH of 68% at 313 ± 3 K). Thus, the temperature changes were found to be a result of the initial reaction and the subsequent transformation processes.

Subsequently, for the reaction beyond 5 min, the peak intensities of $K_2CO_3{\cdot}1.5H_2O$ and $K_4H_2(CO_3)_3{\cdot}1.5H_2O$ in the XRD pattern were markedly enhanced (Figure 4.21(b)). For a longer reaction time (20 min), the dominant peaks were assigned to the $K_4H_2(CO_3)_3{\cdot}1.5H_2O$ phase. However, the peak intensities of $KHCO_3$ increased only slightly with an increase of the CO_2 concentration. Furthermore, in comparison with the patterns after 5 min, the XRD patterns after 20 min showed that the peak intensities of $K_4H_2(CO_3)_3{\cdot}1.5H_2O$ increased considerably, while the peaks of the $K_2CO_3{\cdot}1.5H_2O$ phase almost disappeared. These changes indicate that $K_2CO_3{\cdot}1.5H_2O$ was converted to $K_4H_2(CO_3)_3{\cdot}1.5H_2O$ rather than $KHCO_3$. Accordingly, it can be concluded that $K_2CO_3{\cdot}1.5H_2O$ is the initial product in the bicarbonate formation process of K_2CO_3.

For a reaction time of 40 min, the peak intensity of the $KHCO_3$ phase increased significantly, whereas the peak intensity of the $K_4H_2(CO_3)_3{\cdot}1.5H_2O$ phase decreased (Figure 4.21(d)). This result shows that $K_4H_2(CO_3)_3{\cdot}1.5H_2O$

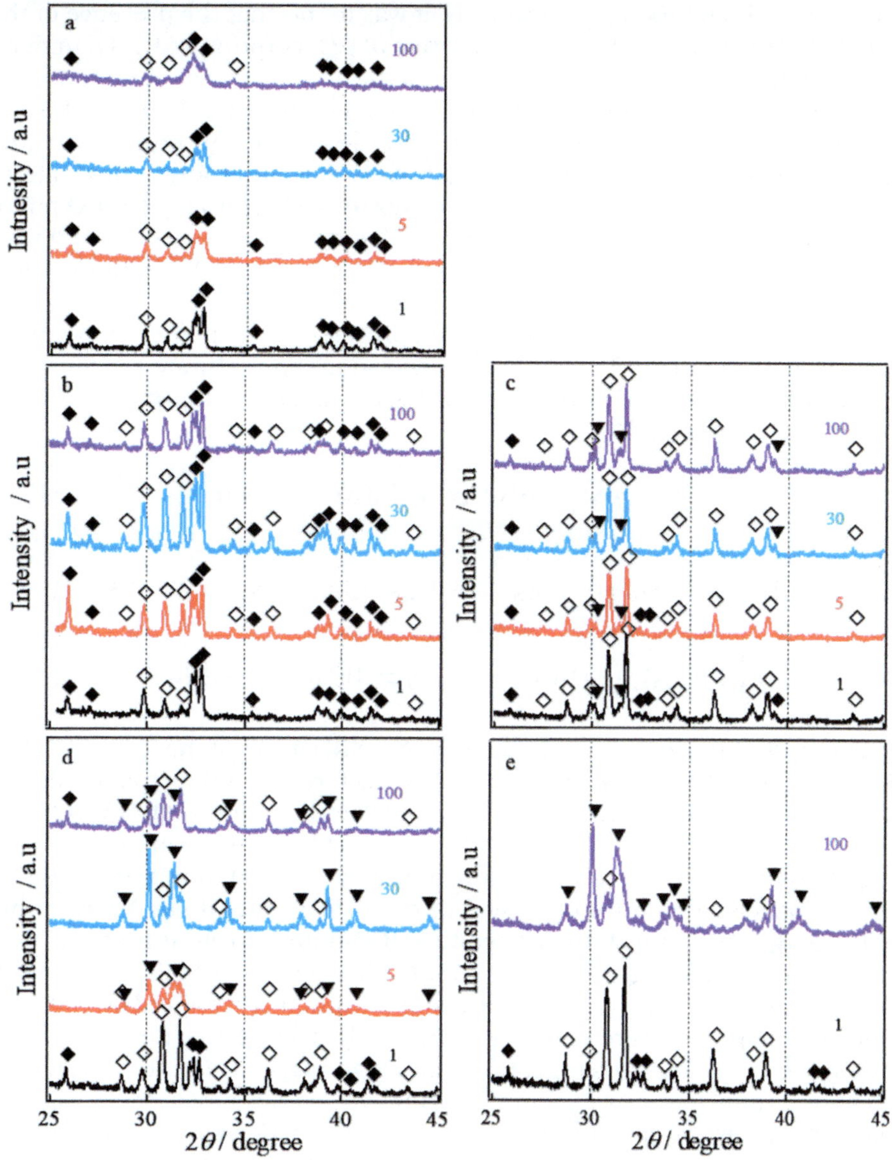

Figure 4.21 XRD patterns after CO_2 occlusion of K_2CO_3 at various CO_2 flow rates (cm^3 min^{-1}), which are indicated with numbers, at reaction times: (a) 1 min, (b) 5 min, (c) 20 min, (d) 40 min, and (e) 60 min. ▼: $KHCO_3$; ◇: $K_4H_2(CO_3)_3{\cdot}1.5H_2O$; ◆: $K_2CO_3{\cdot}1.5H_2O$. Reproduced from ref. 20 with permission from the American Chemical Society, Copyright 2015.

changed to $KHCO_3$ through reaction (4.19). At a low CO_2 flow rate (1 cm^3 min^{-1}), the $K_2CO_3{\cdot}1.5H_2O$ phase remained. This suggests the unhindered transformation of $K_2CO_3{\cdot}1.5H_2O$ from $K_4H_2(CO_3)_3{\cdot}1.5H_2O$ according to the reversible reaction in a dilute CO_2 concentration and saturated water vapor at a mild temperature. Inversely, it was found that the presence of the $K_2CO_3{\cdot}1.5H_2O$ phase induced the slow rate of CO_2 sorption of K_2CO_3 under a high CO_2 flow of 100 cm^3 min^{-1}.

For a reaction time of 60 min, peaks assigned to the $K_4H_2(CO_3)_3{\cdot}1.5H_2O$ and $KHCO_3$ phases appeared in a CO_2 flow rate of 100 cm^3 min^{-1}, while the $K_2CO_3{\cdot}1.5H_2O$ and $K_4H_2(CO_3)_3{\cdot}1.5H_2O$ phases formed at a CO_2 flow rate of only 1 cm^3 min^{-1} (Figure 4.21(e)). The peaks of the $K_4H_2(CO_3)_3{\cdot}1.5H_2O$ phase decreased at a flow rate of 100 cm^3 min^{-1}. These results were consistent with those of TGA and the final crystal structure observed at different CO_2 flow rates.

From the results of the transformation processes and kinetics of CO_2 sorptivity of K_2CO_3 under different CO_2 concentrations and saturated water vapor, the bicarbonate formation pathways of K_2CO_3 can be summarized as follows instead of reaction (4.15):

$$K_2CO_3(s) + 1.5H_2O(g) \rightleftarrows K_2CO_3{\cdot}1.5H_2O(s) \quad (4.20)$$

$$2K_2CO_3{\cdot}1.5H_2O(s) + CO_2(g) \rightleftarrows K_4H_2(CO_3)_3{\cdot}1.5H_2O(s) + 0.5H_2O(g) \quad (4.21)$$

$$K_4H_2(CO_3)_3{\cdot}1.5H_2O(s) + CO_2(g) \rightleftarrows 4KHCO_3 + 0.5H_2O(g) \quad (4.22)$$

The crystallite sizes were estimated for the XRD peaks of the $K_2CO_3{\cdot}1.5H_2O$, $K_4H_2(CO_3)_3{\cdot}1.5H_2O$, and $KHCO_3$ phases for each reaction time (panels (a)–(d) of Figure 4.21) and plotted at the reaction times in Figure 4.22. It was found that the $K_2CO_3{\cdot}1.5H_2O$ and $K_4H_2(CO_3)_3{\cdot}1.5H_2O$ phases grow more slowly for high CO_2 flow rates (30–100 cm^3 min^{-1}) than for low CO_2 flow rates (1–5 cm^3 min^{-1}) (panels (a) and (b) of Figure 4.22). Furthermore, it can be seen that $K_4H_2(CO_3)_3{\cdot}1.5H_2O$ was transformed to $KHCO_3$ more slowly at high CO_2 flow rates (30–100 cm^3 min^{-1}) than at a low CO_2 flow rate (5 cm^3 min^{-1}), as shown for 20–40 min in Figure 4.22(c). Thus, both the initial process and overall reaction of CO_2 sorption were advanced more slowly at a high CO_2 flow rate (above 20 cm^3 min^{-1}). Both low temperatures and high CO_2 concentrations can shift an equilibrium toward CO_2 sorption. However, as discussed below, the initial and final exothermic reactions and the CO_2 sorption rate of K_2CO_3 under different CO_2 flow rates cannot be understood well from XRD data alone.

4.2.2.2.4 Sample Morphologies with Reaction Time. The morphological variation of K_2CO_3 was examined *via* SEM observations before and after CO_2 sorption for different reaction times. The SEM images for 5, 30, and 100 cm^3 min^{-1} CO_2 atmospheres are shown in Figures 4.23–4.25. The K_2CO_3 particles, which were formed from the decomposition of $KHCO_3$ in pure N_2 gas, had a macroporous surface (Figure 4.23(a)).

For a CO_2 flow rate of 5 cm^3 min^{-1} (Figure 4.23(b)), which is milder for the reaction than the 100 cm^3 min^{-1} flow rate (panels (a) and (d) of Figure 4.20), the diameters of the macropores on the surface decreased slightly after

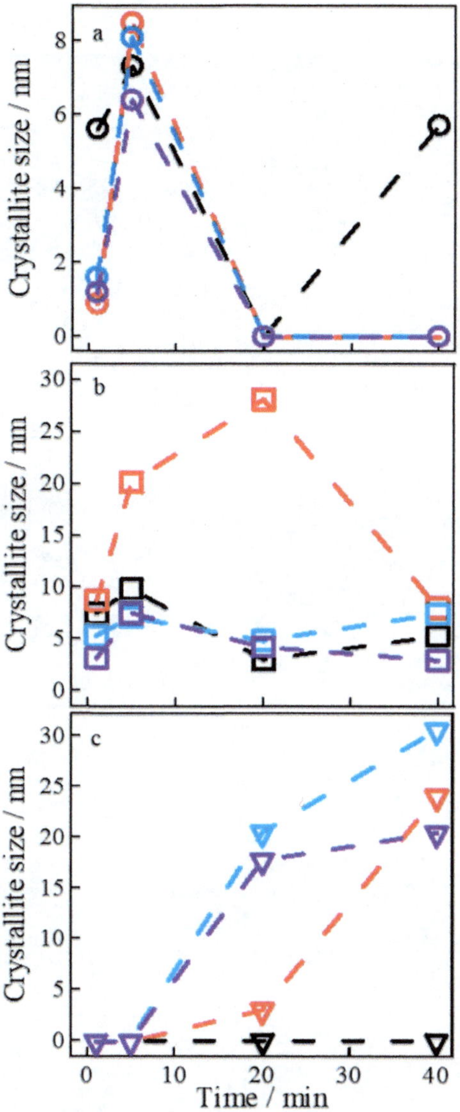

Figure 4.22 Crystallite size of K_2CO_3·$1.5H_2O$ (a), $K_4H_2(CO_3)_3$·$1.5H_2O$ (b) and $KHCO_3$ (c) at different CO_2 flow rates over reaction times obtained from XRD patterns in Figure 4.21(a)–(d) by using Jade software according to the Scherrer equation. The range of Bragg angle: 29–35°. CO_2 flow rate: black (1 $cm^3\ min^{-1}$), red (5 $cm^3\ min^{-1}$), azure (30 $cm^3\ min^{-1}$), and violet (100 $cm^3\ min^{-1}$).

the initial 5 min. The pores on the surface almost disappeared after 20 min (Figure 4.23(c)). This corresponds to the formation of products consisting of major $K_4H_2(CO_3)_3$·$1.5H_2O$ and minor $KHCO_3$ (Figure 4.21(c)). Large granules of $KHCO_3$ developed on the external surface (Figure 4.23(d)) with the reaction of $K_4H_2(CO_3)_3$·$1.5H_2O$ with CO_2 and H_2O for 20–120 min (Figures 4.19

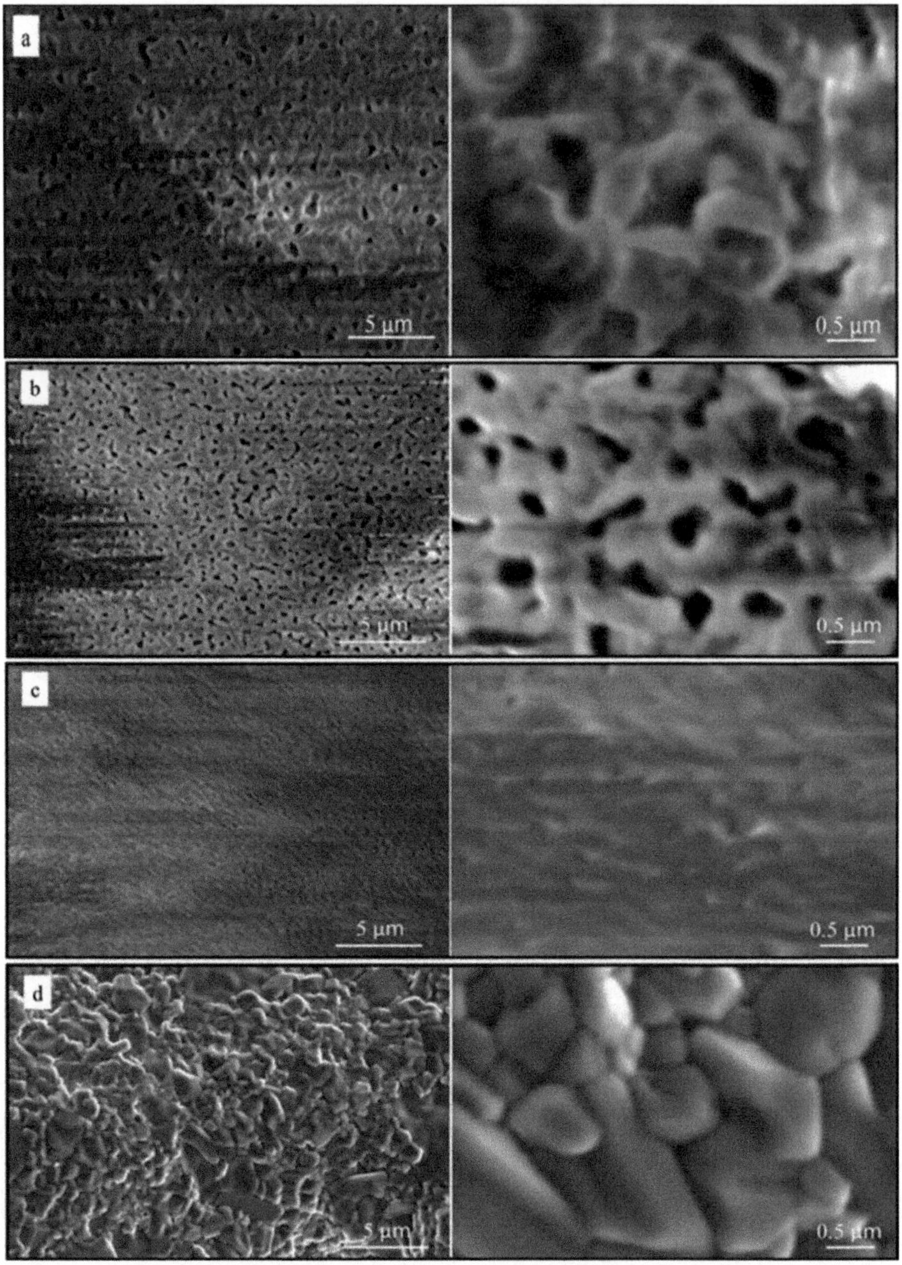

Figure 4.23 SEM images of K_2CO_3 before and after CO_2 occlusion at a CO_2 flow rate of 5 cm^3 min^{-1} (a)–(d) ((a): 0 min, (b): 5 min, (c): 20 min, and (d): 120 min).

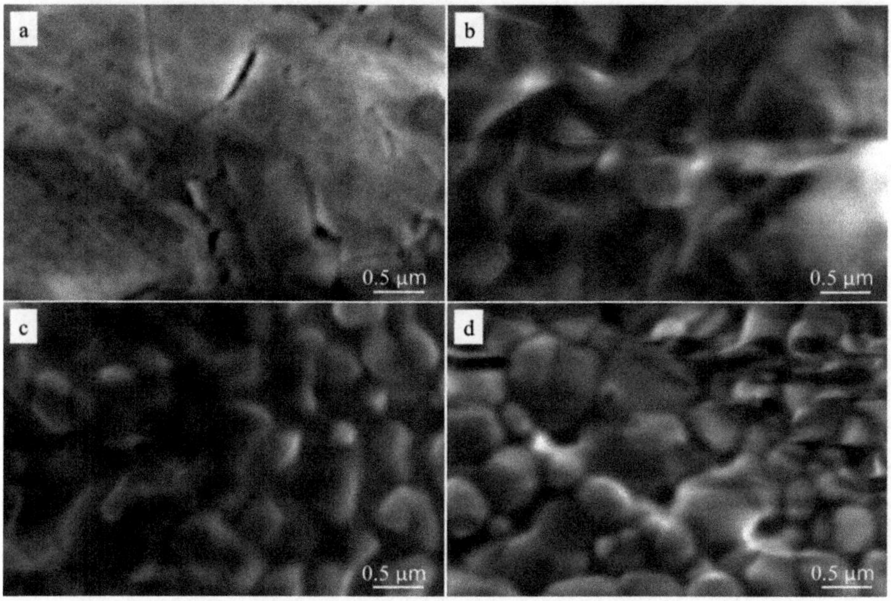

Figure 4.24 SEM images after CO_2 occlusion by K_2CO_3 at a CO_2 flow rate of 30 cm^3 min^{-1} (a)–(d) ((a): 5 min, (b): 20 min, (c): 40 min, (d): 120 min).

and 4.21(d)). CO_2 sorption reached a plateau in about 40 min (Figure 4.18), without any second peak in the DTG and DTA curves during this transformation period (Figure 4.20).

At a CO_2 flow rate of 30 cm^3 min^{-1} (Figure 4.24), the macropores on the surface showed a significant decrease after the initial 5 min and almost disappeared, and then the wrinkled surface formed after 20 min. The granule initially formed on the surface between 20 and 40 min. Finally, smaller granules developed on the surface after 120 min than those for 5 cm^3 min^{-1}. These results show the definite development process of granules on the surface. At a CO_2 flow rate of 100 cm^3 min^{-1}, the initial formation of a wrinkled surface within the initial 5 min (Figure 4.25) was observed, followed by restricted formation of $K_2CO_3{\cdot}1.5H_2O$ and $K_4H_2(CO_3)_3{\cdot}1.5H_2O$. The initial granule growth on the surface was accompanied with the disappearance of the pores at 20 min (Figure 4.25(f)), causing the slow bicarbonate formation of $K_2CO_3{\cdot}1.5H_2O$ and $K_4H_2(CO_3)_3{\cdot}1.5H_2O$ between 5 and 20 min ((b) and (c) of Figure 4.21). The surface resistance was enhanced because of the formation of the wrinkled surface, which should cover the initial product during this transformation period. As a result, it was observed that the $K_2CO_3{\cdot}1.5H_2O$ phase remained at 100 cm^3 min^{-1} and the formation of $KHCO_3$ from $K_4H_2(CO_3)_3{\cdot}1.5H_2O$ was restricted at 40 min (Figure 4.21(d)). Accordingly, there were steep drops in the temperature, DTA, and DTG curves between 20 and 40 min (Figure 4.20), exhibiting that the CO_2 sorption rate at this CO_2 flow rate was slower

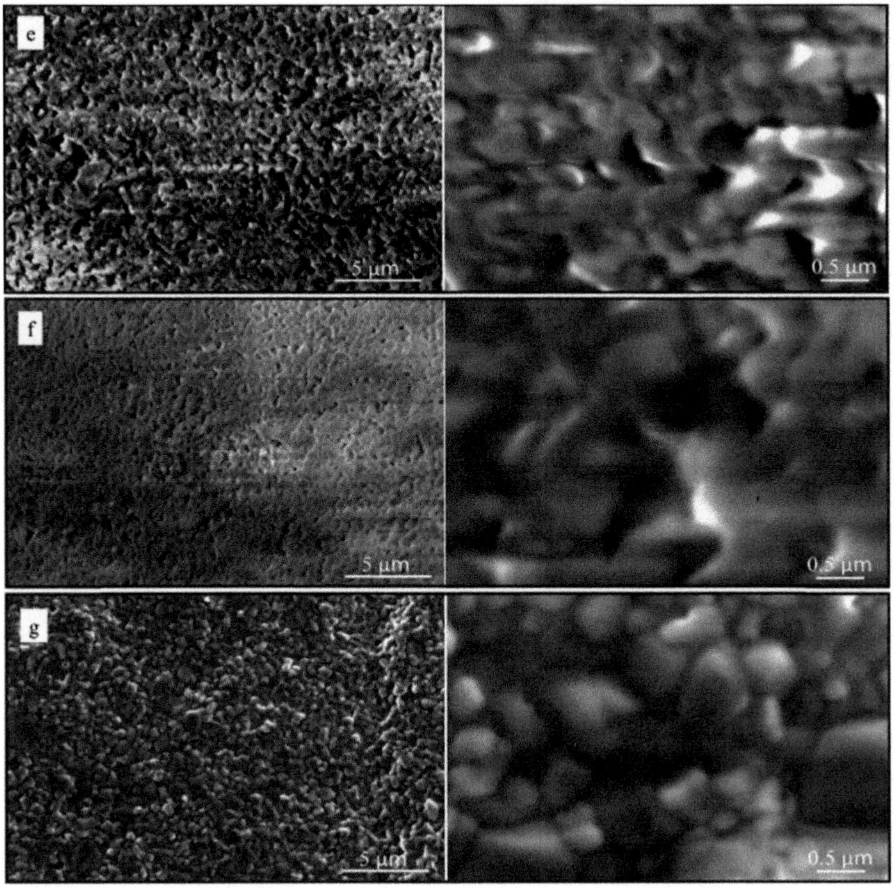

Figure 4.25 SEM images of K_2CO_3 after CO_2 occlusion at a CO_2 flow rate of 100 cm^3 min^{-1} (e)–(g) ((e): 5 min, (f): 20 min, and (g): 120 min).

than those for other CO_2 flow rates (except for 1 cm^3 min^{-1}). For the development of larger granules along with a further reaction of $K_4H_2(CO_3)_3 \cdot 1.5H_2O$ between 40 and 60 min, the wrinkled surface should be broken, followed by a decrease of the CO_2 sorption rate and temperature according to the DTG and DTA curves in Figure 4.20. From the morphology (Figure 4.25(f)–(g)) and XRD results (Figure 4.19 and 21(e)) at a CO_2 flow rate of 100 cm^3 min^{-1}, it was found that smaller granules developed on the external surface due to bicarbonate formation from $K_4H_2(CO_3)_3 \cdot 1.5H_2O$ after 20 min. Owing to promotion of the chemisorption and gas diffusion processes by smaller granules,[43,44] the acceleration of reaction (4.21) between 60 and 80 min (final stage) possibly induced the sharp peaks in the DTA and DTG curves in Figure 4.20.

To understand the second peaks for higher CO_2 flow rates (20–100 cm^3 min^{-1}) during the final stage in Figure 4.20, morphology changes were observed at 120 min for CO_2 flow rates of 30 cm^3 min^{-1} and 100 cm^3 min^{-1}. The results exhibited the development of smaller granules of $KHCO_3$ on the surface at 120 min. Furthermore, the CO_2 sorption rate and temperature change exhibited a transient and sharp increase during the final stage. By contrast, the morphology change in the initial stage of CO_2 sorption must be important in the overall reaction rate. This can explain the decreases in the temperature and the peak intensities in the DTA and DTG curves, as shown in panels (d)–(f) of Figure 4.20, with an increase of the CO_2 flow rate except for 1 $cm^3\ min^{-1}$.

4.2.2.2.5 CO_2 Capture of K_2CO_3 Under Various H_2O Concentrations. As mentioned in the CO_2 capture of K_2CO_3 under different CO_2 concentrations, the bicarbonate formation of K_2CO_3 proceeds *via* three stages. The formation and transformation of $K_2CO_3{\cdot}1.5H_2O$ and $K_4H_2(CO_3)_3{\cdot}1.5H_2O$ at the initial stage have a profound influence on the overall CO_2 occlusion rate of K_2CO_3 along with morphology variation. To identify the special effect of H_2O concentration on the CO_2 capture of K_2CO_3 for a higher CO_2 concentration under moist conditions, a study on the CO_2 capture of K_2CO_3 under different humidities is essential. The CO_2 occlusion of K_2CO_3 under different H_2O concentrations with a CO_2 flow rate of 30 $cm^3\ min^{-1}$ and XRD patterns of the products after the CO_2 occlusion were exhibited. Figure 4.26 shows that the CO_2 sorption amount had almost no increase while the occlusion rate significantly increased with an increase in the humidity. This result indicates that the higher H_2O concentrations accelerate the CO_2 capture of K_2CO_3.

XRD patterns after CO_2 occlusion of K_2CO_3 under different H_2O concentrations showed that major $KHCO_3$ existed in the phases of peaks of K_2CO_3 after CO_2 occlusion under different humidities, as can be seen in Figure 4.27. With an increase in the H_2O concentration, the peak intensity of $K_4H_2(CO_3)_3{\cdot}1.5H_2O$ increased. This result illustrated that the bicarbonate formation of $K_4H_2(CO_3)_3{\cdot}1.5H_2O$ was inhibited under the higher H_2O concentration.

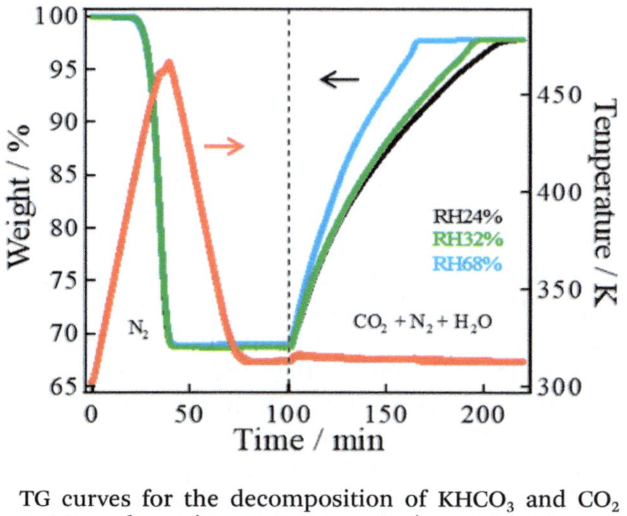

Figure 4.26 TG curves for the decomposition of $KHCO_3$ and CO_2 occlusion of K_2CO_3 under various H_2O concentrations at 313 K.

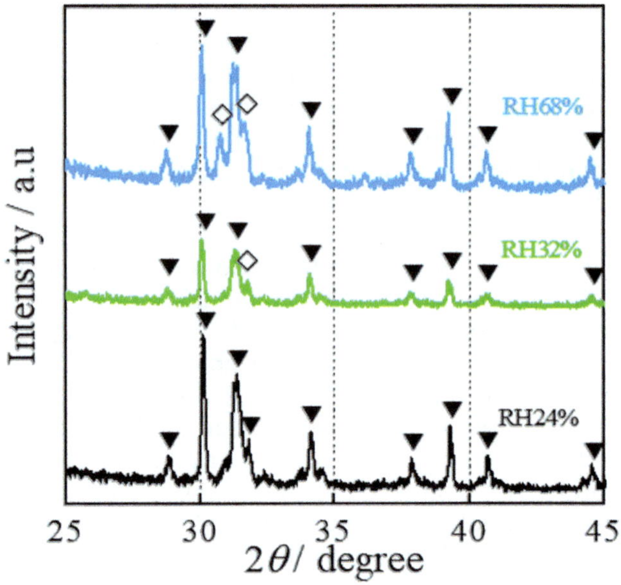

Figure 4.27 XRD patterns after CO_2 occlusion for various H_2O concentrations. ▼: $KHCO_3$, ◇: $K_4H_2(CO_3)_3{\cdot}1.5H_2O$.

The kinetic properties of the CO_2 occlusion of K_2CO_3 can be tracked by monitoring the temperature under different humidities as the bicarbonate formation of K_2CO_3 leads to a temperature change of the exothermic reactions. Figure 4.28 showed that both DTA and DTG curves showed a marked similarity, showing a series of peaks at the initial stage and final stage corresponding to exothermic reactions at the initial and final stages of the CO_2 occlusion of K_2CO_3.

Similarly, there was a temperature rise at the initial stage of the reaction with the H_2O concentration increasing from RH24% to RH68%. After the initial temperature rise, the temperature decreased moderately in all curves (middle stage). A second obvious increase at the final stage was observed for the higher H_2O concentration (RH68%) when the CO_2 occlusion was close to equilibrium. These results demonstrated that the most thermal energy during the CO_2 occlusion process was obtained from the reactions at the initial stage. The peak (final stage) had an analogue peak in the DTA and DTG curves. Particularly, the CO_2 occlusion rate for the higher H_2O concentration, as shown in the DTG curves, had a sharp rise at the final stage. This implied that the higher H_2O concentration was favorable for the initial and final stages of the CO_2 occlusion process. Furthermore, these results displayed that both the initial stage and the overall reaction proceed to readily reach the CO_2 occlusion equilibrium while the H_2O concentration is increasing. Also, these results indicated that the overall reaction rate under different H_2O concentrations depended upon the CO_2 occlusion rate at the initial and final stages. Especially, there was a slower CO_2 occlusion

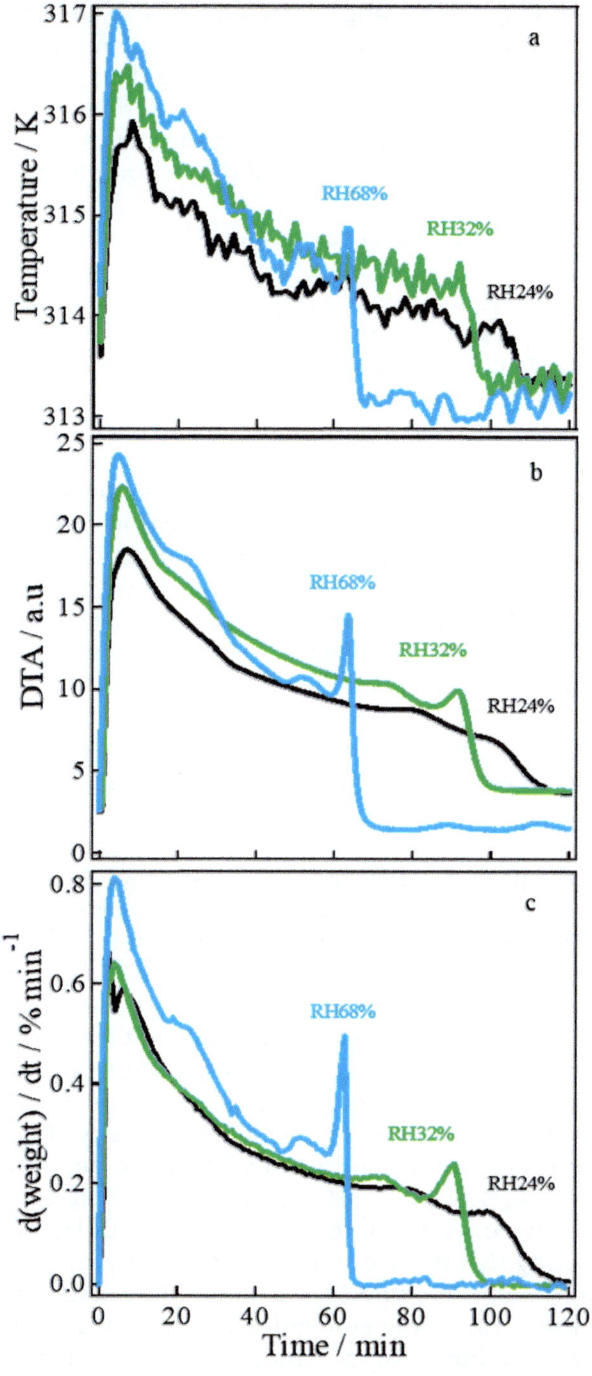

Figure 4.28 Temperature changes (a), DTA (b) and DTG (c) curves for various H_2O concentrations at a CO_2 flow rate of 30 cm^3 min^{-1}.

rate at the final stage according to the analyses of the CO_2 occlusion of K_2CO_3 under different CO_2 concentrations. Thus, this indicated that the rate-determining step was derived from one of the reactions at the final stage. These results and the rate-determining step for different H_2O concentrations can be understood together with the analyses of the reactions at the initial and final stages on the basis of the crystalline structural change, which are discussed later.

XRD measurements on CO_2-occluded-K_2CO_3 based-materials under different humidity conditions with a CO_2 flow rate of 30 cm^3 min^{-1} were carried out to determine the rate-limiting step of the bicarbonate formation of K_2CO_3. Numerous studies concerning the effect of H_2O concentration on the CO_2 occlusion of K_2CO_3 have been reported. Park *et al.* reported that the CO_2 occlusion capacity increased with the H_2O flow rate from 0.009 to 0.077 cm^3 min^{-1}.[29] Zhao *et al.* reported that the CO_2 sorptivity of pure K_2CO_3 was considerably improved with the H_2O concentration from 0% to 21%, especially between 3% and 15%, with 15% CO_2 at 338 K at 0.1 MPa.[39] Particularly, it was claimed that the hydration reaction of K_2CO_3 was considered as the rate-determining step for the CO_2 occlusion process so that the overall reaction rate and CO_2 uptake increased with an increase in the H_2O concentration.[37,39] However, our results indicated that there was a different possibility for the rate-determining step. Besides, detailed information on the transformation of the crystalline structure at the initial stage has not been revealed so far.

The results of XRD patterns at different reaction times under different H_2O concentrations can be seen in Figure 4.29. With the reactions occurring over 5 min (Figure 4.29(a)), the XRD patterns showed the presence of $K_2CO_3{\cdot}1.5H_2O$ and $K_4H_2(CO_3)_3{\cdot}1.5H_2O$ for a H_2O concentration between RH32% and RH68%. Also, the peak intensity of $K_2CO_3{\cdot}1.5H_2O$ and $K_4H_2(CO_3)_3{\cdot}1.5H_2O$ in the XRD pattern increased significantly with an increase in the H_2O concentration from 32% to 68%. This result indicated that the higher H_2O concentration contributed to the occurrence of reactions (4.20) and (4.21). Thus, a maximum in the temperature is a result of the released heat from two reactions at the initial stage, although we could not claim that the heat was mostly released from either reaction (4.20) or reaction (4.21).

As these two reactions proceeded for over 20 min for H_2O concentrations between RH32% and RH68%, the peak intensity of $K_2CO_3{\cdot}1.5H_2O$ in the XRD pattern decreased significantly while the dominant peaks were derived from the $K_4H_2(CO_3)_3{\cdot}1.5H_2O$ phase. As a result, the $K_2CO_3{\cdot}1.5H_2O$ phase was mostly converted into the $K_4H_2(CO_3)_3{\cdot}1.5H_2O$ phase *via* reaction (4.21). During this transformation period between 5 and 20 min corresponding to the observed decline at the middle stage of DTA and DTG curves in Figure 4.28, it was concluded that $K_2CO_3{\cdot}1.5H_2O$ transformed from K_2CO_3 *via* reaction (4.20) provided the main heat for the CO_2 occlusion process. On the other hand, for a H_2O concentration of RH68%, the peaks of the $KHCO_3$ phase transformed from the $K_4H_2(CO_3)_3{\cdot}1.5H_2O$ increased slightly through reaction (4.22). In this case, there was still an

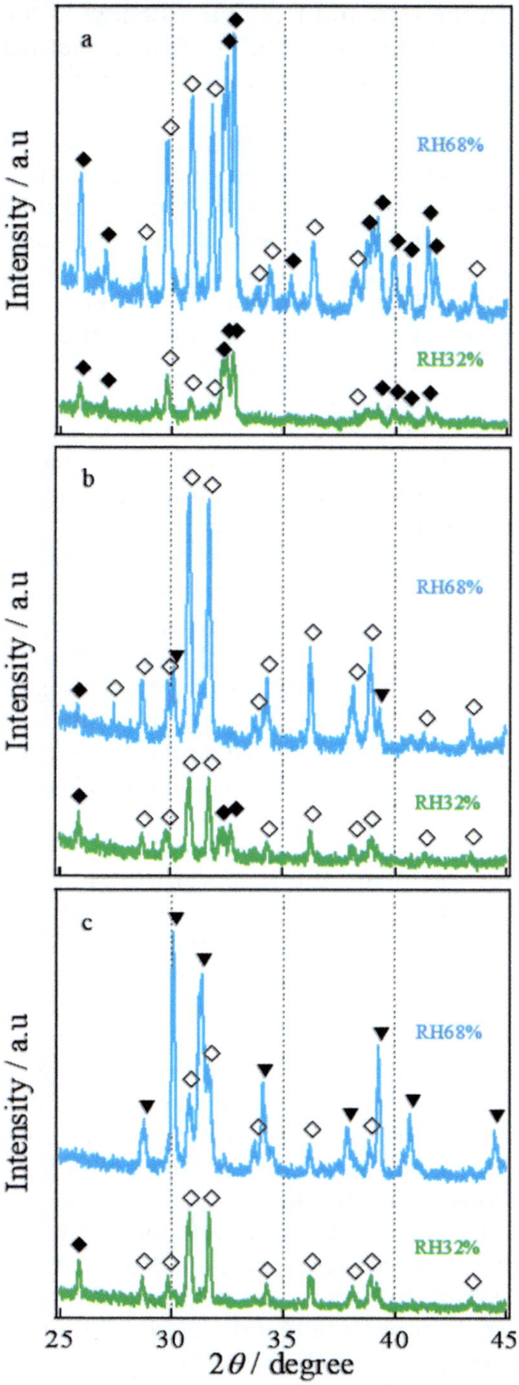

Figure 4.29 XRD patterns after the CO_2 occlusion of K_2CO_3 for various H_2O concentrations at the reaction times: (a) 5 min, (b) 20 min, and (c) 40 min. ▼: $KHCO_3$; ◇: $K_4H_2(CO_3)_3{\cdot}1.5H_2O$; ◆: $K_2CO_3{\cdot}1.5H_2O$.

obvious descent in the DTA and DTG curves between 5 min and 20 min. According to these changes, it can be confirmed that reaction (4.20) is a highly exothermic initiation reaction rather than reaction (4.21) for the CO_2 occlusion of K_2CO_3.

When the reactions proceeded for over 40 min, for a H_2O concentration of 32%, the major $K_4H_2(CO_3)_3{\cdot}1.5H_2O$ phase existed in the XRD pattern while the dominant peaks originated from the $KHCO_3$ phase and $K_4H_2(CO_3)_3{\cdot}1.5H_2O$ phase for a H_2O concentration of 68%. In addition, it was observed that the $K_2CO_3{\cdot}1.5H_2O$ phase remained and no information on the $KHCO_3$ phase appeared for a H_2O concentration of 32%. During the transformation period between 20 and 40 min, the CO_2 occlusion process proceeded through reaction (4.21) for a low H_2O concentration and reaction (4.22) for a high H_2O concentration, respectively, as the intensities of the DTA and DTG curves continually descended. Consequently, it was concluded that the exothermic property of reaction (4.20) was much stronger than that of reactions (4.21) and (4.22). Also, the slower CO_2 occlusion *via* reaction (4.22) indicated that reaction (4.22) was the rate-determining step.

With the reactions terminating at 120 min, the peak intensity of the $K_4H_2(CO_3)_3{\cdot}1.5H_2O$ phase in the XRD pattern increased with an increase in the H_2O concentration. Compared with the XRD patterns between 40 and 120 min (Figures 4.27 and 4.29), the bicarbonate formation of $K_4H_2(CO_3)_3{\cdot}1.5H_2O$ mainly occurred during this transformation process corresponding to a rise in the second peak intensity in the temperature, DTA and DTG curves in Figure 4.28. This result is in accordance with the stronger temperature dependence on reaction (4.19) rather than reaction (4.18) in Section 4.2.2.2.1.

The morphology change of K_2CO_3 after the CO_2 occlusion for different reaction times was also examined with SEM. The SEM images for RH32% and RH68% H_2O concentrations are shown in Figures 4.30 and 4.31, respectively. There was a slight difference with the formation process for the granules on the surface *via* three stages for H_2O concentrations between RH32% and RH68%. For the H_2O concentration of RH32%, the macropores on the surface decreased significantly at the initial 5 min and then remained after 20 min. The initial formation of large-sized granules proceeded on the surface between 20 and 40 min. Finally, small-sized granules developed on the surface after 120 min. Comparing the morphology at 120 min for the H_2O concentration of RH68% with that for the H_2O concentration of RH32%, it was found that the wrinkled surface was formed at 20 min and smaller-sized granules developed on the surface between 40 and 120 min for the H_2O concentration of RH68%. As mentioned in Section 4.2.2.2.1, the morphology variation from the wrinkled surface with the porous surface disappearing to smaller-sized granules on the surface led to the rate changes of reaction (4.22). These variations caused the observed reduction and then the rise in temperature variation and the intensity of the peaks of the DTA and DTG curves in Figure 4.28 between 20 and 120 min. Thus, the morphology at the middle and final stages of

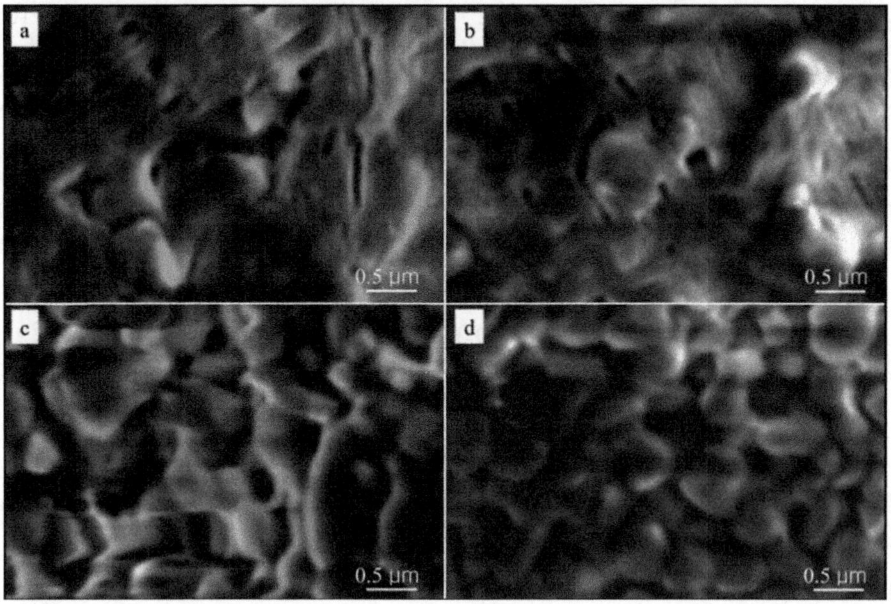

Figure 4.30 SEM images after CO_2 occlusion for a H_2O concentration of RH32% at a CO_2 flow rate of 30 cm^3 min^{-1} at 313 K (a)–(d) ((a): 5 min, (b): 20 min, (c): 40 min, (d): 120 min).

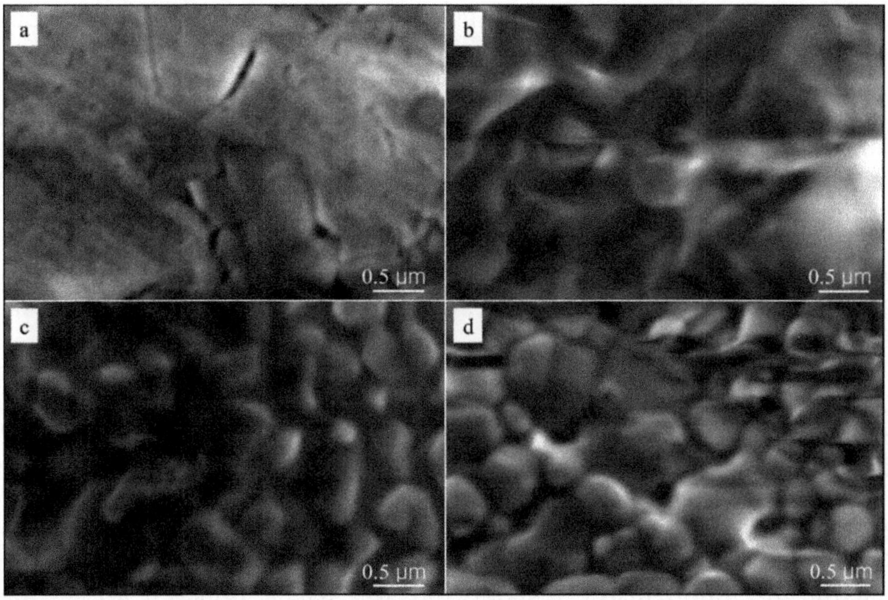

Figure 4.31 SEM images after CO_2 occlusion for a H_2O concentration of RH68% at a CO_2 flow rate of 30 cm^3 min^{-1} at 313 K (a)–(d) ((a): 5 min, (b): 20 min, (c): 40 min, (d): 120 min).

CO_2 occlusion played an important role in the rate of reaction (4.22). With reaction (4.22) advancing, the residual $K_4H_2(CO_3)_3{\cdot}1.5H_2O$ still existed in the XRD pattern (Figure 4.27) because the $KHCO_3$ granules formed on the surface increased the surface resistance. In contrary, there was not a significant variation of the morphology on the surface within 5 min. Consequently, the CO_2 occlusion rate at the initial stage depended upon reactions (4.20) and (4.21). These results of the initial and final stages for a higher H_2O concentration were promising for the CO_2 occlusion of K_2CO_3 at low temperature.

4.2.2.2.6 CO_2 Capture of $K_4H_2(CO_3)_3{\cdot}1.5H_2O$ Under Different H_2O or CO_2 Concentrations. The CO_2 occlusion of $K_4H_2(CO_3)_3{\cdot}1.5H_2O$ displayed a slower bicarbonate formation process for higher CO_2 or lower H_2O concentrations, as shown earlier. In order to better understand the exothermic property and bicarbonate formation behavior of $K_4H_2(CO_3)_3{\cdot}1.5H_2O$, the CO_2 occlusion was performed under a pure CO_2 atmosphere and different CO_2 concentrations containing saturated H_2O, respectively, as can be seen in Figure 4.32. As both CO_2 concentration and H_2O concentration increased, the CO_2 occlusion amount was significantly enhanced. The CO_2 occlusion amount increased to 2.86 mmol g^{-1} for a CO_2 flow rate of 100 cm^3 min^{-1}, which is about 5% in excess of the theoretical amount of 2.73 mmol g^{-1}. This result indicated that there was an error for the TG data caused by the multiple exchange process from a N_2 to a $CO_2 + H_2O$ atmosphere. With reaction (4.22) proceeding, the CO_2 occlusion rate for a CO_2 flow rate of 5 cm^3 min^{-1} reached equilibrium at 20 min, while it reached equilibrium at 40 min for a CO_2 flow rate of 100 cm^3 min^{-1}.

Figure 4.33 shows that the XRD patterns of $K_4H_2(CO_3)_3{\cdot}1.5H_2O$ before and after CO_2 occlusion were exhibited. Only the $K_4H_2(CO_3)_3{\cdot}1.5H_2O$ phase was exhibited in the XRD patterns of the $K_4H_2(CO_3)_3{\cdot}1.5H_2O$ stored in a vacuum oven and under a pure CO_2 atmosphere. This result demonstrated an error for the TG data caused by the multiple exchange process from a N_2 to a CO_2 atmosphere. For CO_2 flow rates of 5 and 100 cm^3 min^{-1}, the XRD patterns contained the minor $KHCO_3$ phase and major $K_4H_2(CO_3)_3{\cdot}1.5H_2O$, indicating a slower bicarbonate formation because of an increase in the surface resistance, as mentioned before.

Figure 4.34 shows the temperature change, and DTA and DTG curves of $K_4H_2(CO_3)_3{\cdot}1.5H_2O$ over time. With an increase in the concentrations of H_2O or CO_2, there was an increase in the temperature and CO_2 occlusion rate along with the strong exothermic property. There was a high value of approximately 15 a.u. in the DTA curves in Figure 4.34(b), compared to that in Figure 4.20(b). This demonstrated that the exothermic property of reaction (4.19) was weaker than that of reaction (4.16). Figure 4.34(c) showed that the maximum in the reaction rate was 0.53% min^{-1} at a CO_2 flow rate of 5 cm^3 min^{-1}, which was much lower than that at the initial stage in Figure 4.20(c). This result can confirm that the bicarbonate formation of $K_4H_2(CO_3)_3{\cdot}1.5H_2O$ (reaction (4.22)) is the rate-determining step for the

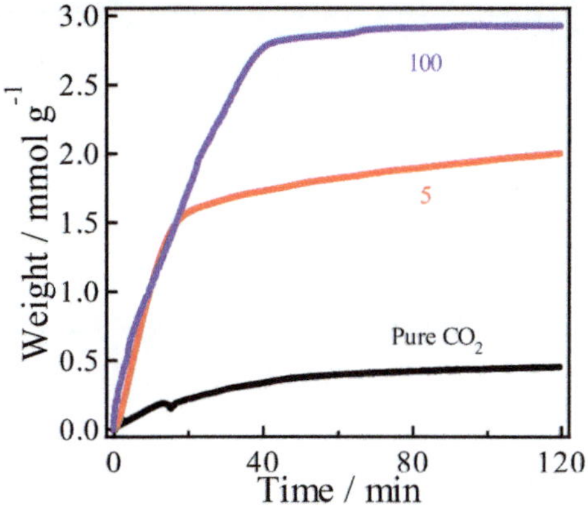

Figure 4.32 TG curves for the CO_2 occlusion of $K_4H_2(CO_3)_3{\cdot}1.5H_2O$ under a pure CO_2 atmosphere, and CO_2 flow rates of 5 $cm^3\ min^{-1}$ and 100 $cm^3\ min^{-1}$ at 313 K.

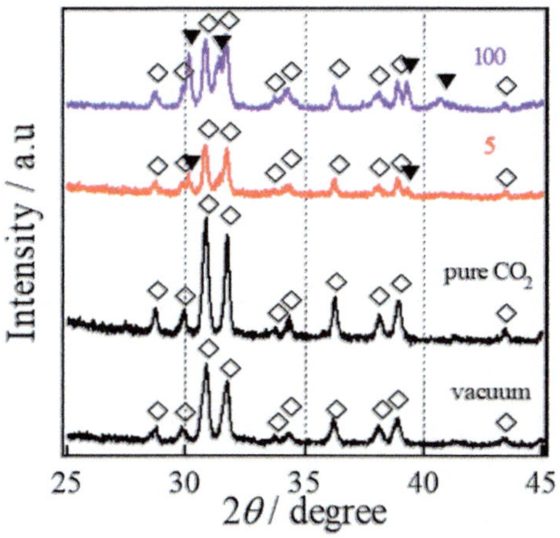

Figure 4.33 XRD patterns of $K_4H_2(CO_3)_3{\cdot}1.5H_2O$ before and after CO_2 occlusion. ▼: $KHCO_3$; ◇: $K_4H_2(CO_3)_3{\cdot}1.5H_2O$.

CO_2 occlusion of K_2CO_3 under moist conditions in the presence of saturated water vapor. Furthermore, with an increase in the CO_2 concentration, the maximum value in the CO_2 occlusion rate decreased, as seen in Figure 4.34(c), which was consistent with those at the final stage under various CO_2 and H_2O concentrations.

The morphology of $K_4H_2(CO_3)_3{\cdot}1.5H_2O$ before and after CO_2 occlusion for CO_2 flow rates of 5 and 100 $cm^3\ min^{-1}$, respectively, are observed in Figure 4.35. Figure 4.35(a) illustrates that the macropores almost disappear to form a compact plane on the surface according to the preparation method of $K_4H_2(CO_3)_3{\cdot}1.5H_2O$ at a higher temperature. Panels (b) and (c) of Figure 4.35 show that the $KHCO_3$ granules developed on the surface.

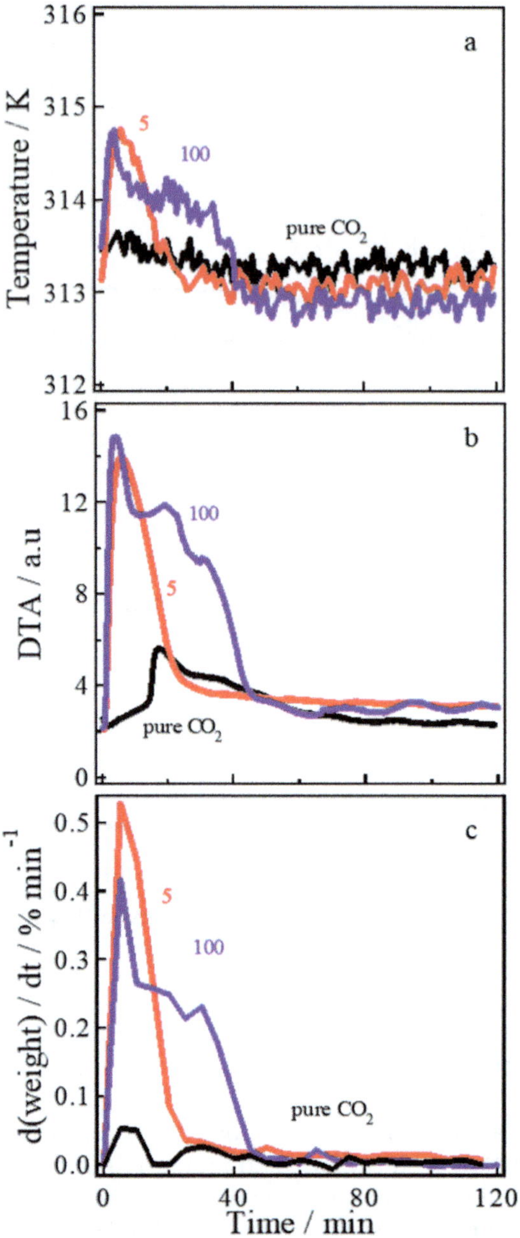

Figure 4.34 Temperature changes (a), DTA (b) and DTG (c) curves for different CO_2 or H_2O concentrations.

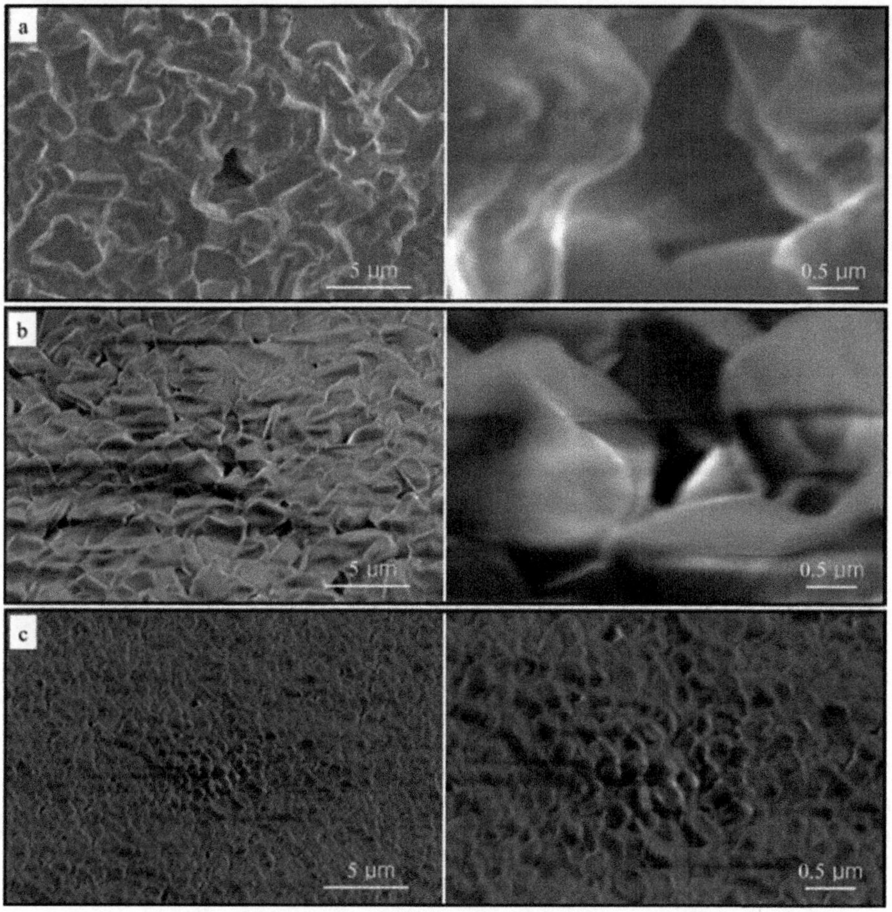

Figure 4.35 SEM images for $K_4H_2(CO_3)_3{\cdot}1.5H_2O$ before (a) and after CO_2 occlusion for CO_2 flow rates of 5 cm^3 min^{-1} (b) and 100 cm^3 min^{-1} (c).

Finally, based on the analyses of these XRD patterns and morphology, reaction (4.22) as the rate-determining step plays an important role in the CO_2 occlusion of K_2CO_3. The granules developed on the surface and the surface resistance increased with the bicarbonate formation. Improving the morphology on the surface to make the bicarbonate formation favorable for application in the CO_2 occlusion of K_2CO_3 should be a requirement.

4.2.2.3 Conclusions

The analysis of changes in XRD patterns with the reaction time of the CO_2 sorption of K_2CO_3 under different CO_2 and H_2O concentrations shows that the bicarbonate formation of K_2CO_3 proceeds *via* three reactions. The changes in exothermic properties, temperature, and reaction rate show that the CO_2 sorption of K_2CO_3 under moist conditions also goes through three stages. The first reaction is the hydrate formation from K_2CO_3 to $K_2CO_3{\cdot}1.5H_2O$, which can be converted to $K_4H_2(CO_3)_3{\cdot}1.5H_2O$ in the initial stage of the CO_2 sorption. The conversion of $KHCO_3$ from $K_4H_2(CO_3)_3{\cdot}1.5H_2O$ occurs slowly at the final stage, although there is a second increase in the CO_2 sorption rate and temperature under high CO_2 flow rates (20–100 $cm^3\ min^{-1}$). The analysis of morphology changes shows that the CO_2 sorption of K_2CO_3 is suppressed at high CO_2 concentrations, but the conversion of $KHCO_3$ from $K_4H_2(CO_3)_3{\cdot}1.5H_2O$ is favorable because of a higher CO_2 sorption amount with an increase in the CO_2 concentration. The morphology of the particles seems to be an important property for the kinetic behavior of the CO_2 sorption of K_2CO_3. TG-DTA and XRD analyses show the dependence of the CO_2 sorption of K_2CO_3 on the CO_2 concentration under moist conditions. Therefore, it is concluded that K_2CO_3 from the decomposition of $KHCO_3$ can be used as a sorbent for CO_2 capture at an ambient atmosphere and temperature.

4.2.3 Improvements of the CO_2 Sorptivity of Na_2CO_3 and K_2CO_3

As described, the advantages and disadvantages of the CO_2 sorptivity of Na_2CO_3 and K_2CO_3 can be summarized as shown in Table 4.1.

Their sorption capacity, cost, cyclability, and recovery rate are good. However, regeneration temperatures are relatively high and reactions are slow, in particular for the CO_2 occlusion of Na_2CO_3 under moist conditions. For the practical use of these salts as CO_2 sorbents, their disadvantages should be improved.

4.2.3.1 Improvement of the CO_2 Sorptivity of Na_2CO_3

As the CO_2 sorption of Na_2CO_3 under moist conditions is very slow, the kinetic aspect must be improved. This was done as follows: as the crystal structure of Na_2CO_3 is very stable, the bicarbonate reaction from Na_2CO_3 to $NaHCO_3$

Table 4.1 Advantages and disadvantages of the CO_2 sorptivity of Na_2CO_3 and K_2CO_3.

	Na_2CO_3	K_2CO_3
Sorption capacity	◎	○
Regeneration temperature	△ > 383 K	X > 423 K
Reaction rate	X	△
Side reaction	△	△
Cost	○	△
Cyclability	○	○
Recovery rate	○	○

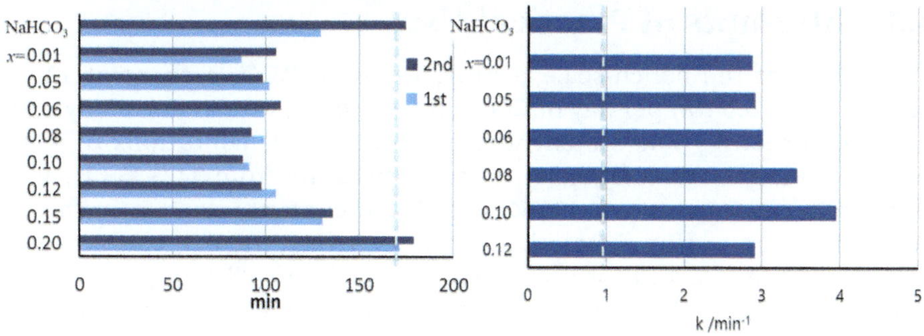

Figure 4.36 Reaction time (left) and rate constant (k min^{-1}) for the CO_2 occlusion of the materials after the heat treatment of $Mg_{x/2}Na_{(1-x)}$ HCO_3 at 473 K under moist conditions. The reaction was repeated twice. The rate constant was evaluated for the 2nd reaction.

is slow. Therefore, the formation of unstable structures of Na_2CO_3 was attempted by introduction of foreign atoms. We tried to introduce Mg^{2+} into the crystal lattice of $NaHCO_3$ as $Mg_{x/2}Na_{(1-x)}HCO_3$ (x = 0.01–0.20). After heat treatment at 473 for 10 min, the CO_2 sorption of these materials was measured under moist conditions. From the results, the reaction time, which is the time from the onset to the saturation, and reaction rate constant, which was evaluated based on the 1st-order reaction of the CO_2 concentration, were estimated, as shown in Figure 4.36. This result showed that the reaction time was shortened and a rate constant with a value four-fold greater was obtained at x = 0.10. Thus, the reaction rate can be improved by this kind of modification of the crystal structure.

4.2.3.2 Improvement of the CO_2 Sorptivity of K_2CO_3

The formation of nanoparticles of K_2CO_3 was attempted for improvement as follows.[45] Nanoparticles of $KHCO_3$ were prepared in mesopores of carbon aerogels, which have mesopores with different average pore sizes of 7 nm, 16 nm, and 18 nm. The contents of $KHCO_3$ prepared in the pores were 19 or 21 wt%. The regeneration temperature decreased from 423 K to 420 K, 400 K, and 390 K for $KHCO_3$ nanoparticles prepared in 7 nm pores, 16 nm pores, and 18 nm pores, respectively. Thus, the nanoparticle formation showed the effect on lowering the regeneration temperature of the reaction from $KHCO_3$ to K_2CO_3, although the temperatures are still higher than that for Na_2CO_3. Also, there was a problem in that the sorption capacity was not good enough for the composite of the K_2CO_3-carbon aerogel because the initially introduced content of $KHCO_3$ in the carbon aerogel was not so high. An improvement for the efficient introduction of $KHCO_3$ into the carbon pores is also necessary.

4.3 Attempts of Practical Use

The results of the bench-scale testing, where a K_2CO_3 sorbent of 120 kg was used and 0.5 ton per day of CO_2 was recovered, were reported by using a K_2CO_3 sorbent.[46] The system was operated under conditions of 1 atm gas containing 7 v/v% CO_2, a sorption temperature from 343–353 K, and a regeneration temperature >453 K. The K_2CO_3 content was 35% in the sorbent with a supporting material. After the recovery, the CO_2 concentration was reduced to 1%, and sorption capacities of 3.5 to 4% CO_2 per g-sorbent were obtained. The capacity is not so high because a large amount of sorbent was used. The regeneration temperature of 453 K is too high to save energy for the regeneration. The temperature should be lowered for the use of waste heat.

Acknowledgements

The authors thank Mr H. Chioyama, Dr G. Yang, Ms A. Nakamura and Ms Nasiman Tuerxun for their experimental support and Dr T. Itoh of the Center for Analytical Instrumentation, Chiba University, for his help with the SEM observations. This work was supported by The Iwatani Naoji Foundation's Research Grant and partially by Grants-in-Aid for Challenging Exploratory Research and for Fundamental Scientific Research from the Japan Society for the Promotion of Science.

References

1. M. Aresta, *Carbon Dioxide Recovery and Utilization*, Kluwer Academic Pub., Boston, 2003, p. 53.
2. T. F. Wall, *Proc. Combust. Inst.*, 2007, **31**, 31.
3. W. C. Yang and J. Hoffman, *Ind. Eng. Chem. Res.*, 2009, **48**, 341.
4. K. J. Fricker and A. H. A. Park, *Chem. Eng. Sci.*, 2013, **100**, 332.
5. M. G. Plaza, S. García, F. Rubiera and J. J. Pis, *J. Chem. Eng.*, 2010, **163**, 41.
6. S. Cavenati, C. A. Grande and A. E. Rodrigues, *J. Chem. Eng. Data*, 2004, **49**, 1095.
7. V. Presser, J. McDonough, S. H. Yeon and Y. Gogotsi, *Energy Environ. Energy Environ. Sci.*, 2011, **4**, 3059.
8. Y. Liang, D. P. Harrison, R. P. Gupta, D. A. Green and W. A. McMichael, *Energy Fuels*, 2004, **18**, 569.
9. S. C. Lee, B. Y. Choi, C. K. Ryu, Y. S. Ahn, T. J. Lee and J. C. Kim, *Korean J. Chem. Eng.*, 2006, **23**, 374.
10. Y. Seo, S. H. Jo, C. K. Ryu and C. K. Yi, *Chemosphere*, 2007, **69**, 712.
11. C. Zhao, X. Chen and C. Zhao, *Energy Fuels*, 2010, **24**, 1009.
12. C. Zhao, X. Chen, C. Zhao and Y. Liu, *Energy Fuels*, 2009, **23**, 1766.
13. J. Hoffman and H. Pennline, *Energ. Environ. Res.*, 2001, **1**, 90.

14. C. Zhao, X. Chen and C. Zhao, *J. Combust. Sci. Technol.*, 2009, **15**, 135.
15. Y. Liang and M. S. Dissertation, *Carbon Dioxide Capture from Flue Gas Using Regenerable Sodium-based Sorbents*, Louisiana State University, LA, 2003.
16. C. Zhao, X. Chen and C. Zhao, *Ind. Eng. Chem. Res.*, 2010, **49**, 12212.
17. C. Zhao, X. Chen and C. Zhao, *Energy Fuels*, 2012, **26**, 1401.
18. H. Luo and H. Kanoh, *J. Energy Chem.*, 2017, **26**, 972.
19. H. Chioyama, H. Luo, T. Ohba and H. Kanoh, *Adsorpt. Sci. Technol.*, 2015, **33**, 243.
20. H. Luo, H. Chioyama, S. Thürmer, T. Ohba and H. Kanoh, *Energy Fuels*, 2015, **29**, 4472.
21. S. C. Lee, B. Y. Choi, S. J. Lee, S. Y. Jung, C. K. Ryu and J. C. Kim, *Stud. Surf. Sci. Catal.*, 2004, **153**, 527.
22. J. B. Lee, C. K. Ryu, J. Baek, J. H. Lee, T. H. Eom and S. H. Kim, *Ind. Eng. Chem. Res.*, 2008, **47**, 4465.
23. H. Knuutila, E. T. Hessen, I. Kim, T. Haug-Warberg and H. F. Svendsen, *Chem. Eng. Sci.*, 2010, **65**, 2218.
24. R. R. Kondakindi, G. McCumber, S. Aleksic, W. Whittenberger and M. A. Abraham, *Int. J. Greenhouse Gas Control*, 2013, **15**, 65.
25. W. Dong, X. Chen, Y. Wu, C. Zhao, C. Liang and D. Liu, *Energy Fuels*, 2012, **26**, 6040.
26. W. Dong, X. Chen, F. Yu and Y. Wu, *Energy Fuels*, 2015, **29**, 968.
27. C. Zhao, X. Chen and C. Zhao, *Chemosphere*, 2009, **75**, 1401.
28. Y. C. Park, S. H. Jo, D. H. Lee, C. K. Yi, C. K. Ryu, K. S. Kim, C. H. You and Y. S. Park, *Energy Procedia*, 2013, **37**, 122.
29. S. W. Park, D. H. Sung, B. S. Choi, J. W. Lee and H. Kumazawa, *J. Ind. Eng. Chem.*, 2006, **12**, 522.
30. Y. Seo, S. H. Jo, C. K. Ryu and C. K. Yi, *J. Environ. Eng.*, 2009, **135**, 473.
31. C. Zhao, X. Chen and C. Zhao, *Energy Fuels*, 2009, **23**, 4683.
32. Y. C. Park and S. H. Jo, *Int. J. Greenhouse Gas Control*, 2010, **4**, 655.
33. H. Hayashi, J. Taniuchi, N. Furuyashiki, S. Sugiyama, S. Hirano, N. Shigemoto and T. Nonaka, *Ind. Eng. Chem. Res.*, 1998, **37**, 185.
34. N. Shigemoto, T. Yanagihara, S. Sugiyama and H. Hayashi, *Energy Fuels*, 2006, **20**, 721.
35. Y. W. Seo, S. H. Jo, H. J. Ryu, D. H. Bae, C. K. Ryu and C. K. Yi, *Korean J. Chem. Eng.*, 2007, **24**, 457.
36. S. C. Lee, H. J. Chae, B. Y. Choi, S. Y. Jung, C. Y. Ryu, J. J. Park, J. I. Baek, C. K. Ryu and J. C. Kim, *Korean J. Chem. Eng.*, 2011, **28**, 480.
37. C. Zhao, Y. Guo, C. Li and S. Lu, *Chem. Eng. J.*, 2014, **254**, 524.
38. C. Zhao, X. Chen, C. Zhao and Y. Liu, *Energy Fuels*, 2009, **23**, 1766.
39. C. Zhao, X. Chen and C. Zhao, *Ind. Eng. Chem. Res.*, 2012, **51**, 14361.
40. Y. Nakahigashi, H. Kanoh, T. Ohba, M. Baba, Y. Hattori, N. Inoue and M. Morimoto, *Adsorpt. Sci. Technol.*, 2012, **30**, 621.
41. C. Zhao, X. Chen, C. Zhao and Y. Liu, *Energy Fuels*, 2010, **26**, 3062.
42. S. C. Lee and J. C. Kim, *Catal. Surv. Asia*, 2007, **11**, 171.

43. M. J. Venegas, E. Fregoso-Israel, R. Escamilla and H. Pfeiffer, *Ind. Eng. Chem. Res.*, 2007, **46**, 2407.
44. F. García-Labiano, A. Abad, L. F. de Diego, P. Gayán and J. Adánez, *Chem. Eng. Technol.*, 2002, **57**, 2381.
45. G. Yang, H. Luo, T. Ohba and H. Kanoh, *Int. J. Chem. Eng.*, 2016, **2016**, 4012967.
46. Y. C. Park, S. H. Jo, S. Y. Lee, C. K. Ryu, J. B. Lee and C. K. Yi, *Korean J. Chem. Eng.*, 2016, **33**, 73.

CHAPTER 5

Application Status of Post-combustion CO_2 Capture

DEEPAK PUDASAINEE, VINOJ KURIAN AND RAJENDER GUPTA*

Department of Chemical and Materials Engineering, University of Alberta, Edmonton, T6G 1H9 Canada
*E-mail: rajender.gupta@ualberta.ca

5.1 Introduction

5.1.1 GHG Emission

Carbon dioxide (CO_2) is a major anthropogenic greenhouse gas responsible for global warming and climate change. The concentration of CO_2 has been continuously increasing since the beginning of the industrial revolution. The atmospheric concentration of CO_2 has increased from 280 ppm, in the mid-1800s, to about 407 ppm in 2017.[1] Global anthropogenic GHG emissions in 2010 have reached 49 ± 4.5 $GtCO_2$-eq per year, 78% of which is contributed by CO_2 from fossil fuel combustion and industrial processes.[2] Due to the increase in anthropogenic CO_2 and other greenhouse gas emissions, the Earth's average surface temperature has risen by 1.1 °C since the late 19th century.[3] Due to global warming and climate change, there has been evidence of a sea level rise, decreasing snow and ice cover, increase in ocean heat content, increase in frequency and intensity of droughts *etc.*[4] In order

Inorganic Materials Series No. 2
Post-combustion Carbon Dioxide Capture Materials
Edited by Qiang Wang

Published by the Royal Society of Chemistry, www.rsc.org

to cope with this condition, governments around the world have targeted the reduction of anthropogenic GHG emissions.

The technological options currently under consideration for capturing CO_2 from combustion and gasification facilities are pre-combustion capture, post-combustion capture, oxy-fuel combustion, and chemical looping combustion. Post-combustion and oxy-fuel combustion can be applied to capture CO_2 from coal- and gas-fired plants. Pre-combustion CO_2 capture is mainly applied to coal gasification plants. Post-combustion, oxy-fuel combustion and pre-combustion CO_2 capture are still not ready for implementation in coal-based power plants due to several reasons: (i) these technologies have not been demonstrated in large plants, (ii) the required parasitic loads to supply both power and steam to the CO_2 capture plant reduces power generation by nearly one-third; and (iii) they are not cost effective at the present developmental stage.[5]

Several CO_2 capture approaches, such as absorption, adsorption, cryogenic, membrane, and chemical looping combustion *etc.*, have been developed. A large number of sorbents to capture CO_2 from combustion and gasification of off-gas has been developed. CO_2 from off-gas can be captured by liquid or solid sorbents. The ideal sorbents should have a high adsorption capacity, high CO_2 selectivity, durability, and relatively fast sorption and desorption kinetics.[6] The most widely used chemical absorption at present is CO_2 absorption using solvents, mainly amine. Solvents used in this process can be regenerated either by increasing the temperature or changing the pressure. The amine-based absorption process is the most mature among the technologies currently available. Due to the several drawbacks of amine solvents currently there has been a search for solid sorbents. Various solid sorbents have been developed and tested for CO_2 capture.[7–9]

Carbon capture and storage (CCS) has been accepted as a primary option to mitigate anthropogenic greenhouse gas emissions. However, CCS adds to the cost of operations and its implementation is a bit challenging in terms of the economic aspect. The integration of a CCS process with oil and gas operations, such as enhanced oil recovery (EOR), may help to overcome the challenge to utilize the captured CO_2. Major CO_2 capture technologies that researchers from all over the world are currently focused on include: pre-combustion carbon capture, post combustion carbon capture and oxy-fuel carbon capture. Figure 5.1 illustrates an overview of these capture technologies processes for a coal-based power system.[10] Table 5.1 outlines the advantages and challenges faced in each carbon capture technology. Chemical looping combustion (CLC) is another emerging technology focused on energy efficiency and carbon capture.

5.1.2 Pre-combustion Carbon Capture

Pre-combustion carbon capture refers to the separation and removal of CO_2 before the combustion of the fuel itself. One such technology is gasification in which the fuel, such as coal, natural gas or biomass, is partially oxidized in the presence of steam and air/oxygen at a high temperature and pressure

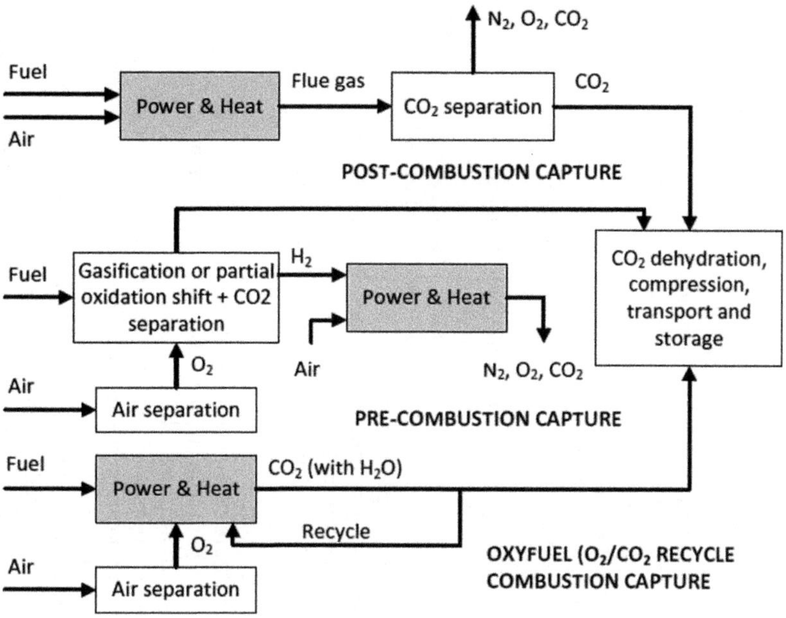

Figure 5.1 Overview of CO_2 capture technologies for a coal-based power system. Reproduced from ref. 10 with permission from Elsevier, Copyright 2011.

to produce synthesis gas. Synthesis gas is a mixture of H_2, CO, CO_2 *etc.* The syngas is then steam reformed to produce a mixture of H_2 and CO_2. The CO_2 can be separated and sequestrated, while H_2 is used as a fuel. The gas for separation is a high pressure (up to 4 MPa) mixture, which contains 15–40% CO_2,[13] much higher in concentration compared to the CO_2 lean flue gas used in post-combustion carbon capture. Therefore, pre-combustion carbon capture is more efficient but the gasification process is expensive compared to pulverized coal power plants. The main research on pre-combustion carbon capture is concentrated on advanced solvents, solid sorbents and membrane systems for the separation of H_2 and CO_2.

5.1.3 Post-combustion Carbon Capture

In post-combustion capture, CO_2 is captured from combustion off-gas after the fuel is burned. The typical concentration of CO_2 in flue gas emitted from a coal-fired plant varies between 12–15%, with the other components mainly N_2, O_2, H_2O and pollutants such as particulates, SO_x, NO_x. In a typical plant, flue gas after combustion goes through air pollution control devices, mainly to control the emission of particulates, NO_x, SO_x and acid gases. Currently, liquid amine solutions have been commonly used in industries to capture CO_2, which has been investigated for decades.[11–14] CO_2 from flue gas is captured in a liquid solvent to form solvent-bound CO_2. This CO_2-rich solvent can be regenerated by increasing the temperature.

Table 5.1 Advantages and challenges associated with carbon capture technologies.[9]

Technology	Advantages	Challenges
Post-combustion capture	• Can be retrofitted to the existing plants • Enables deployment of well-established pulverized coal technology • R&D on improved sorbents and capture equipment may reduce the energy penalty of PCC capture • Sub-scale demonstration of PCC is proceeding, such as 110 MW Boundary Dam project of Saskatchewan power, Canada	• Considerable scale-up of existing commercially available amine processes • Significant space requirement • Most sorbents efficient under pure flue gas (<10 ppm of $SO_2 + NO_2$). Needs optimization before application in commercial plants • Loss of net power output of about 30% and 11% reduction in efficiency with addition of current amine technologies
Pre-combustion capture	• Capture using the water-gas shift reaction and removal of the CO_2 with acid gas removal processes already practiced commercially • Incurs less of an energy penalty (~20%) than current PCC technology (~30%) at 90% CO_2 capture • R&D on improved CO shift catalysts, higher temperature gas clean-up and membrane separation technology can lead to reduction in the energy penalty of capture	• Energy losses with facility addition less than PCC, but still significant • Commercial demonstration of gas turbines firing H_2 has yet to be demonstrated in an IGCC plant with capture • Additional purification may be needed in case of need to vent the CO_2 • Much higher capital costs of IGCC without capture than sub-critical pulverized coal without capture
Oxy-fuel combustion	• Should be able to deploy conventional, well-developed, high efficiency steam cycles without the need to remove significant quantities of steam from the cycle for CO_2 capture • No chemical operations or significant on-site chemical inventory involved in set-up • Ultra low emissions of conventional pollutants can be achieved at no additional cost • Possibility of higher efficiency than air fired power plants	• More expensive technology development than other technologies – requires commitment of whole power plant • Cryogenic ASU and CO_2 compression in capture unit can reduce net plant output by 25% compared to an air fired plant with same capacity • Relaxation in purity requirements of stored CO_2 could lead to reduced operational costs • Air fired combustion during start-ups lead to added flue gas quality controls and hence additional cost

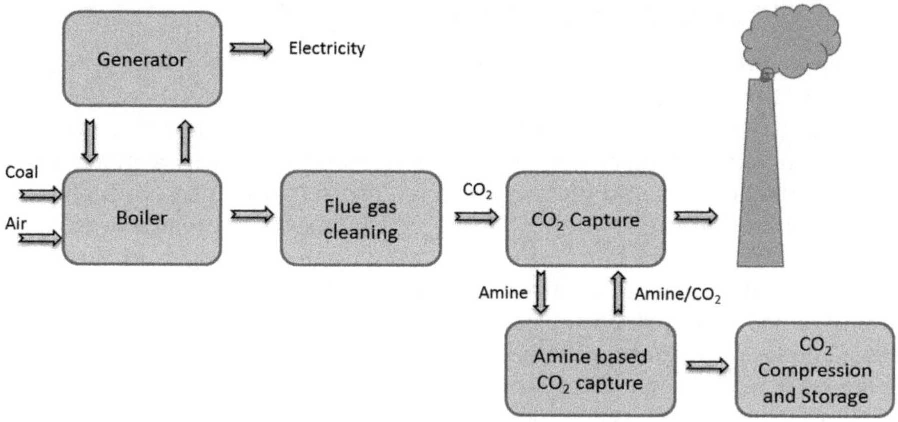

Figure 5.2 Schematic representation of a coal-fired power plant with post-combustion CO_2 capture using an amine-based scrubbing system.

A schematic representation of a coal-fired power plant with post-combustion CO_2 capture using an amine-based scrubbing system is shown in Figure 5.2. Amine solvents used in CO_2 capture can be regenerated. The amine-based CO_2 absorption process is the most advanced among the CO_2 capture technologies. The amine-based absorption process has a high CO_2 adsorption capacity, good CO_2 selectivity, and fast kinetics. The drawbacks of the amine process are that it is costly and highly energy intensive. The major issues with the use of amine solvents for post-combustion capture are: solvent degradation by SO_x and NO_x, decomposition of amine in the presence of oxygen, higher regeneration cost and ammonia loss. In order to overcome these problems, recently, there has been a growing research interest in regenerable solid sorbents.

5.1.4 Oxy-firing

Oxy-firing refers to the combustion of fuel in a pure stream of O_2 to produce flue gas composed of mainly CO_2 and H_2O. The absence of N_2 in the feed stream leads to a reduction in the formation of thermal NO_x and the concentration of CO_2 in the flue gas is higher. H_2O formed during combustion can be condensed and separated easily to get a stream of CO_2, which can be sequestrated. The oxy-firing process provides an opportunity to capture CO_2 by direct physical compression and cooling techniques. The advantages of oxy-firing include: a high concentration of CO_2 that is ready for capture and storage, an opportunity for integrated emission control through reduced flue gas flow and reduced combustor size/volume. But there is a huge energy penalty associated with the separation of air into pure components of O_2 and N_2, when using a standard technique like an air separation unit (ASU). This technology is still at the demonstration phase with research focused on improvements in ASU (includes molecular sieves, membranes) and CLC.

5.1.5 Chemical Looping Combustion (CLC)

In chemical looping combustion, an oxygen carrier is used to transport oxygen from regular air to a solid fuel, and metal oxides (MeO) are the most promising types of oxygen carriers. Metals such as manganese or iron are exposed to air and react with the oxygen present to form a metal oxide through oxidation at a high temperature. The metal oxide is brought into the combustion furnace where no nitrogen is present to undergo a reduction reaction where the carbon in the fuel reacts with the oxygen in the metal oxide to produce CO_2 and convert the metal oxide back to the pure metal. A comparatively pure CO_2 flue gas is produced, which is ready for sequestration. The key advantage of chemical looping compared to oxy-firing is that there is no need for an air separation unit, which causes significant operating deficiency as well as economic losses. CLC requires two reactors, an air reactor and a fuel reactor. In the air reactor, an oxygen carrier will be oxidized by air at temperatures around 700–900 °C. The oxygen carrier (MeO) is then transported to the fuel reactor. The oxygen carrier reacts with a hydrocarbon fuel in the fuel reactor, at around 900 °C, to produce CO_2 and H_2O, while reducing the metal oxide to metal. Two separate flue-gas streams will be exiting the process. The flue gas leaving the air reactor contains mainly N_2 and the remaining O_2, while the stream leaving the fuel reactor contains mostly CO_2 and steam.

5.1.6 Carbon Capture and Storage (CCS)

Carbon capture and storage refers to the whole process of capturing CO_2 from flue gas, compression and transportation to the storage site and storage, mainly in an underground geological formation. CCS includes four major steps: (i) CO_2 capture, (ii) compression, (iii) transport and (iv) storage (injection into a geological formation, in enhanced oil recovery). Figure 5.3 shows the CCS systems with available technical options. Technologies related to each stage are listed in the figure. The solid text represents technologies that are in a commercial improvement phase, the italicized text represents technologies at the pilot plant and demonstration phases and the shadowy text represents a laboratory-scale study and/or evaluation phase. Only post-combustion and pre-combustion capture have been demonstrated and are at a commercial improvement stage. Chemical looping combustion is still at the evaluation phase on a laboratory scale.

Table 5.2 shows the developmental state of CO_2 capture processes, including mature industrial processes and those that require further technical development and demonstration. First-phase industrial applications are more mature than those in the power sector and are ready for use. CO_2 capture technologies related to a natural gas combined cycle are at the pilot plant stage. Post-combustion CO_2 capture in a pulverized coal-fired boiler is the only technology that is at the demonstration level.

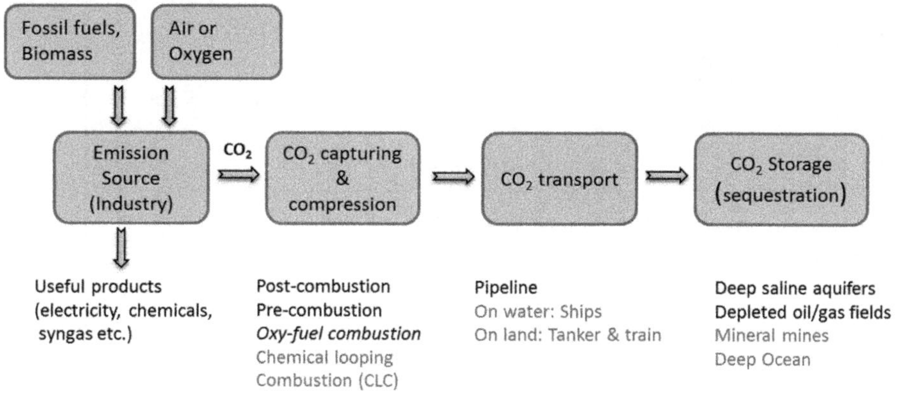

Figure 5.3 Schematic presentation of a CCS system with available technical options.[15]

5.1.7 Natural Gas Combustion

Natural gas combustion produces less CO_2 (less than 400 kilograms per megawatt hour (kg MWh^{-1}) compared to around 800 kg MWh^{-1} for coal).[16] Switching from coal to gas can help in the near-term reduction of GHG emissions. Reduction of 85% or more of CO_2 emissions from gas-fired power plants has been proven to be technically possible in pilot scale projects. The lower capital cost of combined cycle gas plants and lower CO_2 emission in comparison to coal-fired power plants make them attractive.

Globally, in the absence of new policies, natural gas generation is projected to grow 2.7% per year between 2012 and 2040, accounting for nearly 30% of the total worldwide energy generation by 2040, which is nearly equal to the projected coal generation.[17,18] Further, due to worldwide concerns related to CO_2 emissions from coal combustion and lower prices (in some regions), it is foreseen that many countries will shift to natural gas in order to reduce the carbon footprint. As, for example, in the United States of America, energy-related CO_2 emissions from natural gas exceeded those from coal in 2016 for the first time since 1972, which was due to an increase in natural gas consumption and a decrease in coal consumption.[19] Even though natural gas is less carbon-intensive than coal, increasing the natural gas consumption and decreasing the coal consumption over the past decade have resulted in greater natural-gas-related CO_2 emissions than those from coal.

A typical natural gas and combined cycle (NGCC) process with post-combustion CO_2 capture is shown in Figure 5.4. In NGCC, natural gas is combusted with compressed air at a high pressure in a gas combustion turbine, producing mechanical power. A generator converts the mechanical power to electricity. The combustion off-gas leaves at a high temperature, a heat exchanger network is used to capture the heat and the cooled combustion gases are passed through the CO_2 capture system. CO_2 is separated and compressed for transport.

Table 5.2 Routes to CO_2 capture in power generation and industrial applications.[16] Reproduced from ref. 16 with permission from the IEA Copyright 2013.

		Syngas-hydrogen capture	Post-process capture	Oxy-fuel combustion	Inherent separation
First-phase industrial applications	Gas processing	—	—	—	Sweetening[a]
	Iron and steel	Direct reduced iron (DRI),[b] smelting (*e.g.*, Corex)[c]	—	—	DRI[b,c]
	Refining	—	—	—	Coal-to-liquids; synthetic natural gas from coal[a] Hydrogen production[a]
	Chemicals	—	—	—	Ammonia/methanol[a]
	Biofuels	—	—	—	Ethanol fermentation[a]
Power generation	Gas	Gas reforming and combined cycle[e]	Natural gas combined cycle[c]	Oxy-fuel combustion[c]	Chemical looping combustion[e]
	Coal	Integrated gasification combined cycle (IGCC)[c]	Pulverized coal-fired boiler[d]	Oxy-fuel combustion[c]	Chemical looping combustion[e]
	Biomass	IGCC[e]	Biomass-fired boiler[e]	Oxy-fuel combustion[e]	Chemical looping combustion[e]
Second-phase industrial applications	Iron and steel	Hydrogen reduction[e]	Blast furnace capture[c]	Oxy-fuel blast furnace[e]	—
	Refining	Hydrogen fuel steam generation[e]	Process heater and combined heat and power (CHP) capture[c]	Process heater and CHP oxy-fuel[e]	—
	Chemicals	—	Process heater, CHP, steam cracker capture[c]	Process heater and CHP oxy-fuel[e]	—
	Biofuels	Biomass-to-liquids[e]	—	—	Advanced biofuels[e]
	Cement	—	Rotary kiln[c]	Oxy-fuel kiln[c]	Calcium looping[e]
	Pulp and paper	Black liquor gasification[e]	Process heater and CHP capture[c]	Process heater and CHP oxy-fuel[e]	—

[a]Commercial.
[b]Capture approach is dependent on DRI technology used.
[c]Pilot.
[d]Demonstration.
[e]Lab or concept.

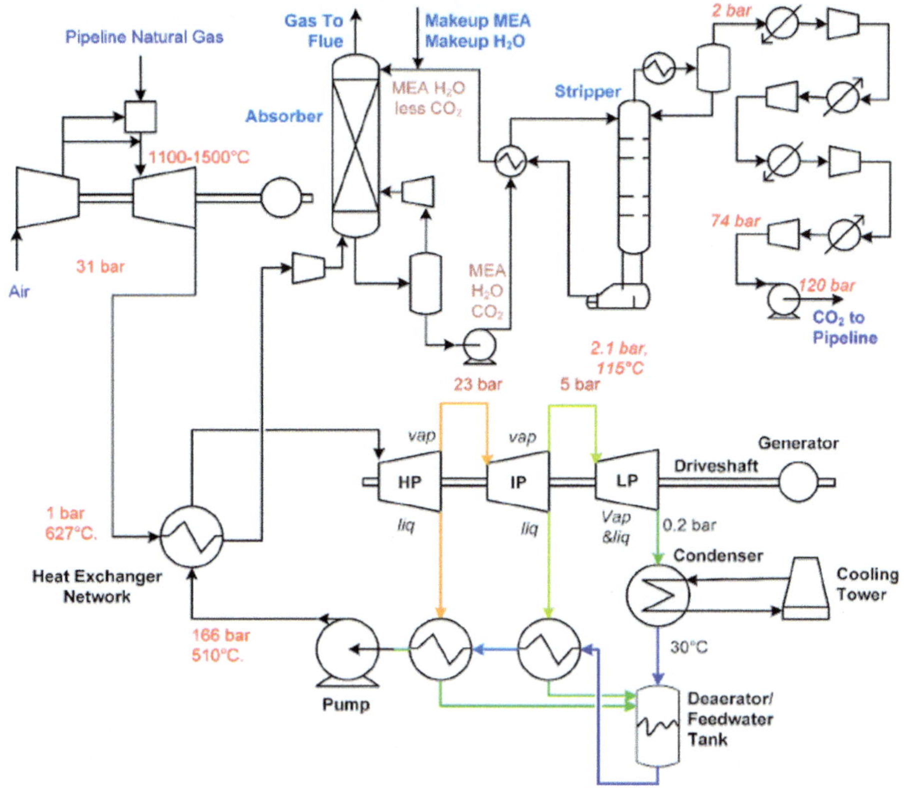

Figure 5.4 A typical natural gas combined cycle (NGCC) process with post-combustion carbon capture using monoethanolamine (MEA) as a solvent. Reproduced from ref. 20 under the terms of the CC BY 4.0 licence, https://creativecommons.org/licenses/by/4.0/, © 2017 The Authors.

5.2 Current Status of CCS Projects

The CCS application dates back to 1972 when several natural gas processing plants in the Val Verde area of Texas began to capture CO_2 for EOR operations. Today, CO_2-EOR also remains a key driver for CCS operations. In 2017, globally, there are 17 large-scale CCS facilities in operation, capturing more than 30 Mtpa of CO_2. In addition, there are 15 smaller-scale CCS facilities in operation and under construction; these facilities can capture about 2 Mtpa of CO_2. Today, the CO_2 captured from large and small facilities equates to more than 8 million motor vehicles being taken off the roads.[21]

Today, more than 6000 Km of CO_2 pipelines are operational with an excellent safety record. There, CO_2 has been safely injected into the geological formation with no evidence of leakage into the atmosphere. The large-scale CCS facilities in operation at present are (i) Petra Nova Carbon Capture, Texas, USA; (ii) SaskPower Boundary Dam- CCS; (iii) Kemper County Energy Facility (IGCC + CCS); and (iv) Callide – Oxy-fuel combustion and carbon storage demonstration plant.[21]

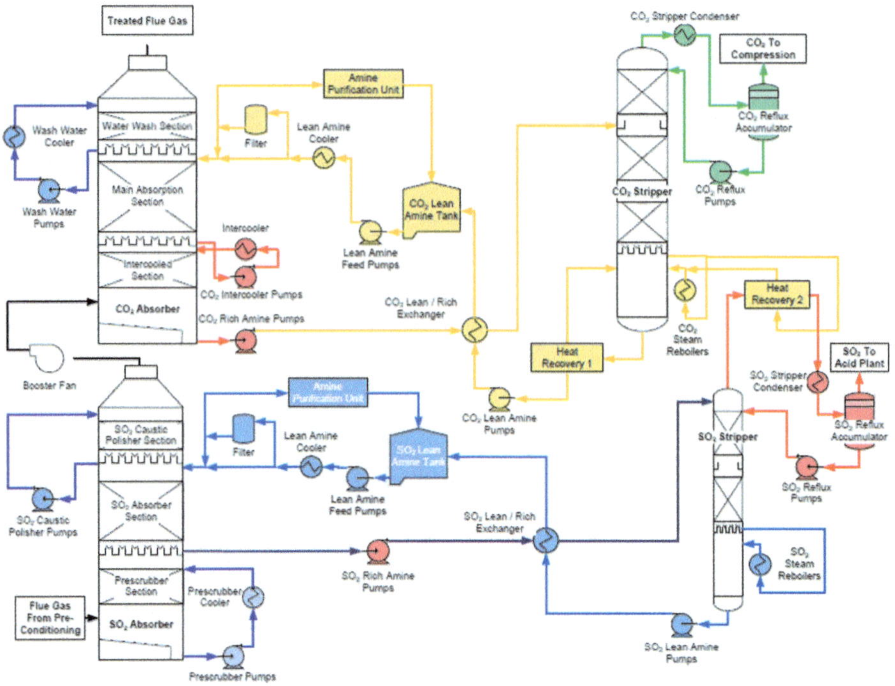

Figure 5.5 Cansolv Process configuration of the SaskPower BD3 CCS Project. Reproduced from ref. 23 with permission from Elsevier, Copyright 2014.

5.2.1 Large-scale Projects

5.2.1.1 SaskPower Boundary Dam- CCS

In October 2014, the Boundary Dam coal-fired power station, Estevan, Saskatchewan, Canada, demonstrated that carbon capture and storage (CCS) can be implemented on a large scale. It became the first commercial-scale post-combustion coal-fired CCS demonstration project to capture CO_2 from flue gas, compress gas and store it underground.[22] The facility uses technology from Shell Consolv to capture CO_2 and SO_2 from the power plant (Figure 5.5). The plant captures more than 1.3 million tonnes of CO_2 every year, which is equivalent to taking 350 000 cars off the roads (http://Saskpower.com). The process also reduces up to 100% of SO_2 emissions from coal combustion and up to 90% of CO_2. The captured CO_2 emissions are compressed and transported by pipelines to Cenovus Energy where CO_2 is used for EOR activities in the Weyburn oil field. Most of the captured CO_2 is used in EOR projects. The captured SO_2 is used as a feedstock for H_2SO_4 plants.

SaskPower and Petra Nova are retrofits, projects that demonstrate that CCS is appropriate for existing coal power plants. Flue gas from the coal

Figure 5.6 Petra Nova Carbon Capture, Texas, USA. Reproduced from ref. 40 with permission from Elsevier Copy right 2017.

combustion is carried to an adjacent carbon capture facility, a 52 meter-high column filled with amines, which absorb 90 percent of the carbon dioxide. The cleaned gas in vented out and the CO_2-rich amine solution is sent to a heater where the CO_2 is removed from the amine and CO_2-free amine is brought back for adsorption. The separated CO_2 is converted into supercritical fluid, most of which is injected into an oil field 65 km away; some is injected into SaskPower's saline aquifer 3.4 Km below the ground.[24]

5.2.1.2 Petra Nova Carbon Capture, Texas, USA

Petra Nova Carbon Capture, Texas, USA, is the world's largest post-combustion capture system at present (Figure 5.6).[21] The plant has been operational at the W. A. Parish plant since January 10, 2017. Production unit 8 of the W. A. Parish power plant, Texas, was retrofitted with a 1.4 Mtpa post-combustion CO_2 capture facility.[21] Parish unit 8 provides a capture capacity equivalent to a 240 MW gross unit. The plant captures more than 5000 ton per day of CO_2, which is used for enhanced oil recovery (EOR) at the West Ranch oil field.[25] The project has boosted the oil production from 300 barrel per day when it began operation to about 4000 barrel per day.[26]

NRG Energy Inc. (NRG) spent \$1 billion to retrofit its existing plant into a post-combustion carbon capture project. NRG's Petra Nova project received a \$190 million clean coal grant from the U.S. Department of Energy (DOE). NRG and Japanese partner JX Nippon Oil and Gas Exploration Corporation (JX Nippon) both contributed \$300 million.[27]

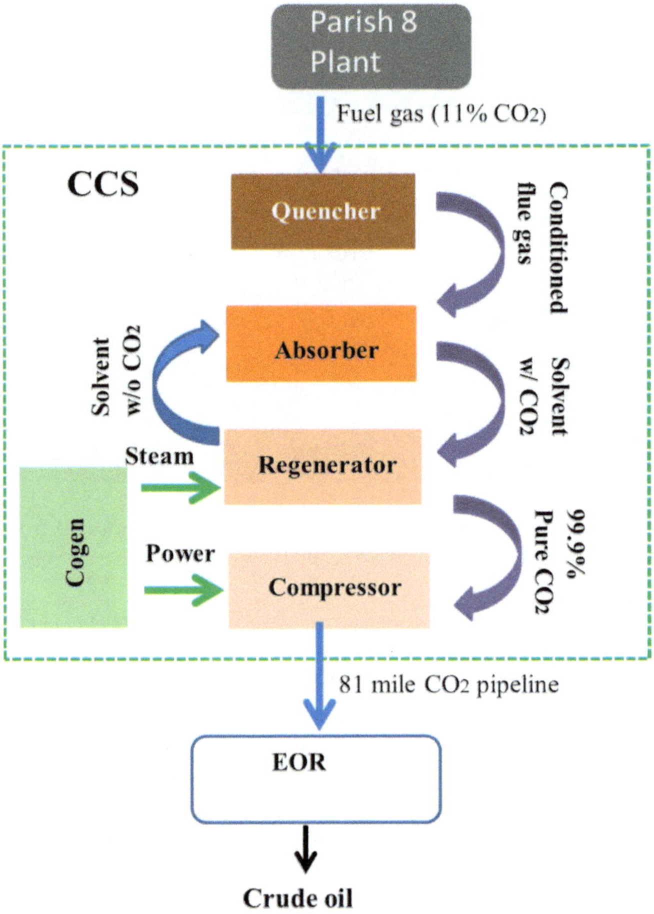

Figure 5.7 CO_2 capture process at Petra Nova.[27]

Petra Nova uses the improved Kansai Mitsubishi carbon dioxide recovery process (KM CDR Process™), which is an amine-based CO_2 capture process, with a low-pressure, oxygen-containing stream (Figure 5.7). Petra Nova (240 MW) has more than double the power capacity of SaskPower's Boundary Dam project and it is designed to capture 60% more CO_2 or about 1.6 million tons per year.[28]

The process uses the KS-1 high-performance solvent for CO_2 absorption and desorption, which was developed by MHI and Japanese utility Kansai Electric Power Co. (KEPCO). In this process, flue gas with CO_2 from a power plant passes to a quencher for cooling, moisture and SO_2. The cooled gas is then sent to an absorber column where it reacts chemically with amine-based sorbents. The clean gas is vented into the atmosphere. The CO_2-rich solvent passes to the regenerator, where CO_2 bound with amine is broken down by heat, and the solvent can be reused. The captured CO_2 is sent to a compressor, where it is compressed into a supercritical stage for pipeline transport for enhanced oil recovery (EOR). The recovered solvent is brought back into the absorber column. In this way, the whole process is regenerable.

5.2.1.3 Kemper County Energy Facility (IGCC + CCS)

Kemper County energy facility is 582 megawatt electric power plant in Kemper County, Mississippi, USA (Figure 5.8). The construction of the plant started in 2010. If it becomes operational, the Kemper project will be the first electricity plant to employ gasification and pre-combustion carbon capture technologies on a large scale.[29] The initial estimated cost of the project was 2.4 billion dollars and it was aimed to be completed by 2014. However, it is not in service as of June 2017 and the cost has risen to 7.5 billion dollars.[30]

Kemper power plant is an IGCC power plant, which is planned to gasify lignite coal to produce synthetic gas, and capture CO_2 for enhanced oil recovery. The plant has been running on natural gas for years, and Kemper's gasifiers have successfully produced electricity from synthetic gas.[29] The Kemper project will have less sulfur dioxide, particulate and mercury emissions compared to conventional coal power plants. After gasification of lignite, three by-products *i.e.*, carbon dioxide, ammonia and hydrogen sulfide, will be removed from the gas stream. The Kemper project will capture at least 65 percent of the CO_2 from off-gas.[30,31] The plant uses transport integrated gasification (TRIG™) technology to be used for capturing 3.5 million tonnes of CO_2 per year.

Figure 5.8 Aerial view of Kemper County Energy Facility. Reproduced from ref. 31 with permission from Elsevier, Copyright 2015.

5.2.1.4 Callide – Oxy-fuel Combustion and Carbon Storage Demonstration Plant

Callide oxy-fuel project (Stage 1 – oxy-fuel combustion and CO_2 capture) is the only oxy-fuel demonstration project located in central Queensland, Australia. Callide power station has four 30 MWe unit and unit No.4 was modified for oxy-fuel combustion.[32] It comprises 2 × 330 ton per day air separation units, a 30 MWe oxy-fuel boiler and a 75 ton per day CO_2 capture plant.[33] The plant was commissioned in 2012. The project demonstrated more than 85% of CO_2 capture from the oxy-fuel flue gas stream and produced a high quality CO_2 suitable for geological storage. Removal of toxic gaseous emissions (such as SO_x, NO_x, particulates, and trace elements) from the flue-gas stream has been demonstrated. The Callide oxy-fuel project is the only CO_2 capture technology demonstrated on this scale in Australia.

The Callide oxy-fuel boiler has three mills to pulverize coal, two of which are needed for normal full-load operation (Figure 5.9).[32,33] Under oxy-firing conditions, the coal is dried and transported by recycled flue gas (consisting mainly of CO_2, H_2O and N_2). This consists of a dehydration system in the primary gas duct to remove SO_3 and to prevent low-temperature corrosion. A flue-gas low-pressure heater at the boiler outlet reduces the temperature of the flue gas to around 145 °C prior to entering the fabric filter. O_2 from the ASUs is mixed with recycled flue gas after the secondary air heater; in addition, O_2 can be directly supplied to the burner flame area. A part of the flue gas is introduced to the CO_2 purification unit (CPU) and purified, compressed and cooled to capture almost pure CO_2. Flue gas is recycled during oxy-fuel operation from the outlet of an induced draft fan (IDF) by a gas recirculation fan (GRF).

Besides coal combustion applications, two noticeable CCS projects were launched[34] in 2016. First, Abu Dhabi CCS Project, Phase 1 the Emirates Steel Industries (ESI) CCS Project was launched as the world's first application of

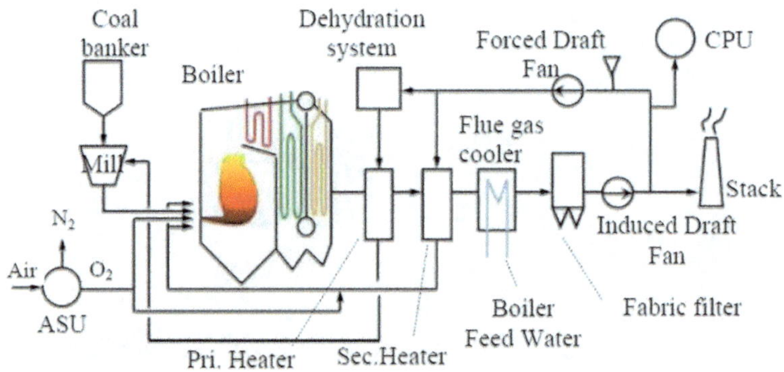

Figure 5.9 Callide oxy-firing process (ASU: air separation unit, CPU: CO_2 purification unit). Reproduced from ref. 32 with permission from Elsevier, Copyright 2014.

CCS to iron and steel production. This captures about 0.8 Mtpa of CO_2 from the direct reduced iron process used and it is used for enhanced oil recovery (EOR). Second, the commencement of CO_2 injection at the Tomakomai (hydrogen production facility) CCS demonstration project. The capture system is processing CO_2 at a rate of at least 100 000 tonnes per annum; this CO_2 is then injected into deep geological formations.

5.2.1.5 NET Power Clean Energy Large-scale Pilot Plant

NET Power clean energy large-scale pilot plant, Texas, USA, a natural-gas-fired demonstration power plant, is the first of its kind. The plant is 50 MW thermal (25 MW electric) and the data from the project will be used to develop a larger 295 MW (590 MWth) commercial power plant.[35] The technology uses the oxy-combustion process, in which natural gas combusted with pure oxygen generates high-pressure CO_2, which is further used to drive a turbine to produce electricity.[35] The plant utilizes proprietary Allam Cycle Technology, which uses supercritical CO_2 as a working fluid to drive a combustion turbine. The working fluid is then cooled and water is separated from it creating a CO_2 stream, which is pressurized and recycled back to continue the cycle. The process inherently captures CO_2, without the use of an add-on carbon capture unit and eliminates NO_x, SO_x, and produces pipeline-ready CO_2. It has been claimed by NET Power that the benefits of the Allam Cycle Technology concept include near-zero air emissions (>97% carbon capture) and low capital cost (comparable to NGCC (natural gas combined cycle) without carbon capture at US$900–1200 per kW).[14]

5.2.2 Small-scale Projects

Small-scale post-combustion capture (PCC) projects are mostly tested in slip streams in coal-fired power plants. Most of the PCC projects to be discussed are solvent-based and use ammonia or proprietary amines as a solvent. A typical 500 MWe coal-fired power plant generates about 12 500 metric tonnes of CO_2 per day, and about 11 300 mt CO_2 per day would have to be captured to yield a 90% capture rate, which gives an insight into the size of a small-scale CO_2 capture facility. Proven techniques of PCC include scrubbing with amines and scrubbing with chilled aqueous ammonia. The main challenges faced are the high energy requirement, solvent degradation and corrosion, therefore, the objectives of the following projects are mainly to mitigate energy usage, and solve corrosion and degradation issues.

5.2.2.1 Fluor Econamine FG Plus^{SM}

Fluor Econamine FG^{SM} is an amine-based technology for post-combustion CO_2 capture. Unlike other amine systems, Econamine FG^{SM} solvents (MEA-based) do not degrade rapidly in the presence of oxygen (up to 15 vol%) and can recover CO_2 from low-pressure oxygen-containing flue gas streams.[36] The

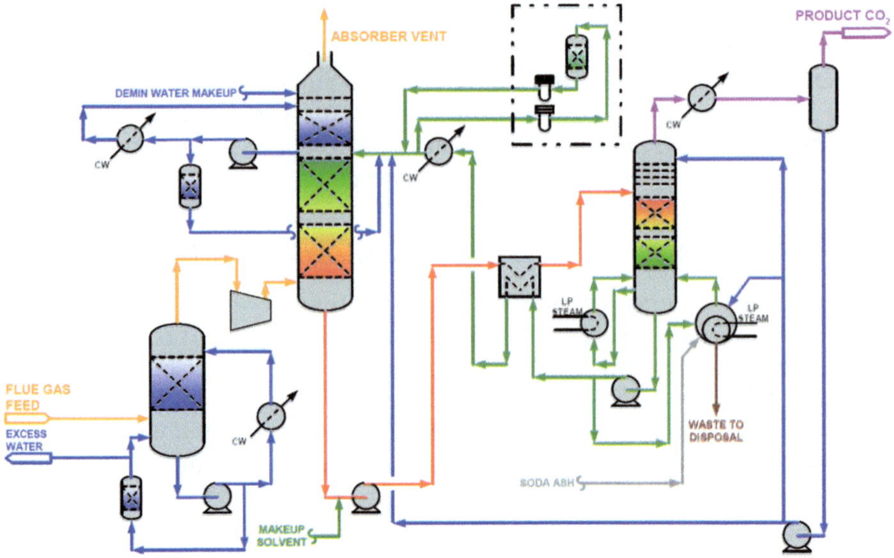

Figure 5.10 Typical Econamine FG^{SM} flowsheet (IEAGHG, 2012).[37] Reproduced from ref. 37 with permission from IEAGHG, Copyright 2012.

Econamine FG^{SM} flow sheet is similar to a generic MEA technology as shown in Figure 5.10.[37] The Econamine FG^{SM} process was successfully installed in pressure-sensitive sources such as gas turbine exhausts and steam-methane reformer (SMR) flue gas lines without disturbing the upstream pressure. In one of the Fluor licensed plants, 330 metric tonnes per day of CO_2 is recovered from the flue gas of a natural gas fired power plant located in Bellingham, MA, USA, which is owned by Florida Power and Light.

With the experience gained from the continuous operation of the Bellingham plant, Fluor has developed an improved Econamine FG^{SM} process through solvent and flow sheet enhancements to lower both the energy consumption and solvent loss. The important features of this Econamine FG Plus^{SM} include:

- Improved solvent formulation
- Absorber intercooling
- Lean vapor compression configuration
- Advanced reclaiming heat integration with the power plant

This improved Econamine FG^{SM} process is said to reduce steam consumption by over 30% compared to a general MEA technology. Proprietary inhibitors used in this process, to control corrosion, solvent degradation, and thermal degradation, allow the use of MEA at higher concentrations to improve the efficiency of CO_2 recovery. The MEA concentration can go up to 30–35% compared to the previous 18–20%.[38] The improved solvent formulation in Econamine FG Plus increases the CO_2 absorption rate, resulting in a smaller absorber and a lower reboiler steam flow. The Econamine FG Plus^{SM} process uses absorber intercooling and a split flow configuration to achieve the optimum absorber temperature and to help reduce energy requirements. Fluor has also implemented advanced solvent reclamation technologies and heat integration strategies. Three heat integration strategies include a flue gas reboiler, vacuum condensate heating and supplementary duct firing. Any combination of these enhancement features can be customized based on the need to optimize the CO_2 capture operation. The flue gas processed by Econamine FG^{SM} was mainly produced by combustion of natural gas or natural gas steam reformers and none of the units processed coal-derived flue gas. The improved Econamine FG Plus^{SM} is targeting coal-based power plants.

5.2.2.2 MHI KM-CDR Process

The Kansai Mitsubishi carbon dioxide recovery process (KM-CDR process) was developed through cooperation between MHI and Kansai Electric Power Company, Inc. (Kansai), which uses an advanced hindered amine solvent called "KS-1". The KM-CDR process demands upstream polishing of the flue gas because the KS-1 solvent requires low levels of SO_x and NO_x (typically 1 ppm) as absorber input requirements.[39] Compared to normal amine plants, KS-1 solvent has improved absorption and regeneration properties with lower corrosiveness and degradation. These features contribute towards lowering the energy and solvent consumptions and low corrosion. As MHI claims, the KM-CDR circulation rate is 60% that of (unspecified) MEA and the regeneration energy is 68% of MEA.[38] The solvent loss and degradation are 10% of MEA. The flow sheet of the KM-CDR process is shown in Figure 5.11.

MHI collaborated with Southern Company to construct a fully integrated CCS demonstration facility for CO_2 capture from the flue gas of a coal-fired plant (with a capacity of 500 tonnes per day) at Alabama Power's James M. Barry plant. CO_2 is recovered from a slip stream of flue gas from the Barry Power Station and it is compressed, dehydrated and transported to a nearby oil field, by pipeline, for injection and permanent storage in a geological storage formation deep underground. Through this demonstration plant, the reliability of a coal-fired plant of a size equivalent to 25 MWe was proven. MHI continues with its technological development for the deployment of large-scale CO_2 capture plants.

5.2.2.3 Alstom Chilled Ammonia Process (ACAP)

Since 2006, Alstom has been working on the development of the chilled ammonia process (CAP), aiming to commercialize it with pilot testing and commercial demonstrations in the future.[41] In this process, the flue gas is in direct contact with a CO_2 lean ammoniated solution at temperatures below

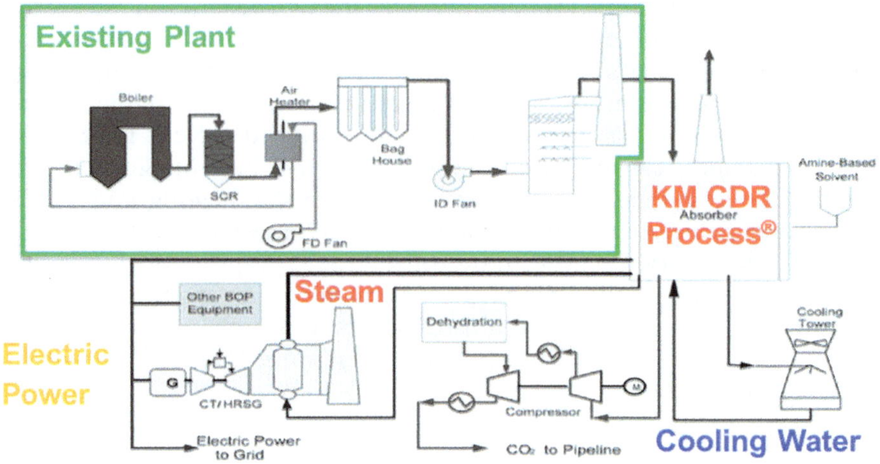

Figure 5.11 Flow sheet of existing power plant with KM-CDR process (Miyamoto *et al.*, 2017).[40] Reproduced from ref. 40 under the terms of the CC BY 4.0 licence, https://creativecommons.org/licenses/by-nc-nd/4.0/, © 2017 Published by Elsevier Ltd.

20 °C, and ammonium carbonate reacts with CO_2 in the flue gas to form ammonium bicarbonate, which precipitates as a solid. This precipitated bicarbonate is concentrated and sent to the regeneration unit where the CO_2 is released by the application of heat. The regenerated lean solution is reused in the absorber to capture CO_2 again.[42] The flow sheet of the Alstom chilled ammonia process is shown in Figure 5.12. So, ACAP utilizes the changes in chemical equilibrium of ammonia, ammonium carbonate, and ammonium bicarbonate with a change in temperature to capture and release CO_2 from the flue gas stream. The low-cost solvent used in ACAP resists degradation with elevated temperatures and with exposure to SO_x and NO_x, but a relatively low reaction rate and CO_2 loading of the solvent demands a larger absorber size.

The ACAP technology development and testing was completed at We Energies Pleasant Prairie Power Plant in Wisconsin and the major achievements of the 1.7 MWe field pilot facility include a CO_2 capture efficiency approaching 90% and a CO_2 purity of about 99.5%. The field pilot plant attained a sustained regeneration at 300 psig with minimal solvent degradation.[43] Following the field pilot capture, a commercial demonstration of 20 MWe equivalent, 100000 ton-CO_2 per year capture and storage pilot began operation in September 2009 at AEP's Mountaineer Power Plant with a 80–90% capture efficiency and at 99.9+% purity. In addition to the improvements made to the We Energies Field Pilot CO_2 capture unit, the key advancement of AEP's Mountaineer Power Plant was the integration of CO_2 capture with compression and storage into a complete CCS demonstration unit.[36] In this system, the captured CO_2 is compressed to nearly 3000 psig using a booster pump

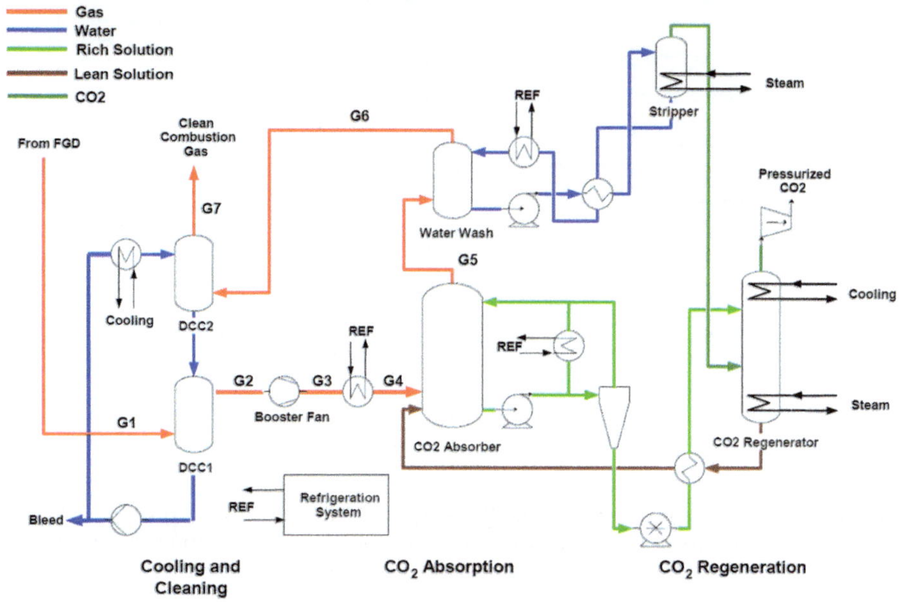

Figure 5.12 Flowsheet of chilled ammonia process (Kozak *et al.*, 2009).[42] Reproduced from ref. 42 with permission from Elsevier, Copyright 2009.

and supercritical CO_2 is transported *via* a pipeline approximately 1200 feet to injection wells. The key objective of the storage project was validating CO_2 injection and storage in geologic reservoirs.

5.2.2.4 Powerspan ECO_2™ Process

ECO_2 is an ammonia-based (not amine) carbon capture process developed by Powerspan and the CO_2 is released for enhanced oil recovery or other form of geological sequestration. The system uses an absorption column and thermal stripper and is interconnected with Powerspan's ECO multi-pollutant control system where the NO_x, SO_2 and particulates are removed from the flue gas prior to CO_2 capture. In the ECO_2 process, aqueous ammonium carbonate solution is used to scrub with flue gas containing CO_2 to form ammonium bicarbonate, which is removed from the flue gas stream and the solvent is regenerated by applying heat to release CO_2 while forming ammonium carbonate for reuse. The liberated CO_2 is collected, compressed and prepared for sequestration. Figure 5.13 illustrates the integration of the ECO_2 system with the ECO multi-pollutant control system. The existing ECO system employs a barrier discharge reactor, wet scrubber and wet electrostatic precipitator to remove NO_x, SO_2, mercury, and fine particulate matter. The wet scrubber uses aqueous ammonia for SO_2 scrubbing, producing an ammonium sulfate fertilizer co-product.[43]

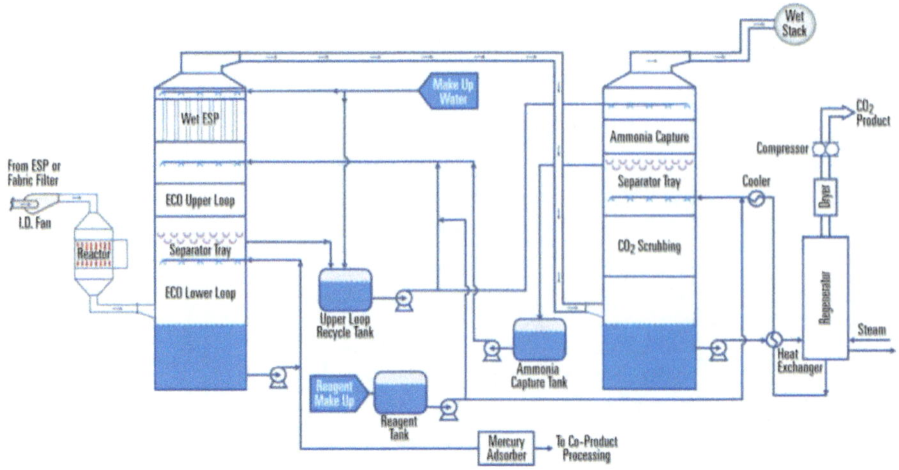

Figure 5.13 Incorporation of the Powerspan ECO_2 scrubbing process with the multi-pollutant control ECO process (McLarnon, 2009).[43] Reproduced from ref. 43 with permission from Elsevier, Copyright 2009.

Pilot-scale testing of a 1 MWe equivalent (~20 stpd) slip stream at FirstEnergy's R. E. Burger Plant in Ohio ran from December 2008 to 2010. According to an independent assessment at Powerspan, the ECO_2 technology is ready for scale-up, with a cost of less than \$40 per ton of CO_2 captured and compressed, based on a 220 MWe net retrofit project size.[44] The energy consumption for solution regeneration is lower compared to amine-based CO_2 capture processes because of the reduced energies of absorption and desorption reactions for an ammonia-based scrubbing solution.

5.2.2.5 *Cansolv*

The Cansolv CO_2 capture system (Figure 5.14), which is very similar to a well-known amine treating process for the removal of SO_2 and CO_2 from refinery streams and natural gas, comprises the following major components: a CO_2 absorber including inter-stage cooling and a water-wash section, regeneration tower and amine purification unit (APU). Cansolv is a subsidiary of Shell Global Solutions International B.V. The Cansolv absorbent is used to absorb CO_2 in the flue gas in the absorber tower with a multi-stage counter current contacting to achieve a maximum loading of the carbon dioxide into the absorbent solution.[45] Cansolv tested their technology in a CO_2 capture test facility at a capacity close to 120 tons of CO_2 capture per day by removing 90% of the CO_2 from the flue gas generated by a natural gas fired boiler. CO_2 was captured using a counter current exchange with the Cansolv DC-103 solvent, using a pre-scrubber followed

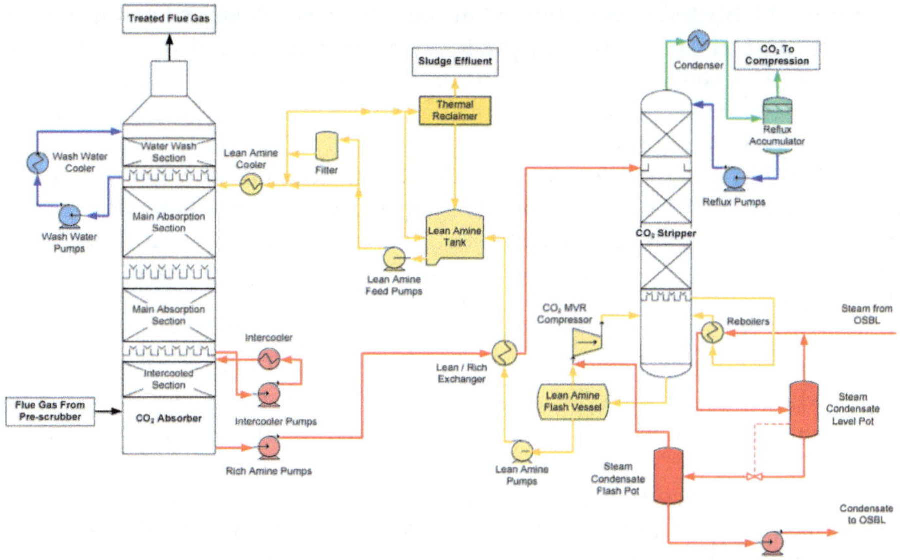

Figure 5.14 Cansolv process line-up (Singh, 2014).[46] Reproduced with permission from ref. 46 with permission from Elsevier, Copyright 2014.

by an absorber. The solvent was regenerated in the stripper and the average CO_2 capture was maintained at more 90%.[46]

A joint project was undertaken between Cansolv Technologies Incorporated (CTI) (part of the Shell Group) and RWE to establish a capture facility at RWE Npower's Aberthaw Power Station in South Wales. This post-combustion process was designed to capture 50 tonnes of CO_2 per day, which is equivalent to the emissions from 3 MW of Aberthaw's 1.5 GW capacity using a slip stream. Operation of the plant since January 2013 was an opportunity for plant familiarization and optimization and also improved the understanding of the performance of capture processes.[47]

5.2.2.6 *Aker Clean Carbon*

Aker Solutions, the parent company of Aker Clean Carbon, has been developing carbon capture technologies and designed, fabricated and continuously operated a mobile test unit (MTU) of a 'Just Catch Process' for a slip stream of PC plant flue gas. The MTU is a custom-built mobile test CO_2 capture facility, which can be used to capture CO_2 from different industrial flue gases to verify new design features and solvents, and operates for long-term testing. The design is based on a conventional amine absorption–desorption process with full packing height columns. MTU was first built in 2009 at the gas research centre in Risavika in Norway and Longannet in Scotland. In June 2011, the MTU performed tests at the National Carbon Capture Center in

Alabama, US, hosted by Southern Company. Scottish Power's Longannet station announced that testing of process improvements and low-energy solvents had achieved a reduction in the energy requirement of about one-third compared to a reference plant. Reduced solvent emissions are a claimed benefit of the just catch process.[48]

The MTU is designed to process 2400 lb per hours of flue gas with a CO_2 capture rate of up to 400 lb per hours equipped with a novel emission control design to test at Alabama. It consists of an anti-mist design and polishing step for pH control.[49] Aker successfully tested the performance with more than 2500 hours of operation with CO_2 capture, to investigate the factors influencing the specific reboiler duty, solvent CO_2 loading, and emissions. The results showed a significant reduction in emissions of solvent components when using Aker's low emissions design and reduced energy consumption using the Aker CC6 solvent compared to 30 wt% MEA solvent for 90+% CO_2 capture.[50]

Aker Solutions constructed an amine plant in CO_2 Technology Centre Mongstad (TCM DA), which is one of the world's largest test facilities for CO_2 post-combustion capture technology. The process flow diagram of this plant at TCM is shown in Figure 5.15. The main objective of the test runs was to verify the Advanced Carbon Capture™ process with two proprietary solvents, S21 and S26. The tests were completed using a slip stream with relatively low (~3.5–4.0 vol%) CO_2 content from a combined heat and power (CHP) plant at Mongstad. The results showed very low levels of produced nitrosamines, very low emissions to the atmosphere and very low degradation rates for solvents S21 and S26 with respect to environmental issues. Emission of solvent

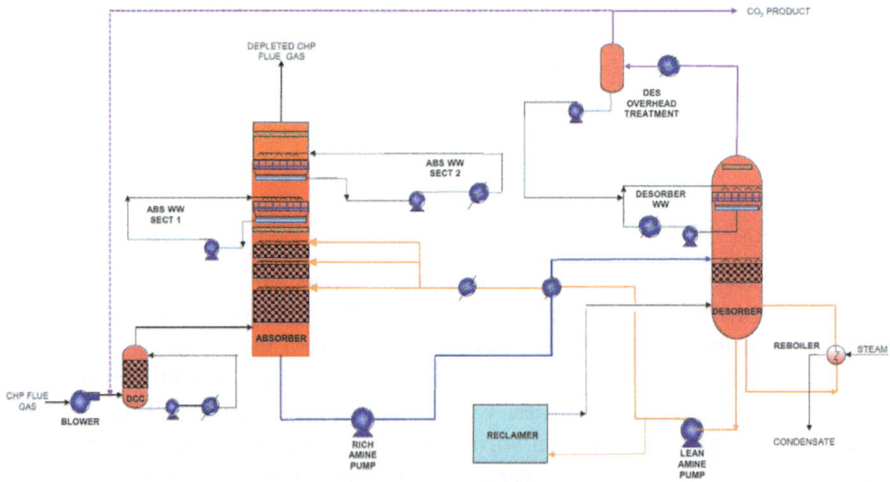

Figure 5.15 Process flow diagram of the amine plant at CO_2 Technology Centre Mongstad for the treatment of CHP flue gas (Gorset *et al.*, 2014).[51] Reproduced with permission from ref. 51 with permission from Elsevier, Copyright 2014.

amines to the atmosphere was in the range 10–90 μg Nm^{-3}, and emission levels of individual nitrosamines and nitramines were typically around 0.1 μg Nm^{-3}. Aker Solutions' emission control design equipped with the MTU is operated with the CHP flue gas, excellent results were obtained for S26 solvent at TCM DA.[51]

5.2.2.7 Alstom Advanced Amine Process

The advanced amine process (AAP) technology is based on the ability of amine solution to absorb CO_2 at low temperatures and release it at moderately elevated temperatures. CO_2 reacts with aqueous amines in an absorption column to form chemical compounds and gets removed from the gaseous stream. AAP developed by Alstom and the Dow Chemical Company relies on an advanced amine solvent, UCARSOL™ FGC 3000, developed by Dow specifically for CO_2 capture from combustion gas. The AAP technology was validated first in a small pilot plant, capable of capturing 5 tonnes CO_2 per day at a Dow chemical plant in South Charleston, West Virginia. This facility was used to develop the operations experience necessary for the design of larger, subsequent plants and establish suitable amine management strategies.[52]

Following this demonstration facility, Alstom and Dow developed the "Advanced Flow Scheme" improving absorber, regenerator and thermal management systems. Alstom established a partnership with EDF Group's Le Havre (France) thermal energy plant, to test and refine the Alstom AAP using coal-derived flue gas. The demonstration unit (shown in Figure 5.16) at Le Havre, France, is designed to capture 25 tonnes CO_2 per day at a 90% capture rate from the flue gas from a hard coal-fired plant. The simplified process flow diagram of the Alstom–Dow advanced amine process is shown in Figure 5.17. The demonstration unit captures around 1900 tonnes of CO_2 with a solvent circulation for 2700 hours. The test run consistently demonstrated a thermal performance of 2.3–2.4 GJ t^{-1} CO_2 captured at a 90% capture rate.[53] Subsequent large-scale demonstrations are proposed and the Dow amine is claimed to capture CO_2 with increased efficiency, a higher tolerance for oxygen and trace contaminants, and decreased solvent degradation.

5.2.2.8 Siemens POSTCAP Amino Acid Salt

Siemens developed the POSTCAP process (Figure 5.18), which is an absorption–desorption process based on a second-generation amino acid salt to capture CO_2 from power plant flue gases. In comparison to amines, the amino acid salt technology has several advantages, such as a fast rate of CO_2 absorption, higher selectivity for CO_2 absorption, an environmentally friendly amino acid salt solvent, good solvent stability against degradation, lower volatility, reduced heat of regeneration, and ease of handling by plant personnel.[54]

Figure 5.16 EDF pilot plant at Le Havre (France) (Baburao *et al.*, 2014).[53] Reproduced with permission from ref. 53 with permission from Elsevier, Copyright 2014.

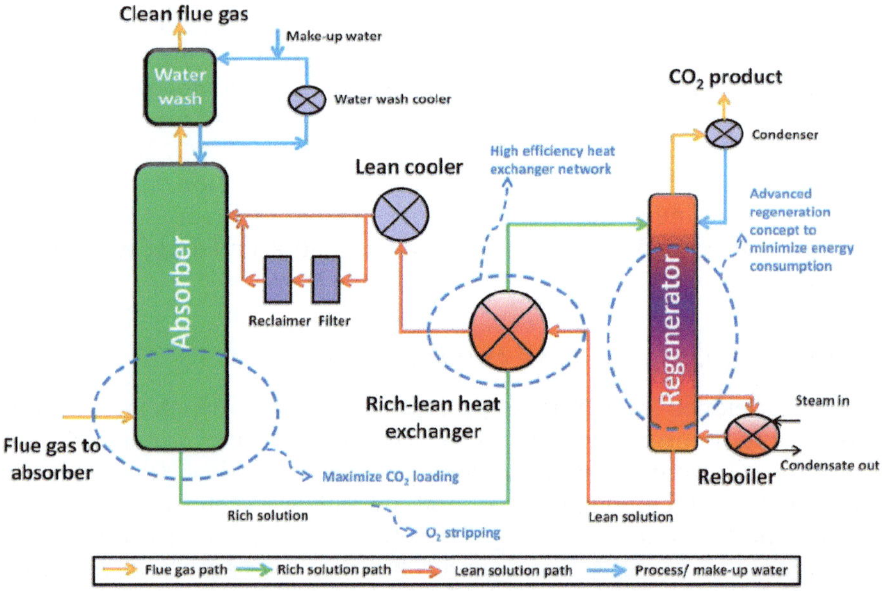

Figure 5.17 Simplified flow diagram of the Alstom–Dow advanced amine process (Baburao *et al.*, 2014).[53] Reproduced with permission from ref. 53 with permission from Elsevier, Copyright 2014.

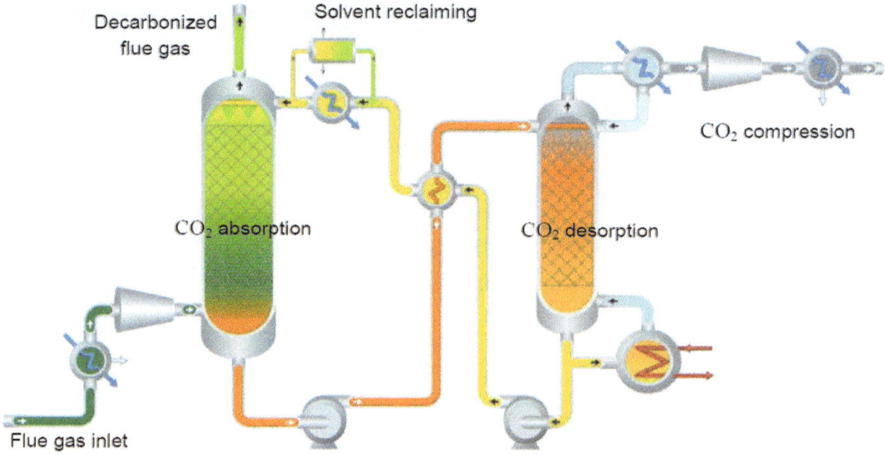

Figure 5.18 Siemens POSTCAP process configuration (IEAGHG, 2012).[37] Reproduced from ref. 37 with permission from IEAGHG, Copyright 2012.

The POSTCAP process is being tested with real flue gas in a slip stream, 1 MWe equivalent pilot unit at E.ON power station Staudinger, Germany. The energy requirement for regeneration of the amino-acid salt solution was reported as 2.7 GJ per ton of CO_2, which compares favorably with the regular amine processes. The results confirmed lower emissions and that the amino-acid salt was stable under thermal conditions and the oxygen environment, which prevail under real operating conditions.[38] Based on the results of the pilot plant operation, it was suggested that a scale-up of the PostCap™ technology to a large-scale demonstration and full-scale projects is possible.

5.3 Environmental and Economic Concerns

A CCS unit involves CO_2 capture, compression, transport and storage. Environmental issues arise in each step. Environmental concerns when adding a CCS unit to an existing power plant should include its impact, such as the release of pollutants into the air, water and land; solid waste generation; and the consumption of resources such as fuel, water, minerals and land. In order to provide a broader perspective of the environmental issues in a post-combustion CCS system, the level of detail involved in the Life Cycle Assessment (LCA) has been presented as an example in Figure 5.19.

Few works have assessed the environmental effect of adding a CO_2 capture unit to combustion facilities. Application of CO_2 capture technologies requires substantial amounts of energy. The amount of fuel used is so high that adding CO_2 capture in such facilities makes the environmental impact as much as 10–67% worse in many other kinds of environmental

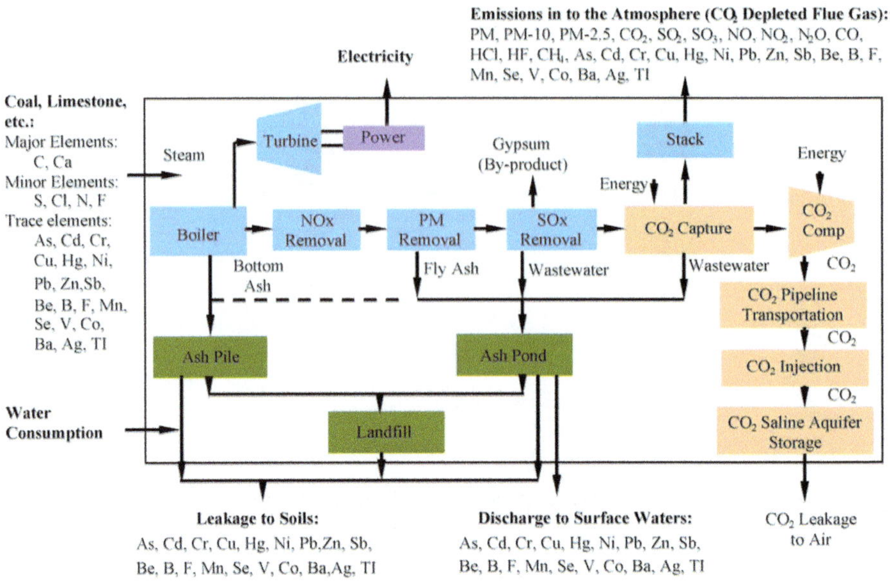

Figure 5.19 The level of detail involved in the LCA of a post-combustion CCS system (Korre *et al.*, 2010).[55] Reproduced with permission from ref. 55 with permission from Elsevier, Copyright 2010.

impact categories.[56,57] In addition to increased fuel consumption, it also increases the use of chemicals, mostly amines, and limestone, which affect the environment more than the plant without carbon capture. It has been shown that adding post-combustion capture to a coal combustion power leads to 38% more fossil fuel consumption, 46% more toxicity impact in humans, 58–59% more harm to aquatic and terrestrial ecosystems, and 67% more freshwater depletion.[20,57] Korre *et al.*[55] have shown that the Life Cycle Impact Assessment results for pulverized coal-fired power generation with post-combustion CO_2 capture can achieve an 80% reduction in greenhouse gas emissions, compared to systems without capture, without a significant increase in other environmental burdens. Using LCA, Clarens *et al.*, 2016,[58] showed that the inclusion of the CO_2 capture unit reduces the emissions of CO_2, particulate matter and terrestrial acidification. Reductions in the climate change impact with the CO_2 capture unit are estimated to be 73%, 66% and 72% for CaO looping, conventional and advanced amine processes, respectively.

Studies on the costs of CCS have estimated that for new coal-fired plants built in the 2020s, after large-scale demonstration has been achieved, all three different routes to CO_2 capture in coal-fired generation have comparable costs using today's technologies.[59] It has been estimated that the costs of coal-fired power generation can be increased 40% to 63% after addition of a CO_2 capture unit; however, this is still comparable to or lower than solar photovoltaic and offshore wind costs (Figure 5.20).[16,60]

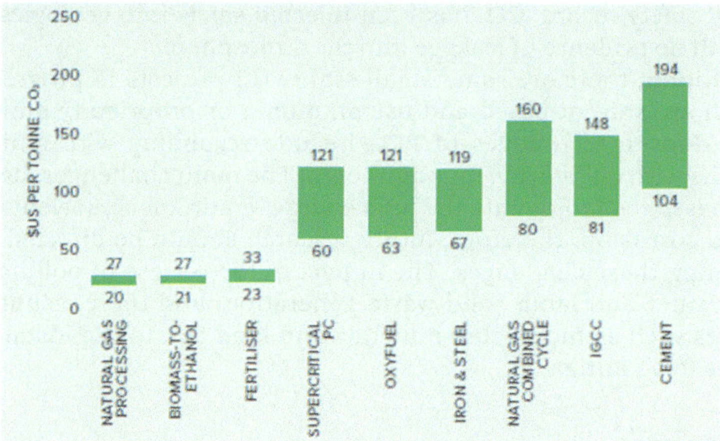

Figure 5.20 The costs of implementing CCS technologies in power plants and other industries. Note: costs defined as the cost per tonne of CO2 avoided (in US$). Reproduced from GCCSI, 2017.[21]

5.4 Conclusion

CO_2 is a major anthropogenic greenhouse gas responsible for global warming and climate change, the emission of which needs to be controlled. Several CO_2 capture approaches, such as absorption, adsorption, cryogenic, membrane and chemical looping combustion, have been developed. CO_2 from off-gas can be captured by liquid or solid sorbents. The amine-based absorption process is the most advanced among the technologies currently available. Due to the several drawbacks of amine solvents, research on solid sorbents has been prioritized. The technological options currently under consideration for capturing CO_2 from combustion and gasification facilities are pre-combustion capture, post-combustion capture, oxy-fuel combustion, and chemical looping combustion. Post-combustion, oxy-fuel combustion and pre-combustion CO_2 capture are still not ready for implementation.

CCS has been accepted as a primary option to mitigate anthropogenic CO_2 emissions. However, CCS adds to the cost of operations and its implementation is a bit challenging in terms of the economic aspect. The integration of a CCS process with oil and gas operations such as EOR may help to overcome the challenge to utilize the captured CO_2.

In 2017, globally, 17 large-scale CCS facilities were in operation, capturing more than 30 Mtpa of CO_2. The CO_2 capture from large and small facilities today equates to more than 8 million motor vehicles being taken off the roads. The large-scale coal-related CCS facilities in operation at present are (i) SaskPower Boundary Dam- CCS; (ii) Petra Nova Carbon Capture, Texas, USA; (iii) Kemper County Energy Facility (IGCC + CCS); and (iv) Callide – oxy-fuel combustion and carbon storage demonstration plant. SaskPower and Petra Nova demonstrate that the CCS is appropriate for existing coal power plants. Today, more than 6000 Km of CO_2 pipelines are operational with an excellent safety record. CO_2 has been injected safely into geological formations with no evidence of leakage into the atmosphere.

In addition, there are some small-scale PCC projects in progress, most of which are solvent-based and use ammonia or proprietary amines as a solvent. Proven techniques of PCC include scrubbing with amines and scrubbing with chilled aqueous ammonia. The main challenges faced by an amine-based technology are the high energy requirement, solvent degradation and corrosion; therefore, future research should be directed towards overcoming these challenges. The impact of the release of pollutants into the air, water and land; solid waste generation; and the consumption of resources such as fuel, water, minerals and land due to the installation of CCS, has to be mitigated.

References

1. NASA, *National Aeronautics and Space Administration. Carbon Dioxide*, 2017, https://climate.nasa.gov/vital-signs/carbon-dioxide/, accessed November 28, 2017.
2. IPCC, *Climate Change 2014 Synthesis Report Summary for Policymakers*, 2014, http://ar5-syr.ipcc.ch/ipcc/ipcc/resources/pdf/IPCC_SynthesisReport.pdf.
3. NASA, *National Aeronautics and Space Administration. Global Climate Change Vital Signs of the Planet, Global Temperature Rise*, 2017, https://climate.nasa.gov/evidence/, accessed November 28, 2017.
4. IPCC, *Climate Change 2001: The Scientific Basis. Contribution of Working Group 1 to the Third Assessment Report of the Intergovernmental Panel on Climate Change*, ed. J. T. Houghton, Y. Ding, D. J. Griggs, M. Noguer, P. J. van der Linden, X. Dai, K. Maskell and C. A. Johnson, Cambridge University Press, Cambridge, UK, and New York, USA, 2001, p. 881.
5. DOE/NETL, *Carbon Dioxide Capture and Storage R&D Roadmap*, U.S. Department of Energy, December 2010.
6. A. Samanta, A. Zhao, G. K. Shimizu, P. Sarkar and R. Gupta, *Ind. Eng. Chem. Res.*, 2012, **51**, 1438.
7. N. K. Sandhu, D. Pudasainee, P. Sarkar and R. Gupta, *Ind. Eng. Chem. Res.*, 2016, **55**, 2210.
8. A. Sayari and Y. Belmabkhout, *J. Am. Chem. Soc.*, 2010, **132**, 6312.
9. N. K. Sandhu, Steam regeneration of amine impregnated silica based sorbents for post combustion CO_2 capture, M Sc Thesis, University of Alberta, 2014.
10. J. C. M. Pires, F. G. Martins, M. C. M. Alvim-Ferraz and M. Simões, *Chem. Eng. Res. Des.*, 2011, **89**, 1446.
11. K. Bennaceur, CO_2 capture and sequestration, in *Future Energy*, ed. T. M. Letcher, Elsevier, Boston, 2nd edn, 2014, ch. 26, pp. 583–611.
12. J. Kemper, G. Ewert and M. Grünewald, *Energy Procedia*, 2011, **4**, 232–239.
13. M. Wang, A. Lawal, P. Stephenson, J. Sidders and C. Ramshaw, Post-combustion CO_2 capture with chemical absorption: a state-of-the-art review, *Chem. Eng. Res. Des.*, 2011, **89**(9), 1609–1624.

14. B. Zhao, Y. Su, W. Tao, L. Li and Y. Peng, *Int. J. Greenhouse Gas Control*, 2012, **9**, 355.
15. E. S. Rubin, H. Mantripragada, A. Marks, P. Versteeg and J. Kitchin, The outlook for improved carbon captures technology, *Prog. Energy Combust. Sci.*, 2012, **38**, 630.
16. IEA, *Technology Roadmap Carbon Capture and Storage*, International Energy Agency, 2013, p. 63.
17. EIA, *International Energy Outlook 2016*, Energy Information Agency, May 11, 2016, https://www.eia.gov/forecasts/ieo/electricity.cfm.
18. US DOE, *Carbon Capture Opportunities for Natural Gas Fired Power Systems*, 2016, https://www.energy.gov/sites/prod/files/2017/01/f34/Carbon%20Capture%20Opportunities%20for%20Natural%20Gas%20Fired%20Power%20Systems_0.pdf, accessed 2018 March.
19. EIA, *Energy-related CO_2 Emissions from Natural Gas Surpass Coal as Fuel Use Patterns Change*, Energy Information Agency, 2016, http://www.eia.gov/todayinenergy/detail.php?id=27552.
20. T. A. Adams, L. Hoseinzade, P. B. Madabhushi and I. J. Okeke, *Processes*, 2017, **5**, 44.
21. GCCSI, *Global Carbon Capture and Storage Institute*, The Global status of CCS, 2017, http://www.globalccsinstitute.com/sites/www.globalccsinstitute.com/files/uploads/global-status/1-0_4529_CCS_Global_Status_Book_layout-WAW_spreads.pdf.
22. Saskpower.com, http://www.saskpower.com/our-power-future/carbon-capture-and-storage/capturing-carbon-and-the-worlds-attention/, accessed August, 2017.
23. K. Stéphenne, *Energy Procedia*, 2014, **63**, 6106.
24. P. Fairley, J. Ackerman and T. Gruber, *Technol. Rev.*, 2015, **118**, 84–86.
25. OGEM, North America: petra nova carbon capture project successfully completed, *Eur. Oil Gas Mag.*, 2017, **43**(2), 62.
26. *DOE-supported Petra Nova Captures More than 1 Million Tons of CO_2*, http://www.energy.gov, https://energy.gov/fe/articles/doe-supported-petra-nova-captures-more-1-million-tons-co2, accessed November 2017.
27. AIChE, 2017, https://www.aiche.org/chenected/2017/01/worlds-largest-carbon-capture-project-launches-texas, accessed August, 2017.
28. S. Goldenberg, *Can Kemper Become the First US Power Plant to Use 'clean Coal'? the Guardian*, 12 March 2014, accessed July 2014.
29. K. Swartz, *Southern Co.'s Clean Coal Plant Hits a Dead End*, 2017, https://www.eenews.net/stories/1060056418, accessed November, 2017.
30. *Mississippi power Plants, Generating Energy: Our Diverse Portfolio Benefits You*, http://www.mississippipower.com, http://www.mississippipower.com/about-energy/plants/, accessed November, 2017.
31. Q. Zhuang, M. Biondi, S. Yan, K. Bhagat, R. Vansickle, C. Cheng, H. Tan, Y. Zhu, W. You and W. Xia, *Fuel*, 2015, **152**, 103.
32. A. Komaki, T. Gotiu, T. Uchida, T. Yamada, T. Kiga and T. Spero, *Energy Procedia*, 2014, **63**, 490–496.
33. GCCSI, *Callide Oxyfuel Project– Lesson Learned*, Global CCS Institute, 2014, p. 52.

34. GCCSI, *Global Carbon Capture and Storage Institute*, 2016, https://www.globalccsinstitute.com/projects/petra-nova-carbon-capture-project, accessed August, 2017.
35. http://www.power-technology.com.
36. Fluor, *Fluor's Econamine FG Plus^{SM} Technology for Post-Combustion CO_2 Capture*, 2008, http://www.fluor.com/SiteCollectionDocuments/FluorEFG-forPost-CombustionCO2CaptureGPAConf-Feb2008.pdf.
37. IEAGHG Report, *Gaseous Emissions from Amine Based Post Combustion CO2 Capture Processes and Their Deep Removal*, May 2012.
38. GCCSI, *CO2 capture Technologies: Post Combustion Capture (PCC)*, 2012, http://hub.globalccsinstitute.com/sites/default/files/publications/29721/co2-capture-technologies-pcc.pdf.
39. T. Kamijo, T. Yonekawa, M. Inui, T. Hirata, T. Tsujiuchi and O. Miyamoto, *Mitsubishi Heavy Ind. Tech. Rev.*, 2015, **52**(1), 24.
40. O. Miyamoto, C. Maas, T. Tsujiuchi, M. Inui, T. Hirata, H. Tanaka, T. Yonekawa and T. Kamijo, *Energy Procedia*, 2017, 5616.
41. G. Valenti, D. Bonalumi and E. Macchi, *Energy Procedia*, 2009, **1**, 1059.
42. F. Kozak, A. Petig, E. Morris, R. Rhudy and D. Thimsen, *Energy Procedia*, 2009, **1**, 1419.
43. C. R. McLarnon and J. L. Duncan, *Energy Procedia*, 2009, **1**, 1027.
44. Powerspan, *Overview of ECO_2™ for CO_2 Capture*, 2017, http://c.ymcdn.com/sites/www.icac.com/resource/resmgr/Greenhouse_Gas_Controls/PowerSpan_ECO2Overview_06-07.pdf, accessed December, 2017.
45. D. Shaw, *Energy Procedia*, 2009, **1**, 237–246.
46. A. Singh and K. Stéphenne, *Energy Procedia*, 2014, **1**, 237.
47. Rwe.com, http://www.rwe.com/web/cms/en/2756/rwe/innovation/power-generation/fossil-fired-power-plants/co2-removal-uk/, accessed December, 2017.
48. http://www.zeroco2.no/projects/developers/aker-clean-carbon, accessed January, 2018.
49. J. N. Knudsen, O. M. Bade, M. Anheden, R. Bjorklund, O. Gorset and S. Woodhouse, *Energy Procedia*, 2013, **37**, 1804.
50. J. Northington, F. Morton and R. A. Yongue, *Advanced Technology Testing at the National Carbon Capture Center, 29th Annual International Pittsburgh Coal Conference October 15–18*, 2012.
51. O. Gorset, J. N. Knudsen, O. M. Bade and I. Askestad, *Energy Procedia*, 2014, **63**, 6267.
52. *Carbon Capture Journal, Alstom and Dow Progress on Amines*, Nov 04 2012, http://www.carboncapturejournal.com/news/alstom-and-dow-progress-on-amines/3201.aspx?Category=all.
53. B. Baburao, S. Bedell, P. Restrepo, D. Schmidt, C. Schubert, B. DeBolt, I. Haji and F. Chopin, *Energy Procedia*, 2014, **63**, 6173.
54. R. Schneider and H. Schramm, *Environmentally Friendly and Economic Carbon Capture from Power Plant Flue Gases: The SIEMENS PostCap Technology, 1st Post Combustion Capture Conference, 17–19 May 2011*, 2011.

55. A. Korre, Z. Nie and S. Durucan, *Int. J. Greenhouse Gas Control*, 2010, **4**, 289.
56. P. Zapp, A. Schreiber, J. Marx, M. Haines, J. F. Hake and J. Gale, *Int. J. Greenhouse Gas Control*, 2012, **8**, 12.
57. J. Nease and T. A. Adams, *Appl. Energy*, 2015, **150**, 161.
58. F. Clarens, J. J. Espía, M. R. Giraldi, M. Rovira and L. F. Vega, *Int. J. Greenhouse Gas Control*, 2016, **46**, 18.
59. IEA, *Cost and Performance of Carbon Dioxide Capture from Power Generation, IEA Working Paper Prepared by Matthias Finkenrath*, OECD/IEA, International Energy Agency, Paris, 2011.
60. IEA, *IEA World Energy Outlook 2011 Exclusive Summary*, International Energy Agency, 2011, http://www.iea.org/Textbase/npsum/weo2011sum.pdf.

Subject Index

Entries in **bold** represent tables and charts.